PALE BLUE

PALE BLUE

A THRILLER

MIKE JENNE

YUCCA

Yucca Publishing books may be purchased in bulk at special discounts for sales promotion, corporate gifts, fund-raising, or educational purposes. Special editions can also be created to specifications. For details, contact the Special Sales Department, Yucca Publishing, 307 West 36th Street, 11th Floor, New York, NY 10018 or yucca@skyhorsepublishing.com.

Yucca Publishing® is an imprint of Skyhorse Publishing, Inc.®, a Delaware corporation.

Visit our website at www.yuccapub.com.

10 9 8 7 6 5 4 3 2 1

Library of Congress Cataloging-in-Publication Data is available on file.

Jacket design by Haresh R. Makwana
Jacket photo: Gemini Mission courtesy of NASA

Print ISBN: 978-1-63158-084-0
Ebook ISBN: 978-1-63158-091-8

Printed in the United States of America

For Adele, with Love

A sweet smile, a joyful laugh, a kind heart,
a hand to hold, and a life to share.

Author's Note

The year is 1972. Blue Gemini astronauts Major Drew Carson and Major Scott Ourecky have flown a total of seven missions to intercept and destroy suspect Soviet satellites, and one mission to rescue a Navy astronaut from a secret ocean surveillance variant of the MOL (Manned Orbiting Laboratory) that was crippled by a massive solar flare. Only one more mission-ready Gemini-I/Titan II "stack" remains. The Aerospace Support Project's clandestine PDF—Pacific Departure Facility—on Johnston Island has been destroyed by Hurricane Celeste. Depending on the results of the upcoming Presidential election, the Project might be extended for twelve more missions or cancelled outright.

Ourecky is hospitalized after he returns to earth following the MOL rescue mission. After his release from the hospital, he returns home to discover that his wife Bea has left him and is staying with a friend dying of cancer. Although the separation is temporary, he is woefully concerned that his marriage will end in divorce unless he is soon freed from the high demands and airtight secrecy of the Project.

Since his secret space missions will never be reflected on his official personnel records, Carson is still obsessed with flying combat missions in Vietnam.

General Mark Tew, the Project's commander, is anxious to retire and put Blue Gemini behind him but insists on staying on to protect Carson and Ourecky.

After the tragic incident involving his ocean surveillance MOL, Admiral Tarbox schemes to gain more influence over Blue Gemini, with the objective of eventually taking over control of the Project after Tew retires. To this end, he enlists the support of Tew's deputy, Virgil Wolcott, who wants the Project to be extended for another twelve missions.

The Soviets, led by RSVN Lieutenant General Rustam Abdirov, continue to develop the nuclear-armed *Krepost* space station. A two-man crew—Lieutenant Colonel Gogol and Major Vasilyev—are selected for the *Krepost's* first mission. As Abdirov pushes the *Krepost* program toward fruition, he gradually loses his equilibrium and becomes obsessed with unilaterally deploying the *Krepost's* nuclear warhead—the "Egg"— to force a thermonuclear confrontation with the West.

Major General Gregor Yohzin, Abdirov's deputy, has been recruited by American intelligence and has been passing secrets to the West for years. Eager to stymie Abdirov's scheme, he is now providing information about the *Krepost* as well.

1

HIATUS

Aerospace Support Project
Wright-Patterson Air Force Base, Ohio
7:04 a.m., Tuesday, August 29, 1972

Only slightly more than a week after returning from his eighth mission into orbit, Major Scott Ourecky felt less like a seasoned spacefarer and more like an invalid confined to a sanatorium. Arriving at the Project's headquarters, he pulled into a parking place, shut off his car, and then hacked up a blood-tinged lump of phlegm.

He was just returning from the base hospital, where he had endured yet another grueling round of respiratory therapy. Despite the intensive treatments, it was rare when he could sleep through an entire night without having to drape a towel over his head while sucking in medicated steam from a humidifier.

Ourecky's chest ached as he wheezed for breath; his lungs felt like rough burlap sacks jammed with coarse gravel. Although the doctors assured him that his lingering bronchitis was healing, his agonizing ordeal was far from over; they also cautioned that he should expect to cough up bits and pieces of dead lung tissue for at least another month.

Despite his frail state, his crushing workload had not subsided even an iota. He and Drew Carson spent most of their duty days consumed in seemingly endless rounds of post-flight interrogations about last month's MOL rescue mission. In addition to the usual array of Project officers who conducted the exhaustive debriefings, a trio of Navy engineers also participated, seemingly intent on gleaning evidence concerning the MOL's communications failure prior to the massive solar flare. It was terribly frustrating for Ourecky; even though he had gone aboard the MOL to rescue Ed Russo and shut down the MOL's nuclear reactor, there was little he could share with his inquisitors that might shed light on why their station's radios had malfunctioned at such a crucial juncture.

Likewise, there was no change to his marital disharmony; he and Bea were still separated. Ourecky thought of her often. As much as he longed to see her, he decided to grant her at least a few more weeks to cool off and collect her thoughts. After all, as it was often said, absence makes the heart grow fonder, so maybe the more time that they spent apart, the more inclined she might become to return home. Besides, his chronic cough was just one more thing that he didn't want to explain to her.

Although he missed her immensely, and missed Andy as well, he wanted to wait until the time was right to see them. She and Andy were still in Dayton, staying with her friend Jill and her mother. He and Bea talked on the phone almost every night, but the conversations always seemed lukewarm. Politely evading the subject of their separation, they could have been chatting from opposite ends of the earth instead of being merely a few miles apart. More often than not, she talked about her friend Jill's medical issues. Only recently, Jill's oncologist had pronounced that her ovarian cancer was terminal. The last time they talked, Bea tearfully related that they didn't expect her to survive past Christmas. The way that Bea described it, Jill's mother was on the verge of becoming a basket case. Losing a child is always traumatic, so witnessing her daughter's demise in slow motion was probably more than she could effectively bear.

Bea made it clear that she intended to stay with Jill until the very end, whenever that might be. He and Bea agreed that it was better if Andy believed that Ourecky was gone on another one of his extended TDY trips. After all, the toddler was accustomed to his father's prolonged absences; that was certainly easier to explain than why his parents could not currently live under the same roof.

The more often they talked, Ourecky came to suspect that Bea was either also a strict adherent to the "absence makes the heart grow fonder" belief, or she just wasn't in an almighty hurry for them to reconcile. Although she had indicated that he was always welcome to visit, she never actually invited him to drop by. Perhaps she was just waiting for him to broach the issue. Sighing, he trusted that their fractured relationship would eventually be mended, if granted adequate time, much as he hoped that his scarred and aching lungs would someday heal.

7:52 a.m., Tuesday, August 29, 1972

As he waited for the weekly intelligence update, General Mark Tew referred to an index card that listed his hour-by-hour regimen of prescribed medications. His hands quaked and his left eyelid twitched. Willing his hands to be steady, he poured a glass of water from the pitcher on his desk and then sorted through the collection of bottles and vials in the desk drawer, which was now his personal pharmacy. He swallowed seven pills of various sizes and colors, chased them with water, and closed the drawer.

Granting himself a few minutes to relax as the medicines took effect, Tew perused the newspaper as he waited for the others to arrive. He was fascinated with the story of a champion American swimmer—Mark Spitz—who was an odds-on favorite to win several gold medals at the Summer Olympics, which had commenced this week in Munich. A lifetime ago, Tew had been a varsity swimmer in college—winning

All-State in the freestyle and breaststroke—but his athletic career was cut short when he drafted into the Army Air Corps.

The newsprint blurred as he found it increasingly difficult to focus. One or more of his prescriptions was apparently wreaking havoc on his eyesight. He no longer drove, because he just didn't feel safe behind the wheel, and could only read for a few minutes at a time.

Tew set the paper aside, closed his eyes, took a deep breath, and listened to the clock on the wall. *Tick…tick…tick…tick.* It reminded him that his life—what remained of it, given his current circumstances—was metered out in precious seconds.

Blue Gemini had exacted a grim toll on Tew; his health was deteriorating at an almost exponential rate. Each visit to his cardiologist yielded a more dismal pronouncement of his life expectancy. He was concerned that at some point in the relatively near future the doc would simply declare that he had already died and had missed his own funeral. As of last week, according to the physician, he shouldn't expect to be above ground for more than a year at most. He was no longer seriously considered as a viable candidate for a heart transplant, because the rest of his circulatory system was so damaged that his arteries and veins probably couldn't handle a new ticker.

Despite his health woes, and they were legion, he was diligently trying to stay at the Project until the very last mission was flown, if for no other reason than to protect Carson and Ourecky.

In his opinion, he had something in common with the pair: from his perspective, all three of them were living on borrowed time. He had personally assured them that he would not bow out until they were transferred—permanently—away from the Project and safely out of the clinging grasp of Wolcott and Tarbox. When that time came, Ourecky would be enrolled at MIT to complete the doctorate that had long been promised to him. After the last flight, Carson was slated to enter a career "holding pattern" as the commander of a Germany-based fighter squadron until it was deemed safe for him to fly in Vietnam, provided that the war was still being prosecuted in Southeast Asia.

And although Virgil Wolcott had repeatedly sworn that he would promptly release Carson and Ourecky after the last mission had flown, or at least grant each an opportunity to determine their own fate, Tew just couldn't shake the infuriating notion that Wolcott would not hold true to his word.

In truth, Tew was far less concerned about Carson's fate than he was with Ourecky's; he felt confident that Carson could fend for himself, but Tew was very aware that Ourecky's marriage was on the verge of failure. So, determined to stretch out the process as long as possible, in order to protect his two veteran astronauts, Tew was decisively engaged in an endurance contest that could likely end in his death.

Tew's concerns about Ourecky might all be for naught if Blue Gemini folded after the last mission, since the engineer would certainly be released from the Project forthwith, but there was still considerable potential that the secret space program might be extended. If that happened, Carson would probably elect to stay on, and if given ample opportunity, he would likely convince Ourecky to remain as well. After all, when it came to flying the Blue Gemini missions, the two men were like the wheels on a bicycle: one was useless without the other.

As much as Tew respected their friendship, he didn't want Ourecky's marriage to further suffer, so he looked for any and all opportunities to keep the two men physically separate. If there were any practical way to keep them on opposite ends of the country, Tew would do just that, just to stifle Carson's influence.

In any event, he desperately wanted the last mission to fly sooner than later, particularly since his influence was on the wane. The Pentagon leadership had decreed that the two military manned space programs—Blue Gemini and the Navy MOL—which had previously competed for scarce resources and funding, would be combined into a joint enterprise.

Once the two programs were joined, Tew would step aside, and Tarbox would take the helm. Unless he elected to retire, Wolcott would remain in his current assignment, serving as Tarbox's civilian deputy. Once openly antagonistic, the two former adversaries had forged a synergistic alliance. Although the merger was still very much subject to the winds of politics, it was accepted as a foregone reality by most of the Project's workers; many were already looking to Tarbox for guidance and leadership.

Although Tarbox's transition was theoretically still pending, he was almost a regular fixture at the Project's offices, as he sought to cultivate relationships and curry favor with the Air Force senior leaders who would make critical decisions on the future of military manned space programs. It would be months before the Air Force and Navy formally combined their programs, but the transformation was already well underway, as the politically savvy Tarbox moved swiftly to assert his influence. Tew grudgingly endorsed the merger between the Air Force and Navy programs and had ceded most of his day-to-day responsibilities to Wolcott. But as much as he welcomed a relief, Tew resented Tarbox's rapid encroachment.

The consolidation would not be without casualties. Roughly three quarters of the Project's civilian employees would lose their jobs. At the conclusion of the last Blue Gemini mission—whenever that might be— the bulk of the Wright-Patterson operations would be shifted to the Navy's more modern facilities in California.

As it was, Blue Gemini was theoretically at its culmination point. There was only one stack left to fly, and it was earmarked to attack the Soviets' elusive *Krepost* orbital bombardment system. The Project's headquarters and mission control facilities were still operational at Wright-Patterson, but their launch options were severely limited, since the PDF—Pacific Departure Facility—on Johnston Island had been decimated by Hurricane Celeste shortly after the last mission. If the Project was extended, most of the anticipated missions probably could be executed with launches from either Vandenberg or Cape Kennedy, but there was also some discussion that the PDF would be rebuilt, possibly as an even more expansive facility.

Although there was considerable support for Blue Gemini to continue into a second phase, momentum had stalled. Nothing could happen until after the Presidential elections in November. If Nixon was reelected, then it was virtually a sure bet that an extension of Blue Gemini would be immediately approved and secretly funded. The Project's prospects weren't nearly as clear for a Democratic win; in all likelihood, McGovern would cancel Blue Gemini as soon as he was briefed on it. In Tew's mind, the election hinged on the potential outcome in Vietnam. To his credit, Nixon was making every effort to end US involvement in the war-torn nation, but he clearly had a long way to go before the electorate was convinced of his sincerity.

The Vietnam War had already contributed substantially to the demise of the Air Force's MOL. In creating a massive program that was in the public eye while simultaneously concealing the secret purpose for its existence, the Air Force had painted themselves into an odd corner. The program's leaders found themselves in the awkward and untenable position of defending the MOL's continuation in two different forums. On the unclassified side of the discussion, the costs of the Vietnam war were already overwhelming the military budget, so it was virtually impossible for the Air Force to publically justify a two-man space station that was—on the surface—little more than a duplication of NASA's current and planned endeavors to explore space. Furthermore, on the classified side, it was equally difficult for the Air Force to justify how the MOL's crew could perform their actual mission—strategic reconnaissance of the Soviet Union—more efficiently than unmanned reconnaissance satellites currently in development.

Even as the Air Force's MOL program abruptly died in 1969, Tarbox was shrewdly able to keep the Navy's ocean surveillance variant a secret. Tew suspected that the "Ancient Mariner" was only successful in this venture because of the unusual accounting lessons gleaned by Russo during his stint at the Project. Blue Gemini's expenditures were scattered across so many other Air Force programs that only Tew actually knew how much money was being spent. Learning from Tew's experience, conveyed by Russo, Tarbox had managed to keep his program flying underneath the budgetary radar.

Besides shrouding his accounts in black budgets, Tarbox's greatest coup was absorbing virtually all of the MOL hardware, less the KH-10 "DORIAN" camera system and several mainframe computers, from the program's contractors after the Air Force program was abruptly cancelled. Forced to lay off thousands of loyal workers in a matter of weeks, the MOL contractors were reeling in shock, and even though they had been directed to simply destroy the parts, they were obviously reluctant to wreck their handiwork. After all, Tarbox's gesture probably lent them at least a glimmer of hope that the defunct MOL might eventually be revived. Consequently, it was relatively simple for Tarbox's scavengers to collect and consolidate the modules at a massive hangar located at North Island Naval Air Station in San Diego.

In one fell swoop, by saving the invaluable components from the scrapyard, Tarbox amassed the key building blocks necessary to assemble the Navy's MOL. Even as the Air Force's MOL effort was blowing apart faster than a high-speed crash on a Formula One racetrack, Tarbox was able to convince key leaders of the pressing need for a manned ocean surveillance platform. The hardest sell was persuading them to allow a nuclear power plant in orbit, but once Tarbox cleared that hurdle, his secret space program was quickly underway.

Although the catastrophic event in August had been a setback, the Navy's clandestine MOL program was still active. A second ocean surveillance MOL was currently in production and would likely be ready for flight by next year. Even though one Navy astronaut had died in orbit and the other—Russo—only rescued as the result of Ourecky's brave actions, the first mission was not considered a failure. After all, most of the systems aboard the Navy's MOL worked exactly as advertised. The station's nuclear reactor performed flawlessly, and the various reconnaissance systems had gathered a treasure trove of data. The communications failure was isolated and fixed, and redundant communications systems were being installed as well. Certainly, it was not a foregone conclusion that the second MOL mission would fly, but it was looking increasingly more likely with every day that passed.

8:00 a.m.

Tew was on the verge of dozing when he heard Wolcott's voice. "Takin' a siesta this early?"

Blinking, Tew replied, "Just giving my eyes a break, Virgil." Wolcott was accompanied by Tarbox and Colonel Ted Seibert, the Project's intelligence officer. The three men had scarcely changed in the past few years. Adorned in his starched jeans, denim shirt, hand-tooled leather belt with turquoise-encrusted silver buckle and white Stetson hat, Virgil Wolcott still looked every bit like an extra in a Roy Rogers movie. Wolcott's weather-beaten face may have acquired an additional crinkle or two, but otherwise he had hardly aged at all.

White-haired and rail-thin, Tarbox still appeared to be an evil gnome. In tweed jacket and button-down Oxford shirt, Seibert seemed comfortable in the guise of a dapper Ivy League professor. Tew suspected that if he were to relax the grooming standards ever so slightly, the intelligence officer would probably grow a modest goatee to complete the transformation. Tew wondered if the intelligence officer would ever be comfortable wearing a uniform again once the Project folded.

"It's a light agenda this morning, General," said Seibert, quickly passing out briefing folders. "Just an update on the *Krepost*. Unfortunately, there are several discrepancies in the reporting, and we need to discuss the implications."

"Discrepancies?" asked Tew.

"We're receiving a lot of contradictory information," explained Seibert. "Particularly concerning the Soviets' readiness to fire a Proton rocket."

Proton? Tew remembered that *Proton* was the Soviets' new name for the massive hypergolic-fueled booster formerly called the UR-500. "I would think that this would be very clear-cut," he noted.

"It should be, General, but it's not. These are national level sources," explained Seibert. He placed two intelligence dispatches side by side and gestured at a string of numbers printed at the top of each document. "These are unique tracking numbers, which do lend us some limited insight into the origin of the information. This material is extensively filtered before it reaches us, but my guys have scoured these documents, and they assure me that the bulk of this information is reported by just two sources.

"Unfortunately, by the time their information is finally funneled to us, it's been entirely scrubbed of any details that might reveal the identity of the sources, who they work for, or where they work. But just for the sake of discussion, let's call them Source One and Source Two."

Seibert continued. "So, here's our tale of two sources. On one hand, Source One emphatically states that there will not be another Proton launch for at least another six months. And on the other hand, Source Two insists that this *Krepost* is in the final processing stages, and that a Proton booster is ready for launch right now."

"So, which do you believe?" demanded Tarbox.

Seibert placed a series of black-and-white satellite reconnaissance photographs on the table. "Admiral, I put my greatest faith in those things that I can see. This is the latest overhead imagery available for Baikonur. These particular shots depict the launch complex for the Proton booster. It's currently being repaired and rehabilitated after the shot in July. If you look closely, you can see that some of the gantry components have been removed entirely."

"So your assessment is?" asked Tew.

Seibert shook his head. "In short, General, the Soviets aren't going to launch a Proton from here, not now or in the relatively near future. These images validate what Source One indicates, that's there's not going to be another Proton launch for at least another six months. Moreover, Source One provided us advance notice of the Proton launch in July, but Source Two didn't make even the slightest peep about it. I don't know where Source Two is, but I have to believe that Source One is in a position to directly observe launch preparations, so I have a lot more confidence in his reports. He's proven to be exceptionally reliable."

Tew nodded and said, "That seems like a very logical assumption."

"And there something else that bothers me about Source Two," said Seibert, lighting his briarwood pipe with a wooden "strike-anywhere" match. As he puffed on the pipe to ignite the fragrant-smelling tobacco, he casually squeezed the matchhead between his fingertips to extinguish its dying flame.

"Another burr under your saddle?" asked Wolcott.

Seibert shrugged his shoulders and said, "Until this past March, we hadn't heard any chatter concerning this supposed *Krepost* orbital bombardment system. There wasn't the slightest trickle, and then suddenly we're drowning in a torrent, and it's all flowing from this one guy. Moreover, he reports like clockwork, once a week, every week, on the button, every time."

"That ain't a good thing, Ted?" asked Wolcott, fanning himself with his Stetson.

"Well, Virgil, if he were a millworker punching a time clock, punctuality would certainly be a positive attribute, but in these circumstances, that kind of regularity is a little suspect."

"How so?" asked Tew.

"The Soviets are absolutely ruthless when it comes to counter-intelligence. They relentlessly monitor anybody and everybody, regardless of rank or position. Anyone who has routine access to this sort of information, the stuff that Source Two is reporting, is going to be under active surveillance.

"You just have to imagine security at Baikonur is airtight. Their counter-intelligence people have to be watching him like a hawk. Moreover, he *has* to know that, unless he's an absolute idiot. So, if he's passing messages through a dead drop of some sort, which is the most likely way that the information is getting out, it's improbable that he could hit every single one of his contact windows, every single week, without getting spooked at least once or twice."

"Then what do you make of the information that he's reporting?" asked Tew.

"I think this intel is bogus, General. I think it's just disinformation specifically designed to get us agitated about a threat that does not exist," said Seibert. "There's absolutely nothing to back it up. I think it would be extremely foolhardy for us to put much stock in it."

"I agree, but as much as we want to assume that this intel is erroneous or that it's an outright lure," said Tew, "there's always the possibility that it's credible. While we might have at least six months to prepare for the last mission, we still have to be ready to execute on short notice."

"Yup," said Wolcott. "The fact is that we could only count on Carson and Ourecky to fly the mission if this danged *Krepost* just dropped in our lap tomorrow."

"As much as that bothers me, I'll concede that point," said Tew.

Tarbox spoke. "Mark, assuming that we will have six months of lead time, maybe I can contribute one of my crews."

Tew grimaced; to him, the admiral's high-pitched voice was like the annoying squeak of a perpetually ungreased wheel. He cleared his throat to immediately decline, as was his habit with anything Tarbox brought to the table, but then fell silent as he contemplated the proposal.

Clearly, Tarbox wanted to snatch at least some of Blue Gemini's glory for himself and the Navy. Grabbing the laurels for the last mission—the critical task that was the very reason for Blue Gemini—would be almost as good as leading the Project from the beginning. Tarbox's seemingly magnanimous offer was a blatant ploy to wedge his feet further in the door, and it was likely only a matter of time before he started to pull the requisite political strings to shove his crew to the front of the line.

As much as Tew despised the notion of allowing a Navy crew to destroy the target that had so far eluded them, there was certainly adequate time to bring them up to speed, if Seibert's assessment was correct. Having a Navy crew execute the hazardous mission would greatly increase the odds that Carson and Ourecky would survive Blue Gemini. As obsessed as he was with destroying the nefarious *Krepost* menace as soon as it appeared, he was equally obsessed with protecting Carson and especially Ourecky. *Consequently*, he thought, *maybe Tarbox's scheme wasn't such a bad idea after all.*

Tarbox spoke again. "Mark, I'm not sure if you heard me, but I can contribute one of my crews..."

"*Excellent* idea, Leon," replied Tew. "I would be delighted for you to bring in one of your crews to back up Carson and Ourecky."

With his mouth agape, Wolcott seemed shocked that Tew would be so receptive to *any* suggestion offered by Tarbox. He cupped his ear, as if he was uncertain of his hearing or that he was listening for the faint sounds of trumpets heralding the Apocalypse. "Are you sure, Mark?" he asked. "After all, we ain't completely played out our own roster. We still have Jackson and Sigler. They can back Carson and Ourecky."

"Correct, but I've sent those two into orbit twice already, and they've failed me twice. At this juncture, I'm open to alternative solutions, and the Admiral is certainly offering us a viable option. Virgil, since we've all but shut down the simulator facility, why don't you take Gunter over there today to make sure we knock down the cobwebs to accommodate our Navy guests?"

"Will do, boss," answered Wolcott.

"Unfortunately, although I'm confident that the Admiral's men can execute this task, there's a compatibility issue that has to be resolved," commented Tew.

Wolcott snorted, lightly slapping the surface of the table. "The *suits*," he exclaimed. "I plumb forgot about the suits."

Since the Blue Gemini repertoire had expanded to include EVA—extra-vehicular activity—the astronauts had to switch from wearing standard Nomex flight suits back to wearing pressure suits. More specifically, they were fitted with a modified version of NASA'S G4C suits, manufactured by the David Clark Company, which were equipped with extra layers and other features to protect the wearer as he operated outside the spacecraft. The Navy crews wore suits developed by Hamilton Standard for the Air Force MOL program. While the Hamilton-Standard suits were far easier to don and doff, and were considerably more flexible, the Navy version didn't offer adequate protection for operating outside the spacecraft.

Wolcott smirked slightly, obviously indicating that he glimpsed Tew's hole card, the one that would swiftly stymy Tarbox's attempt to fold the Navy crew into the Blue Gemini mix.

"The suit issue is challenging but not a show-stopper," noted Tew. "There are sufficient funds in our budget to fit your guys with our suits. Besides, it makes perfect sense to have another crew properly equipped, just in case Blue Gemini is extended into a second phase."

"Excellent point," noted Tarbox. He unlatched his attaché case, flipped it open and riffled through several folders.

Tew had seen this same drill so many times that he suspected that the admiral's attaché was literally stuffed with a multitude of schemes, each prepared for a particularly opportune moment. If nothing else, the Ancient Mariner was incredibly adroit at seizing the initiative and consistently came prepared for almost any contingency.

The admiral retrieved a folder bearing several high-level classification markings, opened it, and handed a photograph to Tew. "Since you suddenly seem amenable to my suggestions, Mark, and since we now apparently have the unexpected luxury of time, I would like to propose a temporary project for Major Carson," he said. "With your permission, of course."

Tew donned his reading glasses to examine the image. Startled, he blinked several times, as it took a moment for him to absorb the incongruous aspects of the picture. The long-winged black spy plane was unmistakable, but it looked entirely out of place in the unusual setting. "That's really a U-2?" he asked.

"It is," confirmed Tarbox. "But to be more exact, it's a U-2 with some special modifications."

"Obviously." Tew handed the photograph to Seibert. "Is this real?" he asked.

"It is, sir," answered Seibert. "They started experimenting with that capability in 1963. It's been operational for a few years now."

"Amazing," noted Tew, looking toward Tarbox. "But as fascinating as this is, just what does this have to do with Carson?"

Tarbox explained his plan, concluding with, "Carson shouldn't be out of pocket for more than a few weeks, plus we can recall him on short order, if need be."

"If nothing else, it sounds harmless enough, but if you're right, it might be just the ticket we're looking for. I think Carson would be the perfect fit for this, but we obviously have to ensure that he doesn't draw any undue attention to himself."

"Agreed. I'll make sure that he remains incognito, although we might have to take some extraordinary measures."

"Clearly."

"Then I have your approval, General?" asked Tarbox.

"You do, Admiral. I concur wholeheartedly."

Evidently confounded by Tew's behavior, Wolcott's weather-beaten face bore a quizzical expression. "Are you feelin' all right, Mark?" he asked. "You ain't missed any of your medications today, have you?"

"I'm fine," replied Tew curtly. "I don't know why you would think otherwise, Virgil. Now, let's move on with this discussion."

Simulator Facility
Aerospace Support Project
10:45 a.m., Wednesday, August 30, 1972

Seated beside Gunter Heydrich in the simulator hangar, Wolcott observed as several men restarted and tested the simulation systems. Feeling the onset of a migraine, he shook two aspirin tablets from a flat metal tin, swallowed them, and chased them with a slug of cold coffee from a paper cup. Leaning back in his chair, he opened a foil-lined Red Man pouch and jammed a thick lump of chewing tobacco into his mouth.

Nudging his elbow, Heydrich said, "Virgil, it looks like you have a guest."

Wolcott swiveled his head to glimpse Tarbox beckoning him from behind the last row of desks and consoles. He removed his headset, donned his white Stetson, slowly climbed out of his chair, stretched, and joined the gaunt admiral at the back of the hangar. Standing next to the bank of refrigerator-sized computers that ran the simulator, the two men spoke just loud enough to be heard over the whir of spinning tape reels.

"That was a mighty big bomb you dropped back there in the intelligence briefing," said Wolcott. "*Your* folks backing up ours?" Shucks, Leon, this is a mite confoundin'. For the life of me, I can't quite understand how and why Mark could be so danged quick to agree."

"Does it matter?" asked Tarbox. "Actually, Virgil, I'm not interested in my crew backing up Carson and Ourecky. I want them to fly the mission, and I want your assistance in convincing Mark to do just that."

Snorting, Wolcott laughed so hard that he lost the wad of chewing tobacco tucked in his lower lip. Regaining his composure, he wiped tears from his eyes before bending over to retrieve the brown lump and flick it into a nearby trash can. "You want your Navy boys to *fly* the mission? And you want me to persuade Mark to do that? Don't you s'pose that's a bit of a stretch, Leon? Mark will *never* bless off on that scheme."

"Hear me out," snapped Tarbox. "For starters, Ourecky might not even be physically ready when and if this mission flies. Why waste simulator time on him and Carson if there's even the remotest possibility that they won't be able to go?"

"Good point. I strongly doubt that Mark will chomp on that lure, but it is a good argument."

"Virgil, this is all a matter of timing," asserted Tarbox. "Mark is obsessed with flying this last mission, but after it's over, he's planning to fade into the wings. If you and I don't seize the initiative, this whole joint effort could fall apart even before it begins, regardless of whether Mark's last mission is successful or not. There are a lot of things at play right now. The war in Vietnam is all but over. NASA will wrap up Apollo after the last mission flies in December. Most importantly, the Presidential election is in November; if Nixon wins, then we continue on. If McGovern wins…"

"We're *done*," groaned Wolcott, contemplating the potential outcome at the polls. "Hell, if this last mission ain't flown before he's inaugurated in January, McGovern could scrub that also."

Tarbox nodded. "And that's why we need to fly it as soon as possible, as soon as we know the *Krepost* is overhead, regardless of whether it's occupied or not. Sure, we'll have to accept some risk, but it's no riskier than what we order men to do every single day overseas. But even if we're willing to accept risk, we *can't* risk Carson and Ourecky."

"Why?"

"Because Carson and Ourecky are genuine heroes. I know that, and plenty of other people know it also."

Wolcott gazed upwards at the hangar's curved ceiling. "Heroes? Well, shucks, I certainly agree, but..."

"Virgil, if we really want to move forward, we'll *need* heroes. More specifically, we'll need *live* heroes. At this point, those two have given enough to this effort, much more than anybody should ever ask of them. The truth is, they're like money in the bank to us, and I don't care to squander that capital on this *Krepost* mission, particularly when there could be so much risk associated with it."

"So you're comfortable with offerin' up your own men for sacrifice?" asked Wolcott, twisting the silver tip of his bolo string tie.

"I'm comfortable with offering up my men to execute a dangerous mission," countered Tarbox gruffly.

"Okay, but no matter how you try to sell this horse, Mark will never buy it."

"And that's just one of the reasons that Mark Tew needs to *go*," said Tarbox candidly. "Face it, Virgil, he may be your friend, but he's standing in the way of progress."

Wolcott removed his Stetson and shook his head. "Mark has to *go*? I thought this was about jumpin' your crew to the front of the line, but this is a whole 'nuther issue," he said. "I ain't too comfortable about goin' behind Mark's back. I like you, Leon, and I'm downright anxious to work for you, but Mark and I have a long history, and I ain't too keen to betray him."

"I'm not asking you to betray him. Right now, when the time comes, all I'm asking is that you help me convince him to fly the mission with a Navy crew. Maybe later, you can eventually persuade him to step aside, for the sake of everyone involved. After all, the transition plan to fuse our programs has already been approved, Virgil. The only obstacle is Mark Tew. Instead of putting the transition on the back burner as we focus on this threat, I think it makes more sense to *accelerate* it."

"Okay, just for the sake of the argument, let's assume you can entice Mark into flyin' your crew. What's in it for Carson and Ourecky? What can you do for them?"

"Plenty," answered Tarbox. "Carson still wants to fly in Vietnam, doesn't he?"

Reflecting on the number of times he had denied the tenacious pilot's request, Wolcott chuckled. "Yep, that goes without sayin', but we've discussed this already. Carson's records are flagged so that ornery varmint can't fly anywhere even remotely near a combat zone, so it ain't going to happen. That's straight from the Chief of Staff of the Air Force. I know you had a scheme, Leon, but..."

"It's already done," averred Tarbox. "I've already laid the ground-work. I can make it happen regardless of whether his records are flagged or not. One phone call, and he'll be on his way."

"Okay, let's assume that you sway Carson. How about Ourecky?"

Smiling, speaking softly in his squeaky voice, Tarbox enthusiastically outlined the second part of his ambitious plan for the pair's future. He concluded by saying, "And it should be obvious that it would not only benefit them, but we also stand to gain as well."

"Whew. *Very* interesting," noted Wolcott, kneading his weathered hands as he contemplated the idea. "If nothing else, I have no doubt that you can hook those two. Carson and Ourecky would jump on *that* opportunity like buzzards on a day-old dead calf."

"Perhaps, but I certainly can't tell them about the second phase until after the election. I don't want to get their hopes up until we're sure that it's actually going to happen."

"I concur, but you can't broach *any* of this with Mark Tew. He would never agree."

Tarbox nodded. "So, when the time comes, will you help me convince him to put my Navy boys in the game, Virgil?"

Displaying his crooked teeth, grinning like a hungry possum in a henhouse, Wolcott replied, "Yep. I'm your huckleberry, Leon."

Aerospace Support Project Headquarters
12:05 p.m., Wednesday, August 30, 1972

Strolling across the parking lot toward his new Corvette convertible, Carson was on his way to grab lunch at the BX snack bar when he spotted Wolcott waiting in ambush. Wolcott was attired in his customary outfit of pressed jeans, a white cowboy shirt with pearl buttons, topped by his trademark white Stetson.

"New 'Vette?" asked Wolcott, running his pale hand along the fabric convertible top.

"Brand new. I just bought it last week."

"Quite a mount. How do you like it?"

"It's okay. Handles well, but the engine is only a 350 cubic inch small block, and my old one had the big block 427. It's taking me a while to get used to."

Admiring the sports car, Wolcott nodded. Their conversation was drowned out by a C-5 Galaxy cargo plane making a low fly-over.

Removing his Stetson, Wolcott squinted as he gazed skyward in awe. A thin string of brown-tinged saliva dribbled from the corner of his mouth. As the massive transport receded in the distance and the noise diminished, he spat and then said, "Durn, that thing is *huge*. I s'pose I'll never get used to these new big aircraft. After Germany and Korea, it's just too hard for me to comprehend flyin' in anything that has more leg room than a shoebox."

"I know just what you mean, Virg," replied Carson, thinking of the Gemini-I's tight fit.

"I s'pose you do, son."

"So, Virgil, am I to assume this is not a social visit?" asked Carson, extricating his keys from the pocket of his tan chinos.

Wolcott replaced his hat, thumped the wide brim, and answered, "You would be assumin' correct, pard. I come on behalf of Admiral Tarbox. I'm here to ask you for a favor."

"A favor?" asked Carson. "Wouldn't an order suffice?"

"What he's askin' ain't exactly somethin' we can order you to do, particularly considering that you work for General Tew, at least until Mark finally surrenders the reins to Tarbox, and he doesn't appear to be in any big hurry to do that."

"That's the impression I get, also," observed Carson. "Is that a problem, Virg?"

Wolcott nodded. "Trust me, pard, it's in everyone's best interests that we move ahead with the merger of this project. The faster that Admiral Tarbox takes over this ranch, the better. The problem is that Mark is obsessed with these damned Soviet nukes in space, and he's not willing to step aside until we fly this final mission."

"But why is that even an issue, Virgil?"

"Because Mark ain't thinkin' past this mission. And pard, although we assume that we will fly this mission, the truth is that this whole durned thing could fizzle out before we even get the go, depending on who's elected President in November. I think it's a foregone conclusion that we'll be shuttin' the doors in January if George McGovern lands in the White House."

Arching his eyebrows, Carson asked, "But if that's the case, why would it make any difference whether Tarbox takes over now or later?"

"Because Admiral Tarbox has big plans for this project, and there's a good chance that he can accumulate some momentum behind those plans if he's given half a chance. There's a possibility that we might even survive McGovern if Tarbox has enough of a head start to secure funding and support."

"So, Virg, that's all well and good, but how does that affect me? I assume that you're aware that I am moving on to other endeavors as soon as we're done with this last flight."

"Carson, son, Tarbox's plans include *you*. And Ourecky, too."

"Really?" asked Carson. "Then if you came here to ask me a favor, Virgil, you're barking up the wrong tree. As far as this Project is concerned, once we make this last flight, Scott and I are *done*. Sorry. I don't see how I can help you. Furthermore, even if I could do whatever you're asking, I sure can't envision what you could offer in return."

Wolcott grinned. "Well, son, you're right. Personally, I don't have anything to offer you. On the other hand, Admiral Tarbox has *plenty* to offer you. You and Ourecky both."

"Tarbox? Whatever he's peddling, I'm not interested."

"Shucks, Carson, don't be too hasty to look a gift horse in the mouth. You know, brother, Mark Tew's plan was that you go to Germany to command an F-4 squadron, and when things cool off sufficiently, we'll stick you on the roster to fly in Vietnam. Is that still what you want, Carson?"

"To go to Germany, Virg?" asked Carson.

"No, pard. Do you still want to fly in combat?"

Carson sniffed. "Do I want I to fly in combat? *Yes.* But why are we even discussing this? We've been down this road a thousand times, Virgil, and you've made it abundantly clear that you can't help me."

Wolcott chuckled. "You never change, do you? Let me tell you, son, I have it on extremely good authority that we'll be out of Vietnam before the end of this year, so that clock is tickin' out fast. With that said, what if Admiral Tarbox could make it happen for you?"

"For me to fly in combat?" asked Carson.

"Yep," replied Wolcott. "Trust me, son, all he has to do is pick up the phone and it's all but done. But let me ask you, why is it so danged important for you to fly in Vietnam?"

Carson frowned. "Hell, Virgil, you know *why*. If I don't fly over there, then it's highly unlikely that I'll ever make it beyond full bird colonel, unless we get in another shooting war very soon, and I don't see that happening."

"So, ultimately, you want to be a general, right?"

Carson nodded. "Of course."

"Well, *listen* to me, Carson: Admiral Tarbox can tug some mighty long strings on your behalf. Trust me, we're talkin' about opportunities that you could hardly imagine. Give him half a chance, and you'll eventually see those stars on your shoulders."

"Be specific. What kind of opportunities are you talking about?"

"I can't tell you anything now. You'll just have to trust me, but I assure you, flyin' overseas is small potatoes compared to what Tarbox ultimately has in store for you."

"But, Virg, why would he go to all of this effort?"

"Simple. He's beholden to you, son. He's indebted to you and your buddy, Ourecky, because you pulled his ass out of a king-sized frying pan, and he wants to make good with you."

Carson shook his head and asked, "Really? So am I going to have any choice in what happens, or is everything just going to be foisted on me?"

"If you do what I ask, brother, I *promise* you that we're never going to obligate you to do *anything* that you don't want to do. If you don't want to participate, for whatever reason, you won't. The same deal holds true for your compadre. When the time comes, and we make you two an offer, if Ourecky is still hankerin' to go to MIT, then he'll go. No questions asked. You can write that up on the wall, pard, in big letters, because I'm givin' you my word, and I won't let Tarbox or anyone else do anything that contradicts your wishes. *Savvy?* Are we square, son?"

"We're square, Virgil," answered Carson. "But I have to assume that the Admiral will eventually want something in return."

"At this point, all he wants...all *we* want is for you to keep an open mind about workin' with Admiral Tarbox. He sees a lot of opportunities comin' up on the horizon, and it would help a lot if you rode with us," said Wolcott. "Fair enough?"

"Plenty fair."

"Good," replied Wolcott. "You look like you're in a hurry to get somewhere, son, so don't let me hold you up."

"Thank you, sir," replied Carson, opening the red Corvette's door and sliding in behind the wheel. He started the car and pulled out of the parking lot. As he watched Wolcott slowly shrink in the rear view mirror, he wondered what the Ancient Mariner would want in return for his allegiance, and suspected that it was not a bargain that would come cheaply.

Pre-Launch Processing Facility
Burya Test Complex, Kapustin Yar Cosmodrome
7:17 a.m., Thursday, August 31, 1972

Major General Gregor Yohzin scratched Magnus between the ears before leaving the loyal Alsatian in the care of his driver. He then strolled into the massive horizontal assembly building where the hardware was being prepared for the upcoming inaugural mission of the *Krepost* nuclear weapons space station.

Much had transpired since he had been assigned to the *Krepost* effort over two years ago. At the onset of the project, Yohzin's boss— Lieutenant General Abdirov—had been granted the sprawling old *Burya* testing complex—located at the Kapustin Yar cosmodrome—as a research and testing facility.

Even though the *Krepost* would be designed and built at the *Burya* complex, the Soviet High Command's intent was that all spacecraft associated with the project—the *Krepost* stations themselves, as well as the Soyuz crew vehicles and resupply freighters—would be launched from the Baikonur cosmodrome located at Tyuratam in the Kazakh Republic. Each *Krepost* would be launched on an enormous UR-500 "Proton" booster, originally envisioned by the Chelomei bureau as a super-ICBM, fueled with hypergolic propellants. The Soyuz crew vehicles and freighters would be lofted by smaller R-7 "*Semyorka*" boosters. But Abdirov wanted everything together at one site. Exerting his considerable political influence, he insisted that new facilities be built at Kapustin Yar to accommodate launches of the Proton. But although he desired brand new facilities for his program, a compromise was struck to take advantage of the structures abandoned after the *Burya* project was cancelled in 1960. Construction workers gutted a massive brick building that was previously used to process the experimental ramjet cruise missile designed by the Lavochkin bureau. Its capacious interior was retrofitted with the heavy equipment necessary to assemble the Proton booster. Repurposing the defunct facility yielded an extra benefit: viewed from the outside, or perhaps from the vantage point of a reconnaissance satellite in orbit, the dilapidated structure looked like a derelict relic of a long-cancelled program. An existing rail network was also renovated to support transport of booster components from distant factories to the assembly building, and from the assembly building to the launch pad. Launch facilities for the reliable R-7 already existed at the *Burya* complex and had already been used for practice missions by prospective *Krepost* crews.

Yohzin envied the men laboring to ready the Proton. He wanted to be just like them again, with his hands on the metal, scrambling up tall ladders, dangling from rickety gantries, solving problems on the fly. He disliked being stationary and hated to be cooped up indoors for any extended period of time, but his current assignment entailed just that. To make matters even worse, his time was committed to a project that wasn't even his own, and if truth be told, he wasn't there to actually do anything productive, but to snoop.

Even as its Proton was being assembled, the *Krepost* station waited in a separate processing bay. Save for just a few components to be installed, perishable foods to be loaded, and final checks made on its nuclear warhead, it was ready for flight. In less than a week, it would be mated to the Proton, and then the completed rocket would be transported to its launch pad on a specially built railcar. In addition to the *Krepost* and the Proton, two Soyuz crew vehicles and one resupply freighter—a modified Soyuz—were being processed in their own separate bays in the assembly building.

Yohzin made his way to an isolated work area that was set aside exclusively for personnel associated with *Perimetr*. *Perimetr* was the massive automated network that would eventually control virtually all nuclear weapons in the Soviet inventory. The hardened network would ensure a devastating wave of retaliation against the West, even if the

Americans succeeded in decapitating the Soviet leadership and interdicting the more conventional channels of command and control.

The Soviet High Command had ordered that the *Krepost* stations be eventually linked to *Perimetr*, until the manned stations were eventually supplanted by unmanned *Skorpion* weapons platforms. Now, with one *Krepost* ready to launch and two more in production, the *Perimetr* link was still in development. An interim system—based on a device called the "interlock"—would be installed on this *Krepost*. Since his other pre-launch tasks had long ago been accomplished, Abdirov directed Yohzin to act as a liaison to ensure that the *Perimetr's* interlock system was ready for flight.

Despite his rank, Yohzin waited in line behind the others seeking entrance to the *Perimetr* workspace. When he reached the front of the queue, he handed his credentials to a stern-faced guard and then was subjected to an intensive search. Another guard examined his notebook and rifled through his belongings. They went so far as to open the lunch pail lovingly packed by his wife, Luba, to ensure that no camera or other recording devices were concealed in with the *olivie* potato salad, black bread and pickled herring that would be his noon meal. He knew that at the end of the workday, the process would be reversed. It was extremely demoralizing; he felt like a prisoner in a gulag.

Of course, if the fanatical *Perimetr* leadership learned of Yohzin's true intentions, then he might soon become an actual prisoner in a GRU cell. After all, his current task wasn't even real work, but an act of subterfuge. Abdirov was still very much intent on devising some means to circumvent the fail-safe weapons deployment system intended to prevent an unauthorized deployment of the massive nuclear warhead—now universally known as the "Egg"—that was the centerpiece of the *Krepost's* mission and architecture. The weapons deployment system consisted of two major components: the targeting computer and the interlock system. The targeting computer had been designed and fabricated by a team led by Yohzin, but since the High Command's ultimate intent was that the *Krepost* would be controlled under the *Perimetr* system, the design of the interlock fell under the purview of the *Perimetr* leadership.

Once the *Krepost* crew entered key data concerning a potential target—latitude, longitude, ground elevation and other variables—the targeting computer automatically calculated the angles and timing associated with firing the Egg's braking rockets to initiate reentry into the earth's atmosphere. If all went as planned, and there was no reason that it wouldn't, the Egg would eventually arrive at the appointed location and detonate at the appropriate altitude to ensure the maximum amount of destruction and casualties.

Although the *Perimetr* leadership strived to exercise complete control over all aspects of the *Krepost's* weapons deployment system, they reluctantly ceded development of the targeting computer to Yohzin's team. The underlying reason was simply that Yohzin's men had physical possession of the Gemini guidance computer—stolen from an American museum warehouse by the GRU—that was the basis for the Egg's targeting computer.

Merely possessing the Gemini machine was not the only reason that Yohzin's team maintained the lead role on the targeting computer effort. On orders from the Soviet High Command, Yohzin had demonstrated the Gemini computer to a select team of *Perimetr* engineers and had conducted extensive briefings about its design and capabilities. Early into the session, Yohzin came to realize that the *Perimetr* men could not grasp the digital computer's concepts. As much as they coveted it, the foreign machine was absolutely opaque to them; explaining it to those blockheads was as futile as describing a wristwatch to a mule. So Yohzin held sway over the targeting computer, and the *Perimetr* men kept their interlock.

Simply stated, the interlock functioned much like an ignition switch on a Western automobile. Its main components were a central station and two remote lock stations. Each of the three men possessed a unique key and personal code. Once in orbit, after they received an Emergency Action Message—a directive to deploy the Egg—each man would immediately report to his assigned station, insert his key, and enter his personal code; all three keys had to be turned and the codes entered within a set timeframe—just a few seconds—in order for the interlock to be disabled.

The three stations were sufficiently dispersed throughout the *Krepost*, so that two men could not overpower the third. Once the three personal codes were entered, the final step mandated that the mission commander enter a separate authorization code, which would be sent up from a ground station when an Emergency Action Message was issued. Unlike the personal codes that would remain constant throughout the duration of a mission, the authorization codes changed on a daily basis. The authorization codes were transmitted on a dedicated secure radio network operated by *Perimetr* and automatically uploaded to the interlock interface.

Once the interlock was disabled, the targeting computer passed the instructions to the Egg, at which time the Egg's internal computer took control over the station until the warhead had physically separated from the *Krepost*. The last step was for the mission commander to push an "Arm" button, which finalized the arming process and initiated the actual deployment. After the Arm button was pressed, there was little more for the cosmonauts to do but to hastily grab their belongings before beating a hasty retreat to the Soyuz spacecraft that would return them to Earth. Once the Arm button was pressed, the process could not be halted or reversed, unless a special recall code was issued by *Perimetr* headquarters and entered into the interlock.

In addition to the personal codes, the authorization code and the recall code, there was yet another code to be reckoned with, and that was the Independent Action Code. An Independent Action Code granted full control of the Egg to a single individual aboard the *Krepost*. No input was necessary from the other crewmen, assuming that they were still alive, and no authorization code—issued from a ground station—was required. It was clear to all involved that the High Command would not authorize an Independent Action Code to be issued except in the most exigent of circumstances. When such extreme power was entrusted to one man, it had to be assumed that the *Perimetr* network was on the verge of compromise, and that the entire world was standing at the brink of thermonuclear war.

Thus far, Abdirov's quest had been stymied by the tight veil of secrecy implemented by the *Perimetr* leadership, but the aftermath of last year's Soyuz *Yantar* accident, in which a three-man crew died when their Descent Module was accidently depressurized, afforded a potential avenue towards achieving his secret agenda. The *Krepost* station was configured for a three-man crew, but after the Soyuz *Yantar* tragedy, Soyuz spacecraft crews were limited to two men wearing spacesuits—to deal with the effects of emergency decompression—rather than three men clad in coveralls. Consequently, with the limitation imposed by the Soyuz used to ferry crews to the *Krepost*, the station complement was reduced from three men to two.

Since the original interlock system was expressly designed to accommodate three men, it had to be redesigned. Unfortunately—or fortunately, depending on one's perspective—the first redesigned interlock failed miserably. The redesign had taken the better part of a year, so after the first solution was discarded (and the lead designer sent east to make penance in frigid exile) the remaining engineers struggled to make up for lost time. Only two months ago, they had finally fabricated a working prototype. Normally, any device destined to fly on a Soviet spacecraft would be subjected to an intensive battery of pre-flight certification checks, but the *Perimetr* leadership unilaterally declared the new interface to be flight-ready, despite Abdirov's very vocal complaints to the High Command. After all, no one, but no one, wanted to be seen as the reluctant milquetoast who stalled the *Krepost's* maiden voyage.

Abdirov stood his ground, though, and the *Perimetr* leadership eventually conceded. Even though the interlock device would not be run through the full certification gauntlet, it would undergo a rigorous series of checks under Yohzin's fastidious supervision. And even though the *Perimetr* leadership insisted that Yohzin only see the exterior of the interlock box, Abdirov tenaciously demanded that his deputy be granted full access. So, granted Abdirov's sanction as his sole trusted agent, Yohzin was permitted to scrutinize the detailed schematics as well as the interior of the box.

Operating under the pretense that the interlock might eventually present a hazard to the crew, Yohzin frowned, fussed and fretted as they cycled it over and over and over, under all potential scenarios. With the interlock deactivated, at Yohzin's insistence, the *Perimetr* engineers entered the coordinates of almost every potential target in the catalog, as well as different firing solutions. They entered the coordinates correctly and then intentionally entered them incorrectly, substituting the latitude for the longitude, and vice versa, as well as setting the detonation altitude so that the Egg would theoretically explode thousands of feet below surface level. All the while, with the enclosure removed and the device's guts laid bare, Yohzin used an electronic thermometer to measure heat in the circuits, ostensibly to determine any potential aberrations that might lead to a short circuit or otherwise spark a fire aboard the *Krepost*.

He pretended to be skeptical, but in reality, it was difficult for Yohzin to disguise his admiration for their efforts. Their engineering was rock solid, and the robust device was well constructed, as if it had been assembled by master craftsmen. In theory, when Yohzin declared that he was satisfied, the tedious testing regimen would cease, and the stainless steel box would be hermetically closed with tamper-proof seals. Afterwards, technicians would install it in the *Krepost*.

Of course, the *Perimetr* engineers balked at having an interloper in their midst. Moreover, the haughty *Perimetr* people considered themselves as some sort of chosen elite and typically looked upon all RSVN officers with a very discernible degree of scorn. Even though he substantially outranked all of them, Yohzin did his best to act almost servile. It chafed at him to do so, but he set aside his disgust to gradually establish rapport with them. As he slowly won their trust and gained their confidence, he insinuated himself deeper into the testing process.

Given his unique position, Yohzin eventually realized that *Perimetr* engineers possessed some form of shorthand code that allowed them to swiftly disable and bypass the interface box.

The shortcut was apparently intended to save time in testing. As much as they grew to trust him, they were still cautious not to allow him to watch them enter the shortcut, but even as he turned his back to the box, he was able to count the clicks to determine that the sequence contained eight digits. He also discerned that the special code did not require inserting and turning a key to disable the interlock. He was confident that the special code was the skeleton key that Abdirov sought.

Although he was aware of the special code, gaining access to it was another matter. Although Yohzin had unfettered access to the interlock's schematics and technical documents, the *Perimetr* engineers jealously guarded their notebooks where the special code obviously resided. Circumventing their vigilance was a foray that demanded patience, so he formulated a plan, prepared his arsenal, and bided his time as he awaited an opportunity.

11:45 a.m.

By now accustomed to Yohzin's presence, three of the four *Perimetr* engineers had departed for lunch, leaving Yohzin with the most junior of the group, Aleksey Bogrov, a broomstick-thin bespectacled electrical engineer from Kiev. This had become a common occurrence, since the senior engineers would compel Bogrov to document the morning's sessions in the testing journal. The documentation process was an undesirable task, which entailed transcribing comments from a tape recorder which rarely worked correctly. Denied of his lunch break, Bogrov frequently did not eat until after he departed for the day, and often his stomach audibly growled when the sessions ran late into the evening.

Yohzin lingered in the workspace over lunch because it was such an aggravation to be searched as he left and then searched again as he returned. The *Perimetr* engineers were not subject to such indignities. Thankfully, he was fortunate that he had a loving spouse who saw fit to fix his lunch; as a bachelor who resided in a dormitory, Bogrov lacked that luxury.

Yohzin noshed on a chunk of black bread as he watched Bogrov manipulate the finicky tape recorder. Sensing the opening that he had long anticipated, Yohzin held out the small loaf and casually asked, "Some bread, Aleksey? It pains me that those ingrates leave you here to fend for yourself while they stuff their faces at the cafeteria."

"It's my job," replied Bogrov. "And the nature of the beast. Someday, some poor soul will take my place, and it will finally be my turn to go to lunch."

Yohzin nodded sympathetically.

"But I will have some of that bread, if you don't mind, Comrade General," said Bogrov, reaching out to accept the offering. "Thank you for your kind gesture."

Waving his hands over the other contents of his lunch pail, Yohzin said, "There's plenty here as well. Have some of Luba's *olivie*. It's delicious. Besides, it wouldn't hurt me to cut back a bit, since my trousers have been fitting so snugly as of late."

"Again, sir, *spasiba*," said Bogrov, sparing no time to pick up a spoon to sample the potato salad. "Your generosity is greatly appreciated." As he availed himself of Yohzin's other offerings, the engineer persisted in his efforts to force the tape recorder to function properly.

"It looks like you have your hands full, tinkering with that gadget. I love my Luba, but her bread always comes out of the oven a little dry for my tastes. I'm going to pour myself a glass of tea. Fancy some yourself?"

"Please, Comrade General," replied Bogrov, wiping dark crumbs from his lips. "That would be most kind of you."

Yohzin strolled across the room to the samovar and filled two glasses with steaming tea. He gently tugged a fine wire that removed a wax stopper from the end of a glass pipette concealed in the sleeve of his woolen uniform coat, and carefully dispensed four drops of the phenolphthalein into Bogrov's glass. He had acquired the chemical yesterday from the fuels testing laboratory. Colorless and tasteless, phenolphthalein was a powerful laxative. The surreptitious dispenser was of his own design.

He returned to the bench and handed the glass to Bogrov. The engineer took a brief pause from his chores to enjoy the tea.

Feigning grave concern, Yohzin implored Bogrov with technical questions about how the interlock box connected with *Perimetr's* secure radio network. Bogrov did his best to field the questions that were valid and deflect those inquiries that fell outside Yohzin's purview. All the while, as he gradually amplified the level of his inquisition, Yohzin carefully watched the engineer, confident that the phenolphthalein was taking effect.

Only a few minutes passed until it was blatantly obvious that the inspector was physically uncomfortable. Beads of perspiration appeared on Bogrov's forehead, and his complexion gradually took on an ashen cast. His face was etched in pain as he squirmed on his stool. Yohzin heard the faint sounds of Bogrov's stomach churning and could only imagine the cramps and convulsions that were gripping the engineer's guts. Yohzin's scheme was to pester the inspector right up until the point he reached critical mass.

"But are you *absolutely* sure that this interface will work as it is intended?" asked Yohzin.

Bogrov sipped the rest of his tea, set the glass aside, and said, "Begging your pardon, Comrade General, but this interlock interface is not your concern."

"Granted, Aleksey," replied Yohzin, dunking a scrap of bread into his tea. "That box is your concern, but I have to be confident that it will present our crew with no undue hazards."

"While I understand your skepticism, Comrade General, I can assure you that it has been appropriately certified." And finally, the moment that Yohzin had waited for. The anxious engineer abruptly swiveled around on his stool and blurted, "I am sorry, Comrade General, but I *must* go to the latrine immediately." His hasty departure was rather comedic; bending sharply at the waist, he waddled off like a penguin stricken with severe indigestion.

Yohzin furtively glanced around before casually flipping open Bogrov's notebook. He quickly found the dog-eared page that contained the special code. He opened his own notebook, found a table that he had prepared expressly for this moment, and jotted down the eight numbers—76810723—in the open spaces that awaited them. He confirmed his transcription and precisely replaced Bogrov's notebook on the benchtop. In mere seconds, the deed was all but done.

He sipped the remnants of his tea and smiled, but that smile was soon replaced by a frown as he contemplated the implications of his actions. He envisioned Bogrov squatting over the floor-hole in the latrine, shuddering and grunting as he painfully emptied his bowels, and realized that the final product of the engineer's laxative-fueled diarrhea might literally be the spasm that brought the world to a fiery end.

But there could be no turning back. He was compelled to disclose the code to Abdirov; it was not just a matter of loyalty, but practicality as well. He had known the general for decades, long before Abdirov became the scarred monstrosity that he was today, and was intimately familiar with his methods. With this knowledge, Yohzin could not delude himself into believing that he was the only one surreptitiously pursuing this objective. Abdirov often used a parallel approach in which two men or two teams or even multiple teams were challenged with solving the same problem, without having knowledge of the others' activities. Consequently, Yohzin was positive that someone else might stumble upon a solution, although one not necessarily as elegant and yet simple as the eight numbers now contained in his notebook.

He sincerely hoped that Abdirov just wanted the capability to bypass the interlock, and that he did not actually possess the desire to do so. But all other things aside, although Yohzin could not control the ultimate outcome, being the one to share the share the code might grant him at least some influence over some of the consequences.

As he waited for Bogrov or others to reappear, he remembered that there was another vitally pressing matter: *What should he communicate to the Americans?* He had to compose his weekly message tonight, which he would deliver to the dead drop tomorrow evening. Although he had already decided to inform them—yet again—that the *Krepost* launch was imminent, he was not sure whether he should let them know about this new development. After all, he really had no way to clearly divine

Abdirov's intentions. Moreover, just the mere suggestion that Abdirov might use the code to trigger a thermonuclear war might provoke the Americans to take their own preemptive action.

Yohzin's thoughts were interrupted by Bogrov's return to the workspace. The engineer's posture was considerably more upright now, and he seemed greatly relieved.

"Everything come out okay?" asked Yohzin.

"*Da*, Comrade General, but in the future, should you be so inclined as to offer it again, I feel that I should decline your wife's black bread. It just didn't sit well with me."

Aerospace Support Project Headquarters
10:18 a.m., Thursday, August 31, 1972

Summoned by Wolcott, Carson bounded down the stairs, swooshed past the assistant in the waiting room, and entered the office that Wolcott shared with Tew. "You called, sir?" he asked.

"Come on in, son," said Wolcott. "Admiral Tarbox has something to share with you, pard."

"Is General Tew not here today?" asked Carson, glancing towards Tew's vacant chair.

"Mark's at Walter Reed," answered Wolcott. "Routine check-up."

Routine check-up? thought Carson, taking a seat at the table across Wolcott and Tarbox. Tew's health was anything but routine. Tew's insistence on remaining at the Project was a mystery to all, since his prognosis was far from rosy.

Tarbox slid a black-and-white photograph across the table. Carson examined the image; it depicted a U-2 spy plane on the flight deck of an aircraft carrier. Several members of the carrier's deck crew surrounded the plane, and the carrier's tall "island" was prominently visible in the background.

Tarbox explained. "That's a U-2 aboard the Kitty Hawk in 1963, as part of Operation 'Whale Tail,' a feasibility test to evaluate whether U-2s could take off and land on aircraft carriers."

"Amazing," said Carson. "I had no idea…"

"Whale Tail was a resounding success," interjected Tarbox. "The upshot was that three U-2s—designated as U-2G's—were fitted with tail hooks and other modifications so that they could safely operate from aircraft carriers. Additionally, two groups of pilots were qualified to fly them from carriers."

"Navy pilots, sir?" asked Carson.

"You mean *Naval aviators?* No, Carson, they weren't Navy. The pilots were all civilians. In any event, although it's not public knowledge, the United States has the capacity to launch and recover U-2 reconnaissance aircraft at sea, should the need arise."

Carson nodded and then handed the photo back to Tarbox. "That's all very interesting, Admiral, but I'm not quite sure why you're showing me this."

"As you're aware, we're very concerned about the coverage of contingency landing zones in some areas of the world, particularly those regions which are predominately open ocean," explained Tarbox. "And we're also quite aware that the Gemini-I doesn't usually fare well in water landings, so my staff has proposed a series of tests to determine if an aircraft carrier could be employed for contingency landings, if no land-based sites were readily available."

Landing the Gemini-I on an aircraft carrier? thought Carson. *It doesn't get much more far-fetched than that.* While there might be some slight merit to the idea, Carson immediately recognized it as just another insidious scheme that would enable Tarbox to dig his claws deeper into Blue Gemini. Moreover, as Carson witnessed repeatedly in the past, this approach was Tarbox's tried and true method for a hard sale; the Admiral obviously knew that he would have no success pitching such a notion to Tew outright, so he sold the idea to subordinates first, before approaching the General.

"Honestly, sir, as interesting as this is, I really don't foresee General Tew buying off on this concept," said Carson.

"He already has, buck," noted Wolcott. "As a matter of fact, Mark bit on it hook, line and sinker. It's a done deal."

Dumbfounded, Carson didn't know how to respond. *Tew had agreed to this? It made no sense. And what could this possibly have to do with Vietnam?*

"Son, we're sendin' you to Pensacola to get carrier-qualified," said Wolcott. "You'll leave tomorrow."

"But, sir, I don't think…"

"Don't look a gift horse in the mouth, son," replied Wolcott. "There's a whole lot more to this story that you ain't seen yet. You need to hush up and hear out what the Admiral has to say. Just lend him a minute, and you'll see that you've won the danged jackpot. Savvy?"

"I'm listening, sir," said Carson.

Tarbox cleared his throat and spoke. "In all sincerity, Carson, I'm indebted to you and I'm a man who makes good on my debts. I am very aware that you want to fly in Vietnam. While I don't agree with you about many things, I do concur with your assertion that your Air Force career will suffer if you don't have that experience, because you will be competing against thousands of Air Force officers who do. So, I've developed a plan to secure your wishes.

"Let me make something abundantly clear, Carson: what we will describe stays in this room, between the three of us. If you discuss it with anyone, including Major Ourecky, then the entire deal comes off the table, and we will never make this offer again. Do you understand?"

"I do, sir."

"Good. You should be aware that the next six months are critical for several reasons, but three in particular. First, without delving into a lot of details, we are receiving very mixed messages from the intelligence community about what's happening on the other side of the Iron Curtain, specifically whether the Soviets are ready to launch this new *Krepost*. After extensive analysis, we've reached the conclusion that it will be at least six months before they are ready for another launch attempt."

"So, Ourecky and I have at least a six-month respite?" asked Carson. "With the intelligence that we've received about this *Krepost*, I can't see General Tew wasting the last stack on anything else."

"Correct," replied Tarbox. "Second, mark my words, our nation's direct involvement in the Vietnam War will end within the next six months. The peace talks are ongoing. Ground combat operations have already ceased, and air operations are being curtailed significantly. Air ops will continue to taper off, but it's a foregone conclusion that they will end altogether by Christmas."

"Then as far as I'm concerned, it's now or never, right?" asked Carson.

"Affirmative," replied Tarbox.

"But I still don't see how you can make this happen, sir."

"You'll see in due time, Carson, and that brings us to my third point. Although General Tew has agreed with sending you to Pensacola to undergo carrier quals to support this bogus feasibility study we've cooked up, it's a certainty that he would never agree to the rest of this plan. With that said, your deployment to Vietnam is contingent on his retirement. We are very confident that he will retire within the next six months, so we intend to assemble all the pieces of this plan so that the last phase can be executed on very short notice."

Carson swiveled to face Wolcott and asked, "Do you really think that General Tew will retire that soon?"

"I do, son," answered Wolcott. "Moreover, I sincerely hope that he does. His health is bad, Carson, much worse than you could imagine. He's literally livin' on borrowed time. And as much as I hate the notion that he might miss the last hurrah when we finally get to go up against this danged *Krepost,* he needs to make the most of what life he has left. All this stress is just killin' him, bit by bit."

"But, sir, even if General Tew retires in the next six months, the Chief of Staff of the Air Force has flagged my personnel records to prevent me from flying overseas. I don't see…"

"In due time, Carson," interjected Wolcott. "You'll see in due time."

"When and if you do fly in Vietnam," explained Tarbox, "and I am confident that you will, you will *not* fly as an Air Force officer. During that timeframe, you'll be a Naval aviator. Afterwards, we'll talk to the Chief and have your records amended accordingly, but all of this will be under the pretense that it's easier to beg for forgiveness than to ask for permission. After all, Carson, once this final mission happens, we all know that you can't be readily replaced, so the Chief may rap your knuckles, but there's little that he can do otherwise. And as for me and General Wolcott, there's not a lot that they can do to punish us.

"Major Carson, you should be aware that this will not be a casual endeavor by any means, and an abundance of people are working diligently on your behalf." Tapping his finger on an inch-thick stack of paperwork before him, Tarbox said, "I am working with some of my confederates at the Bureau of Naval Personnel. They're going through a fairly extensive paper drill to create a new set of records for you. Once we execute the final phases of this plan, as far as the Air Force is concerned, you will remain assigned here at Wright-Patterson. On the Navy side, you will be 'sheep-dipped' so that you'll undergo the remainder of this exercise as a Naval aviator."

"Yes, sir," said Carson.

"Provided that you submit to my terms, Major, this venture will take place in four evolutions," he stated concisely. "First, as we've already indicated, you'll proceed to Pensacola Naval Air Station in Florida for temporary assignment with Training Squadron VT-4 to undergo carrier qualifications. During your carrier quals, you will be certified to land and take off from a carrier at sea."

"Yes, sir," replied Carson.

"Before you get too enthusiastic, Major, let me explain a few things. You shouldn't expect any special treatment at Pensacola or anywhere else. At Pensacola, even though you'll be under an abbreviated syllabus, you will be training alongside brand new ensigns in Florida, and won't be treated any differently from them. And just so you're aware, when you report to Pensacola, you'll be in the guise of a Naval aviator, so we can establish your story for later on."

Tarbox continued. "You should be aware that virtually all Naval aviators undergo carrier quals during their initial flight training. In your case, we have worked out a logical exception for this lapse. According to your personnel binder, you graduated from West Point, and then went to Army fixed-wing flight training. After flying OV-1 Mohawks for a year, you transferred to the Navy where you went directly into the P-3 Orion program, and now you're moving from P-3's into fighters.

"The instructors at Pensacola will probably be a trifle curious about your background, but they know enough about Mohawks and P-3's to assume that you were probably involved in some sort of spook work, so they're not likely to ask too many questions. If they do, you'll refer them to my office, and they will be strongly encouraged not to be too inquisitive."

Frowning, the lanky admiral said sternly, "Let me warn you: Pensacola will offer plenty of distractions and opportunities to get neck-deep into trouble. You need to focus on your training. No hot dogging, no womanizing, no colorful antics, no flat hatting, no special treatment, no *nothing*. If you break character, or if you stand out too prominently from the background, the deal is off. Understood, Major?"

Nodding, Carson replied, "Yes, Admiral."

"Good. *If* you successfully complete your carrier quals, you'll come back here. Once General Tew retires, you'll immediately move on to Miramar Naval Air Station in California for the next evolution, where you will attend "Top Gun," the Navy Fighter Weapons School. The Fighter Weapons School is more or less a graduate-level course in air-to-air combat skills."

"Begging your pardon, Admiral," interjected Carson. "I'm not one to toot my own horn, but I do keep my hand in, and I'm already very proficient in air-to-air combat. Now, I understand the necessity of the carrier quals, but since time is of the essence, wouldn't it make more sense for me to skip Top Gun and jump into the fight as soon as possible?"

Shaking his head, Tarbox reached across the table, abruptly snatched the paper from Carson, and crumpled it. He turned towards Wolcott and said, "Did I not tell you that this was just a waste of time, Virgil?" He swiveled his head towards Carson. "The offer is *withdrawn*, forthwith. You are dismissed, Major."

"But, sir," muttered Carson.

"You are *dismissed*, Major," reiterated Tarbox. "If I am not mistaken, and I doubt that I am, '*dismissed*' connotes exactly the same thing in the Air Force as it does in the Navy."

Wearing an expression of shock, Carson slowly rose from his chair and saluted.

"Let's not be too hasty, Leon," said Wolcott. "Couldn't you see fit to afford Carson another pass?"

"Do you honestly think that this headstrong major can be patient enough to even *listen* to all the provisions of this plan before he sees fit to disagree with them?" growled Tarbox.

"Carson?" asked Wolcott. "Tarnations, son, can you not behave? The Admiral is really tryin' to do you a danged favor here. Can you at least hear him out?"

"I can, sir," said Carson meekly, still standing at rigid attention.

"Then have a seat, Major," ordered Tarbox.

As Carson settled back into his chair, Tarbox continued. "As I said, your first evolution is at Pensacola for carrier quals. Later, you will go to Miramar, where you will learn to fly and fight the *Navy* way. *If* you matriculate from Miramar, you'll proceed to a carrier-based fighter squadron, and then—*if* the squadron commander signs off on your aptitude and attitude—you'll be allowed to proceed to the fourth and final evolution: combat air operations in Southeast Asia."

Grinning, Carson replied, "Yes, sir."

"Before you get too excited, let me caution you that if you're expecting to jump into some hot dogfighting action, then you will be sorely disappointed. At this point, you can safely assume that your combat tour will probably be entirely uneventful, because frankly, Carson, the war is *over*. It's merely a formality that the North Vietnamese haven't signed the treaty in Paris, but that could literally happen any day now. In any event, I have it on the highest authority that US combat involvement will conclude on December 31.

"Fortunately for you, even though this boondoggle will be nothing more than a milk run, you will receive legitimate credit for flying in a combat zone. Initially, this will be reflected in your Navy personnel file, but once things settle down adequately, your official Air Force records will be amended to reflect that you served an exchange tour with the Navy, with the appropriate references to flying in an active combat zone."

"Thank you, sir," said Carson. "I promise that I won't disappoint you."

"See to it that you don't," said Tarbox sternly. "And Carson, one more thing..."

"Sir?"

Tarbox opened a folder, removed a black-and-white photograph, and held it out for Carson to view. "Do you recognize this man?"

Carson examined the official portrait from a personnel file and recognized the stern-faced Navy officer as one of his erstwhile dogfighting adversaries from a few years back. "Yes, sir, I know him. That's Lieutenant Commander Steve Billingsley. His call sign is..."

"*Badger*," interjected Tarbox in his squeaky voice. "Correct on all accounts, except one. Billingsley is no longer a Lieutenant Commander; he was just promoted to full Commander." Tarbox reached into his folder, pulled out a document, and slid it across the table to Carson.

Carson glanced at the paper; it was a set of promotion orders, elevating him to the rank of lieutenant colonel. He had been eligible for the promotion since July; it was about time that the orders were formally cut. "Thank you, sir!"

"Don't be too quick to thank me," said Tarbox, reaching over the table to pick up the orders and return them to his folder. "Your promotion will have to wait until you complete this assignment. Badger—*Commander* Billingsley—will be your squadron commander when you deploy overseas. If we promote you now, you two will be of equal rank, although only Badger and you will be aware of that fact. So, in order to make it absolutely clear who is in charge, we're holding your promotion. Understood?"

"Yes, sir," replied Carson. At this point, he would gladly accept a bust back to second lieutenant in exchange for the opportunity being presented to him, so waiting on the promotion to lieutenant colonel certainly wasn't a deal breaker by any means. Besides, an opportunity to fly *beside* Badger? *What could better than that?*" But still, something was missing. Tarbox would never make such a grand offer without expecting at least something in return.

"This is excellent, sir, but I have to believe that you will want something from me in return," said Carson.

"And you would be correct, pard," said Wolcott, twisting the silver tip of his bolo tie. "After you return from Pensacola, you'll go into a holdin' pattern here at Wright-Patt, until the stars align and we can launch you towards Vietnam. Admiral Tarbox here has been kind enough to lend us the services of one of his flight crews, since they are not gainfully employed. As you're waiting, you and Ourecky will work with that crew to bring them up to speed. With any luck, they will be proficient in time to fly the last mission. Savvy?"

"Yes, sir," replied Carson. As much as he dreaded flying yet another mission, he wasn't exactly keen about handing the reins to a Navy crew, since he and Ourecky had done all the hard work to bring them to this place.

"And one more thing," said Wolcott. "I'm sure that you're aware that Blue Gemini might be extended for a second phase. That ain't exactly a sure thing by any means, but if it happens, I want you to stay on. If you do, it's obvious that you'll be flyin' at least one or two more missions, but your main role will be to train the new crews that we've been promised."

"That certainly seems like a fair trade," noted Carson.

"Moreover," said Tarbox, "since Ourecky is such a critical part of this Project, we ask that you persuade him to stay on as well."

And there's the hook, thought Carson. "I'll talk to him," he said. "But you have to know that he wants to leave after the last mission, don't you? General Tew has already promised him that he will be transferred."

"True," observed Tarbox. "But we really don't expect that Mark Tew will still be here then, particularly if we don't get a shot at this target for another six months. As I said, son, I'm a man of my word, and I'll honor Mark's promise to Ourecky, if that's what Ourecky really wants. All I am asking of you is to talk to him, when the time comes, to persuade him to stay on. Is that not fair?"

"I suppose that it is, Admiral." Carson reviewed the developments in his mind. Tarbox was offering an opportunity to fly in combat, in exchange for him remaining at the Project and nudging Ourecky to do the same. It was a no-brainer. *What could he lose?*

Tarbox's high-pitched voice interrupted Carson's thoughts. "I'll need an answer today. Any questions?"

"Just one, sir," replied Carson, anxiously drawing a Skilcraft ballpoint pen from his shirt pocket. "Where do I sign?"

As if by deft sleight of hand, with a speed perhaps only surpassed by Mephistopheles sealing a deal with Faustus, Tarbox produced a contract. "Here," he said. "On this line."

Grinning, Carson abruptly sealed the deal.

"You'll report to Commander Billingsley—"Badger"—at Pensacola tomorrow morning, Carson," said Tarbox, slipping the executed document into a folder. You'll undergo accelerated training for carrier procedures under his supervision, and then you'll fall in with a group of new pilots undergoing carrier quals in the Gulf."

"Thank you, sir," replied Carson. "I can't tell you how much I appreciate this opportunity."

Wolcott cackled. "I admire your gumption, pard. I ain't never seen anyone so anxious to jump into a fight. I just think it's pretty danged ironic that we have to move heaven and earth to sneak you into a war zone when so many young folks are dodgin' the draft and runnin' to Canada to avoid it."

2

WALKING THE DOG

Residential Complex # 4
Znamensk (Kapustin Yar-1), Astrakhan Oblast, USSR
6:45 p.m., Friday, September 1, 1972

Relaxing in his favorite chair, Yohzin lit a cigarette, donned his reading glasses, and perused the document that Abdirov had given him yesterday. The translated report described the ambitious American endeavor to build a space transportation system—a "space shuttle"—to deliver heavy payloads to orbit. The shuttle program was only a few months old—President Nixon had officially announced it in January—and already NASA scientists were making tremendous headway.

Right now, Luba was in the kitchen, clearing the table and washing dishes, while his two sons applied themselves to their academic studies. Yohzin glanced up from his papers to observe Magnus lying by the front door, waiting for his evening walk. Staring at the door, the dog squirmed and fidgeted, as if silently pleading to go out. Yohzin examined his wristwatch as he jabbed his partially smoked cigarette butt into an ashtray.

"Not yet, *hund*," he said sternly. "Ten more minutes. Can you not be patient until then?"

7:03 p.m.

Minutes later, he and Luba ventured out on their evening walk in the large commons area set in the middle of the apartment complex. Except for three other couples, also out for air, and a gaggle of children playing soccer in the fading light, the grassy expanse was unoccupied. Yohzin and Luba paused to speak with a young couple. Pushing a baby stroller bearing their newborn daughter, the husband was a motorized infantry captain who supervised a security force at one of the missile test ranges. As Luba cooed over the infant, Yohzin congratulated the captain and wished the child a long and prosperous life.

Leaving the couple, they walked on. A flock of doves passed overhead; Luba crossed herself, reached into a pocket of her coat, and scattered a handful of breadcrumbs. Yohzin looked at her and sighed; even in her old gray coat and a dark green kerchief covering her auburn hair, she was just as beautiful now as she was on the day that they had met. He considered himself an extremely fortunate man to have found her after his first wife had passed. Looking back, he remembered that those were extremely dark days, a bleak time where he held out little hope for even the slightest semblance of lasting happiness.

After almost half an hour of strolling, Yohzin looked out into the distance, aligned himself with a landmark, and silently counted his steps as he paced out onto an open field. He stopped, pointed his finger at the ground, turned towards Magnus and quietly ordered, "*Hier. Sitz.*"

The impatient canine sat obediently and then softly whimpered as it awaited the next command.

"*Scheisse.*"

Squatting abruptly, Magnus eagerly complied with Yohzin's scatological order. He completed the task in mere seconds. Panting, wearing a canine expression of relief, the black and tan Alsatian whirled about and sat down. Vigorously wagging his tail, he dropped his head slightly, pointing his shiny nose at the steaming deposit as if it were a hard-won trophy offered for his master's approval.

"*Braver hund,*" observed Yohzin in German. "*Sehr gut.*"

Shaking her head in bewilderment, Luba giggled and then said, "I love you dearly, Gregor Mikhailovich, but sometimes you confound me. Why on earth are you so compelled to torture that poor dog like that?"

"*Discipline,*" he answered. "It's the foundation of all effective training."

"I would think it sufficient to bark at him with those German commands, but it's a little much that you insist on even dictating the precise timing of his bowel movements."

Ignoring her, he placed his hands on his hips as he stooped over to examine the fresh scat. He looked at it for over a minute, until he was satisfied, and then slowly stood erect.

"So are you still concerned that he might have worms?" asked Luba.

"Always, dear," he replied. As he dusted himself off and adjusted the brim of his hat, a twinkle of light caught his eye. Just a momentary flicker from the tiled rooftop of an apartment building, he suspected that it was a reflection from the binoculars of one of the clumsy GRU counter-espionage agents who kept a constant vigil on everyone's comings and goings. The goons were always on the prowl, constantly seeking even the slightest evidence of subversive activities. He smiled to himself; after all, there was nothing for the lurking spooks to see but a couple out on their evening stroll, walking the family dog, a clockwork occurrence on any given day.

Krepost Project Headquarters
9:05 a.m., Saturday, September 2, 1972

Yohzin quietly entered Abdirov's office, took a chair, and softly declared, "I have it."

"Have what?" replied Abdirov, glancing up from a blueprint. He had just returned from Moscow the night before, after briefing the General Staff of the High Command concerning the preparations for the *Krepost* launch.

"You had asked for some means to bypass the *Perimetr* system, Rustam. Here it is," said Yohzin, handing the disfigured general a small scrap of paper.

"This?" asked Abdirov. His brow furrowed as his examined the code. "Eight digits? 76810723? How can this be?"

Yohzin explained how he had observed the *Perimetr* engineers entering the special code that effectively disabled the interlock whenever they chose. He concluded by saying "I am confident that this bypass code is permanently hard-wired into the interlock's hardware. It is the key that you seek."

"*Confident?*" asked Abdirov. "*How* confident are you? Are you certain?"

"Honestly, after closely examining the schematics, I am about ninety-five percent confident that it cannot be removed from the interlock's inner workings. In fact, I strongly suspect that this is not just an interim code inserted into the hardware for test purposes, but that it is likely the Independent Action Code as well. If you think about it, there's no practical way for *Perimetr* to relay or otherwise update that code without the risk of compromise, so it has to be hard-wired into the interlock from the very outset."

"But just eight numbers?" said Abdirov. "It could not possibly be that simple."

"But it is," commented Yohzin.

"So, how were you able to accomplish this?" asked Abdirov. "I would have thought that our esteemed *Perimetr* colleagues would have guarded their treasures with a little more care."

"And they do," replied Yohzin. "But as fate would have it, one of the *Perimetr* engineers was stricken with a severe case of diarrhea recently and neglected to pick up his notebook from his workstation when he rushed to the latrine. So, I was able to filch it with no one being the wiser."

"And am I to assume that this man's bowel issues were not an accident?"

Yohzin grinned. "Perhaps."

Abdirov smiled. "That's what I like about you, Gregor Mikhailovich, besides your abundant intellect. I greatly appreciate your willingness to exercise the initiative."

"Thank you, sir. As always, I am glad to be of service."

Abdirov chuckled and then exclaimed, "Such a diabolical caper! Such cunning and guile you possess! Perhaps I should keep a closer eye on you myself, lest you steal *my* secrets."

Yohzin felt his heart stop for a moment. Remaining calm, he replied, "Perhaps you should, friend, because if I can pierce *Perimetr's* security and purloin their secrets, then I could steal yours as well and sell them to the Americans. I would not surrender them cheaply, though; I would insist on top dollar, of course!"

Abdirov laughed uproariously, so much so that tears flowed from his single eye. Regaining his composure, he said, "Hah! Gregor, I fear that you might sell my secrets to your damned German friends, but the Americans, *never*."

And now, time for the moment of truth, thought Yohzin. "So, do you still intend to use this thing to force a confrontation between East and West?" he asked hesitantly. "Do you still plan to just drop the Egg and see what becomes of it?"

"That's a very plausible scenario," stated Abdirov. "But truthfully, I have reconsidered the entire idea."

Yohzin breathed a silent sigh of relief. *Perhaps his old friend Abdirov had finally come to his senses.* "You've reconsidered? How so?"

"As I've always told you, the Egg could certainly be the catalyst that initiates the reaction that breaks this damned deadlock," said Abdirov. "But what good is that, if we inadvertently allow the Americans to gain the upper hand? We would be deceiving ourselves if we chose to believe that they didn't possess a formidable advantage in many areas, particularly nuclear-armed submarines."

"Agreed."

"So, if I'm going to crack this Egg, doesn't it at least make sense to give *our* side at least some modicum of advantage? Assuming that this secret code of yours proves to be valid, and I've directed the crew to use it, then I intend to notify the High Command after the reentry process has been initiated and there's no turning back. That will lend them roughly forty-five minutes to formulate a plan and issue orders to the strategic forces."

Abdirov continued. "That's sufficient warning time to launch the first volley of ICBMs and prepare the rest, as well as enough time to get virtually every bomber into the air and headed towards America. They should also be able to launch every fighter and interceptor so they're positioned to bash the American B-52s out of the skies. Finally, we have a chance to land a knockout blow, even before the Americans get their feet planted under them!"

"Then you've made up your mind?" asked Yohzin.

"I have."

"When?"

"For various reasons, I will wait at least until after the crew receives their first resupply freighter," said Abdirov.

Yohzin nodded but was silent. If Abdirov was willing to wait that long, then perhaps there was still adequate time to persuade him away from this path of insanity. But he knew to tread lightly; even as he coaxed Abdirov to step back from the brink, he had to be cautious not to agitate the old man.

"This is truly momentous," said Abdirov solemnly, gazing at the scrap of paper that contained the code. "I am indebted to you, my brother."

Yohzin took a moment to compose his thoughts. Finally, he spoke hesitantly. "If you are truly indebted to me, sir, may I ask one favor?"

"Of course," replied Abdirov. "Anything."

"Dear friend, you know that you can rely on me to stay at my post no matter what comes," said Yohzin softly, like a condemned prisoner asking a king for dispensation. "But if you would be so kind as to grant me sufficient advance warning, I would like to make arrangements for Luba and my sons to visit her parents, and no one would be the wiser."

Obviously contemplating the request, Abdirov closed his sole eye and was silent for over a minute. Fearful of what ire his mentor might dispense, Yohzin gripped his thighs and willed his pounding heart to slow before it leapt from his chest.

Abdirov opened his eye and spoke in a soft voice that Yohzin could barely hear. "As I recall, Luba's parents live in a rural area southwest of Odessa, not too far from the Black Sea, correct?"

"*Da*," croaked Yohzin. "That is correct."

"And it is a very remote area, far removed from any military bases or industrial complexes, correct?"

"Again, sir, that is correct."

"Then certainly I will grant you that, one week's warning, provided that you keep this secret to yourself. Agreed?"

"Agreed, sir."

"And moreover, Gregor Mikhailovich, since you have effectively finished all of your chores associated with the *Krepost*, I will soon announce that you will be heading up the design team for that damned *Skorpion*. And because I am so pleased with your diligent labor on the *Krepost*, I strongly suspect that I might order you to take a furlough at the same time, so that you might accompany Luba and your boys. Do you understand?"

Swallowing deeply, Yohzin nodded and spoke. "*Spasiba.*"

Abdirov leaned forward and took Yohzin's hands in his. "Please hear me, brother. When that time comes, go and don't look back. After all, who could possibly refuse such a grand opportunity to visit their mother-in-law?"

10:33 a.m., Sunday, September 3, 1972

Reading an urgent dispatch from the Korolev bureau, Abdirov was livid. The High Command had explicitly ordered the exalted aerospace design bureau to directly support Abdirov's effort by providing a steady supply of Soyuz spacecraft to shuttle crews to the *Krepost*, as well as a specially modified version to serve as a freighter to keep the station replenished.

At this point, the Korolev bureau had already delivered the first freighter, which was currently being packed with supplies, as well as the first two Soyuz crew vehicles. The memo before him stated that production of the second freighter would likely be delayed, because some titanium welds had failed inspection. On a positive note, it also stated that production of the additional Soyuz crew vehicles was proceeding apace.

It was an aggravating development, and the manner in which the memo was delivered was aggravating as well. It was an unofficial "back channel" communication that was not on the official record. Theoretically, the Korolev bureau was doing him a favor by granting him advance notice of a potential delay. In reality, it obligated him to either delay the *Krepost* launch, under the assumption that the next freighter would not be delivered on schedule, or launch according to their current plan. In either event, he had to assume risk and also be prepared to accept blame.

As he contemplated the potential impact of the tersely worded memo, he recognized that timely delivery of the crew vehicles was largely irrelevant. In the current scheme, the crews would rotate every forty-two days, so if the crew vehicle production was delayed, or if there was a launch mishap, he could just simply order the crew on orbit to remain aloft. But while they might be able to keep their vigil indefinitely without relief, they certainly wouldn't fare very long without food, water, and oxygen. In order for the *Krepost* to remain operational, the freighters had to be launched at three-week intervals. The only exception was the first mission, with Gogol and Vasilyev; since the *Krepost* itself was so heavy, it would only be stocked with two weeks of supplies, so their overall mission would be thirty-five days.

The freighter, essentially an unmanned truck, had been adapted from the Soyuz spacecraft. A standard Soyuz consisted of three parts, from bottom to top: a Service Module, a Descent Module, and an Orbital Module. The cylindrical Service Module housed life support equipment, electric power supply equipment, various electronic instruments, and communications gear, as well as the propulsion systems—including the associated propellant storage tanks—necessary for maneuvering in orbit and returning to Earth. Two wing-like arrays of electricity-producing solar panels, one on each side, protruded from the Service Module. The Soyuz cosmonauts hunkered in the snug confines of the Descent

Module during launch and reentry, and otherwise occupied the more spacious Orbital Module.

For the freighter variant, the Descent and Orbital Modules had been replaced by a single bullet-shaped Cargo Module that was little more than a titanium shell that encased and protected the goods being ferried up to the *Krepost*. The Cargo Module was internally subdivided into a pressurized "dry" compartment and an unpressurized "wet" compartment. The dry compartment was packed with food and other supplies, while the wet compartment contained bulk water and cryogenic vessels for liquid hydrogen and oxygen. The Cargo Module was a relatively simple vessel, which was topped by a rendezvous and docking pod, which borrowed key technology—the *IGLA* automatic rendezvous system, docking hardware and pressure hatch—from the Soyuz crew vehicle.

Unlike the manned version of the Soyuz, the freighter was intended to be entirely sacrificial. Once the supplies were depleted and the vacant dry compartment filled with trash, the freighter would be undocked from the *Krepost* and would incinerate upon reentry.

Early on in the *Krepost* program, Abdirov had been faced with choosing between the Korolev and Chelomei aerospace bureaus for the design and production of the freighter. Although ordered to do so, neither of the premier aerospace bureaus was enthusiastic about playing second fiddle to Abdirov, and both approached the General Staff of the High Command demanding an opportunity to submit their own design proposal for the *Krepost*. The High Command quickly squelched their objections; obviously recognizing that there was no way to usurp Abdirov, the two bureaus aggressively competed to win the freighter design.

The choice was not an easy one. Since 1965, the Chelomei bureau had been developing its own unique freighter—the TKS—for its *Almaz* military space station. Unfortunately, while the TKS looked promising, no prototype had actually flown yet. Consequently, since betting on the unproven TKS was a wager he just wasn't willing to place, Abdirov elected to stick with the Korolev bureau's proposal, which entailed modifying the existing Soyuz. He had another motivation to stick with the Korolev bureau: they were also refining an even more sophisticated unmanned freighter—the Progress—which was projected to become the logistical mainstay for future Soviet space stations. He felt confident that the Korolev bureau could ultimately shelve the interim Soyuz freighter and switch to the Progress to sustain future *Krepost* missions.

Now, faced with news of the delay at the Korolev plant, he regretted his decision. And there was more for him to be troubled over. He was hearing pervasive rumors that the *Perimetr* engineers were on the verge of perfecting a remotely operated interlock system. If the freighter delay compelled him to wait until the next available launch window, then it was highly likely the *Perimetr* engineers could install the new interlock aboard the first *Krepost* before it was fired into orbit. If that came to pass, then Yohzin's efforts to purloin the secret code would all be for naught, since the *Krepost* crew would have absolutely no role in disabling the interlock in order to drop the Egg.

Abdirov just couldn't shake the notion that the Korolev bureau was colluding with the *Perimetr* faction to stall his program. After all, both entities would be delighted if the *Krepost* foundered even before it left port. Even so, the last thing he wanted to do was to accuse the well-connected Korolev bureau of deliberate sabotage; that would be the equivalent of political suicide.

"Lieutenant Colonel Gogol is here as you ordered, Comrade General," announced his secretary. As usual, even though she had worked for him for over a decade, she averted her eyes slightly, so as to not directly look at him.

Abdirov took a deep breath and calmed himself. Gogol's arrival reminded him that, while aggravating, the production delay was really not a tremendously significant matter in the grand scheme of things. "Send him in," he said. "And ensure that we are not disturbed."

As she pivoted to walk out, he held out the Korolev memo. "And draft a response to this," he said. "I want to make sure that the Korolev bureau is aware that I intend to proceed with the current launch schedule, and I expect that they will fix their problems and resume production of the freighters so that they are delivered on schedule, as they promised."

"*Da*, Comrade General," she replied, peering at the floor. "I will have it ready for you shortly."

Moments later, clearly favoring his left foot, Gogol walked in and reported. Abdirov studied him; the squat cosmonaut reminded him of a less refined version of Nikita Khrushchev, if such a thing was truly possible.

"Are you injured?" asked Abdirov, returning the cosmonaut's salute and gesturing for him to take a seat. "It appears that you are limping."

"I stubbed my toe, Comrade General," replied Gogol. "A moment of clumsiness, sir, but nothing to be concerned over."

Abdirov could scarcely believe that he was prepared to confer the key to Armageddon upon this oafish dolt. It seemed insane that he would be compelled to use such a blunt object to ignite the fires that would purify the world. "How are the preparations for flight proceeding?" he asked.

"Splendidly, Comrade General."

"Good," noted Abdirov. "I summoned you here for a reason, but I must warn you that the things we discuss may never be spoken of to anyone. That is *absolutely* imperative. Do you understand?"

"*Da*, Comrade General. I understand absolutely."

"Do you recall that we discussed your willingness to deploy the weapon if you were issued an Independent Action Code?"

"I do recall that conversation, sir."

"And you still bear no reluctance to deploy the Egg in those circumstances?"

"None whatsoever, Comrade General," replied Gogol without hesitation.

As much as Abdirov valued obedience on the part of subordinates, he wasn't exactly fond of purely blind obedience, particularly given the dire implications of this situation. Surely, any rational man would have at least some pause for thought when faced with the prospect of instantly laying waste to millions.

"Then if that's still the case, there's something that you should know," said Abdirov. "I cannot elaborate, but I have it on excellent authority that the High Command fully intends to issue an Independent Action

Code during your mission, and they will expect it to be executed. Do you understand?"

"I do, Comrade General. Am I to assume that I will be authorized to deploy the Egg on the first available target after receipt of the code?"

Abdirov smiled faintly; Gogol wasn't an absolute dolt. "Drop on the first available target?" he replied. "Not exactly. Although they will likely provide you with the Independent Action Code first, the High Command will still designate a specific target, which you will execute on their order. Obviously, they would prefer to achieve the greatest potential effect."

"Obviously, Comrade General."

"One more thing. This may seem odd, but you should not expect to receive the Independent Action Code or targeting instructions by the usual means, which are controlled by *Perimetr*. Instead, we'll send up the code and other instructions on the intelligence channel."

Besides the automatic communications links for telemetry, there were three dedicated radio channels. The first was a "housekeeping" channel reserved for routine traffic concerning spacecraft operational matters. The second was the weapons deployment channel exclusively managed by *Perimetr*. The third channel was reserved for daily intelligence reports and general news; for brevity's sake, the intelligence broadcasts were encrypted and sent in a compressed burst format. During periods of international crisis, these messages would include specific locations— airfields, ICBM fields, troop concentrations, port facilities—that the cosmonauts should visually monitor as they passed over them. Although not intended as a dedicated reconnaissance platform, the *Krepost* was equipped with a powerful spotting telescope and four portholes.

"Permission to ask a question, Comrade General?" asked Gogol.

"*Da*. Go ahead."

"If the Independent Action Code is to be passed by Channel Three, am I to believe that the High Command does not trust *Perimetr*?"

Abdirov was mildly surprised at the insightful question, as well as Gogol's courage in raising such a sensitive matter. Perhaps he was not nearly the bumpkin that he tried to portray. "Perhaps, but in any event, that is entirely outside your realm, and beyond mine as well, so we should not waste our energies contemplating the thoughts and desires of the High Command. Understood?"

Nodding, Gogol replied, "I understand, Comrade General, and will not speak of it again."

"Good."

"I don't want to offend or anger you, Comrade General, but may I ask another question?"

"Go ahead."

"Why are you telling me this now, Comrade General?" asked Gogol quietly.

"I have two reasons," answered Abdirov. "First, although I am confident that you will execute your duties in a timely manner, I am not as sure about Vasilyev or any of the other cosmonauts. In fact, once the Independent Action Code is issued and the deployment order given, I seriously doubt that Vasilyev will be compliant, so you should be prepared to subdue him, if necessary. I am telling you this now, so you can be prepared to act accordingly."

"And what if the timing is such that I need to exercise lethal force to subdue him, in order to deploy the Egg on the appointed target?"

Again, another tremendously insightful question, thought Abdirov. *Perhaps this is indeed the perfect man for the task.* "Then I trust that you will do what is necessary."

Gogol was silent for a moment, apparently pondering the situation, and then asked, "Comrade General, you said that there was a second reason?"

Abdirov sighed and then said, "Realistically, should this contingency occur as we anticipate, you will be the man who pulls the trigger that ends the stalemate between East and West. It's safe to assume that you will not see me again after you depart this earth, because I'm confident that this place will be reduced to radioactive rubble, as will a large part of our Motherland and ideally, all of America.

"Since you will be in the unique and enviable position to survive the holocaust that you will have wrought, and you will have over a week in orbit to contemplate where you will come back to Earth, I wanted to grant you adequate time to make the appropriate preparations. Fair enough?"

"You can count on me, Comrade General!" vowed Gogol. He seemed almost elated.

"Gogol, do what you need to get ready. Just so you are aware, once you arrive on orbit, I am confident that the order will not be issued until after you have received your first freighter."

"Thank you for granting me such warning, sir. I will study the situation and assemble the necessary materials, but I will do so without attracting any undue attention," declared Gogol. "By your leave, Comrade General? I have some scrounging to do."

"I imagine that you do," answered Abdirov. "You are dismissed."

Dayton, Ohio
5:35 p.m., Tuesday, September 5, 1972

Just yesterday, while they were talking on the phone, Ourecky and Bea had finally decided that it was time to see each other. He pulled into Jill's gravel driveway and parked behind Bea's Volkswagen Karmann Ghia. Just the sight of her red "roller skate," as Carson always called it, made him sigh in longing.

Not sure exactly what to expect, he slowly walked up the front steps like he was ascending the stairs to the gallows. He succumbed to a coughing spell just as he started to push the doorbell. Bent over at the waist, he hacked up a bloody lump and spat it out behind an overgrown boxwood shrub next to the front door.

Gasping and wheezing, he struggled to catch his breath. He had effectively sacrificed his lungs—at least for the time being—but Ed Russo was alive. Now, he would gladly sacrifice every breath he would ever draw if Bea would only come back to him. He was apprehensively sailing into uncharted waters; she had given him no clear indication of when or if she would return, and he was wary of saying anything brash that might sway her decision in a negative way.

He straightened up, regained his composure, and pushed the door-bell. Seconds later, the door creaked open. Barefoot, Bea pushed open the screen door and waved him inside. She wore faded and patched blue jeans, a tie-dyed T-shirt, and had her hair tucked under a blue bandanna. A large silver peace symbol dangled from a chain around her neck.

He was immediately melted by her glamorous smile, just like on the first day they had met. It seemed as if she hadn't changed in the slightest, but he felt like he had aged a thousand years. He hugged her awkwardly, almost like she was a stranger, and drew in her familiar smell.

"That's the *best* you can manage, Ourecky?"

"Well, I just thought…"

"We're *still* married. I just need some time away. Come on in. We need to keep it down. Jill's asleep in the back bedroom. Jill's mom took Andy and Rebecca to the park."

"Rebecca?" asked Ourecky.

"Jill's daughter," she replied. "She's almost a year older than Andy. Anyway, come on in. Want anything?"

"Some water would be nice," he answered, following her inside and taking a seat on the couch. "You look great."

"Thanks," she replied, filling a glass from the kitchen faucet. "Did I hear you coughing outside? You sounded horrible. Are you *still* sick?"

"Yeah. It's the same damned bronchitis. I'm slowly getting better, though."

She sat beside him on the couch and handed him the glass. "Is it catching?" she asked. "I have to be really careful with Jill the way she is, and the kids…"

He sipped the water and replied, "No. It's not contagious. I think I caught it by going from one climate to another too quickly. Too cold to too hot, too fast."

"Yeah. I know how that works."

The television was on, although its sound was turned down so low as to be barely audible. Its monochromatic screen displayed images from yesterday's massacre at the Munich Airport, where several Black September terrorists and the nine members of the Israeli Olympic wrestling team, held hostage by the terrorists, were killed during a shoot-out with German police.

"Sorry," she said, reaching over to switch off the television. "Except for that Watergate break-in stuff, that's just about all that's on the news right now. What a horrible situation. You would think that people would learn to live in peace with each other, especially during something like the Olympics."

He nodded. Uncomfortable with her, he wasn't sure what to say. He nursed the water as he assessed the surroundings. The floor was cluttered with toys, games, and Dr. Seuss books. A typewriter and cassette tape player, the principal tools of Jill's medical transcription work, occupied the small breakfast nook that divided the living room from the kitchen.

He wasn't particularly comfortable with some of the other trappings of Jill's house. For him, the visit was like an excursion into the depths of hippie counterculture. One wall was decorated with Day-Glo posters of Woodstock, Jimi Hendrix, and Janis Joplin. A dog-eared copy of the *Whole Earth Catalog: Access to Tools* and several books on alternative medicine occupied the coffee table. On the end table to his left, shape-shifting blobs of oily goop swam in the cone-shaped glass of a Lava Lamp. Ourecky felt queasy; the undulating blobs were just too reminiscent of the floating spheres of vomit he had dodged during his day aboard the MOL.

If all the psychedelic paraphernalia wasn't disconcerting enough, there was a pervasive scent, a faint lingering odor, something he hadn't smelled since his college dorm years. He sniffed the air and asked, "Is that what I think it is?"

"Pot?" replied Bea. "Marijuana? Yeah, Scott, that's *exactly* what it is."

"Do you…"

"*No.* Jill smokes it sometimes. Her medicines make her really nauseous. The pot helps with the pain and gives her an appetite. If she didn't smoke it, she probably wouldn't eat anything at all."

"Well, I can't believe that you bring our son into this place." As soon as the words left his lips, he knew he had blundered as he watched her expression.

She frowned. "*This* place? Man, aren't you quick to pass judgment? You make it sound like some evil opium den in Chinatown. Look, Jill doesn't ever do it around the kids. I take them outside when she smokes. Jill's *dying*, Scott. It makes her feel better. Can you not grant her that?"

"I suppose," he conceded reluctantly. "I just wish that Andy wasn't around it."

"Well, he *isn't*," she said curtly.

"Okay," he mumbled.

"So have things changed much?" she asked.

"At work?"

She nodded.

Shaking his head, he disclosed, "Things around the office have slowed down a lot, but we know that there's still one more flight..."

"One more flight?" she sniffed. "*Another* test? I suspected that. You'll be flying with Drew Carson again?"

Staring at the wooden floor, he replied, "Yeah."

"When?"

"I'm not sure. Definitely not until after I heal up. Right now, it doesn't look like it will happen for at least another six months, so I'm really not sure."

"Well, *you* may not be sure, but as far as you and I are concerned, *I'm* sure that things can't change until you get out of there, Scott. As much as I want us to be back together, I don't think I can climb back on that emotional roller coaster again. Do you understand that?"

"I do, Bea, but I have to warn you: it might be months before it's over and I have a chance to leave. Is there not any chance that you might come home before..."

"I don't know," answered Bea. "I really don't know if I can handle it. Look, Scott, I feel obligated to stay here with Jill until...well, until that's *over*. After that, we'll just take things one baby step at a time. Can you live with that?"

Ourecky nodded solemnly. At least there was some hope.

3

CALL THE BALL

Over the Gulf of Mexico
10:32 a.m., Wednesday, September 6, 1972

Carson ticked off the final items on his pre-landing checklist and made his "abeam" call as he came alongside the Lexington. Seconds later, he confirmed his power settings and ensured his tail hook was extended before smoothly banking into the tight 180-degree level turn that would line him up for his third "trap" of the day.

Flying roughly twenty nautical miles south of Pensacola, he was at the controls of a T-2C "Buckeye" trainer. The orange and white Navy jet was the slowest aircraft he had flown in recent years, but it was certainly sturdy and nimble. He supposed that it had to be solidly built, since it was expressly designed to be repeatedly smashed onto carrier decks by fledgling Naval aviators.

Carson rolled out of the turn, crossed the trailing wake of the carrier, and verified the "meatball," the Fresnel Lens Optical Landing System, on the carrier's deck. He keyed the mike and stated, "Two-Two-Four, Buckeye Ball, Three Point Six, Qual Nineteen." The meatball's colored light array displayed visual cues to denote if he was on the optimum glide path and centerline to safely bring the jet aboard the carrier. If he was spot on, he would view the illuminated meatball as an amber light hovering between two green bars. If he sank too low, he would see the amber light dip below the green bars and gradually change to red. If he was too high, the amber light would appear to float above the green bars. Right now, he was precisely where he needed to be to securely plant the jet on the deck.

In a patient, unhurried voice, the Lexington's LSO—Landing Signal Officer—replied,

"Two-Two-Four, *roger* ball. The deck is steady. You're in the groove. Fly the ball."

He maintained a disciplined scan, swiftly alternating between his instruments and the glowing meatball. He remembered the Navy instructors' mantra: *Watch the Ball. Watch the Ball. Watch the Ball. Be the Ball.* Fighting years of ingrained training, he double-checked that his feet were *off* the brakes and remembered to resist any urge to flare. He focused intently on the ball, striving *not* to look at the massive gray deck swelling before him.

Carefully monitoring his angle of attack, he kept the aircraft precisely aligned on glide slope and centerline; if he followed the ball all the way to touchdown, the landing would come as a virtual surprise. He had often heard a carrier landing described as a controlled crash, but he found it more like a Zen-like trust exercise, like falling face-first off a stepladder, blindfolded, hoping to be caught in the outstretched arms of waiting comrades.

The plane slammed down on the deck. In anticipation that his tail hook had not snagged an arresting wire, Carson followed procedures to the letter, immediately jamming the throttle to Military Rated Thrust— full power without afterburners—and retracting his speed brakes. If the T-2C wasn't securely trapped, he would immediately "bolter" back into the air for another try.

After confirming that he was safely aboard ship, he directed his attention to the hand signal instructions of an enlisted "yellow shirt" plane director. After he had pulled back to allow his tail hook to disengage from the arresting wire, he followed the yellow shirt's directions past the "foul line" to a deck area where he would undergo a "hot seat" rotation. Since the number of pilots currently undergoing carrier quals exceeded the number of aircraft available to fly—a situation which would undoubtedly change within a day or two—the Buckeye trainers were time-shared.

After climbing down from the aircraft, Carson clambered below decks to a ready room where he would cool his heels while waiting for the LSO's critique of his landing. After the critique, he would join the queue of fledgling pilots awaiting their next turn at the controls. Most were fresh-faced ensigns, kids barely out of college. Carson felt like an ancient amongst them.

As he watched a television monitor that showed carrier landings in progress, Carson contemplated his potential combat deployment. The timing could not be any more critical. With the US anti-war movement in full swing, American involvement in the Vietnam conflict was waning. The last US ground combat troops had departed the country even as he sat beside Ourecky's hospital bed at Lackland in August.

But while the US ground campaign might be over, the air war definitely was not. A massive air operation, Operation Linebacker, was incessantly pounding North Vietnam with the full brunt of US airpower, including B-52 strategic bombers. The air campaign, in response to the NVA's Easter Offensive in the South, was slamming strategic facilities—factories, warehouses, marshalling yards, power plants, railroads, key bridges—in the Communist heartland, particularly in the vicinity of Hanoi and Haiphong.

The question remained how long it would go on and whether it might already be too late for Carson to earn his spurs. It looked like the North Vietnamese were clinging to the ropes, and that it was only a matter of time before they crawled back to the negotiating table.

The nation was woefully tired of war, so it was unlikely that politicians would be overly inclined to allow the military to become embroiled in another conflict, at least for the foreseeable future. Consequently,

Carson wanted to deploy *immediately*, before it was too late, but he also knew that Tew would never approve. His eagerness made the training regimen a difficult pill to swallow. But even though it was a path towards the combat experience that he so desperately desired, the Navy training had also been the most humbling episode of his entire career. Falling in alongside ensigns, he effectively became a nobody, reduced to the lowest common denominator.

Besides being compelled to start at the bottom, Carson's reunion with Badger—who would be his squadron commander overseas—had been very awkward. Shortly after his arrival at Pensacola last week, Badger offered him an off-the-books training hop over the Gulf. Flying an F-4 Phantom against Badger's A-4 Skyhawk, Carson spent a hectic afternoon in full-out, no holds barred engagements. Badger swiftly established dominance in the air, spanking him in seven out of seven encounters. Since their last meeting over the Gulf over two years ago, Badger had undergone two more combat cruises and had served a tour as a Top Gun instructor. To say that he was proficient at air-to-air combat would be a gross understatement.

After issuing Carson his embarrassing comeuppance in the sky, Badger brought him down to earth for a fatherly chat. He informed Carson that the hop was his *last* free ticket; he reiterated Tarbox's assertion that everything, from Pensacola through Vietnam and until his return to the Air Force, had to be legitimately *earned*.

Badger also told him that being known as one of Tarbox's favored children was a rocky path toward making friends and influencing enemies. Besides the latent stigma associated with being a Tarbox protégé, if Carson did eventually matriculate to Yankee Station in the Gulf of Tonkin, he would arrive with two strikes against him. First, Naval aviators were a clannish lot, very reluctant to accept outsiders in their midst.

Second, and probably a much more severe violation, Carson's tentative slot at the prestigious Fighter Weapons School would require that a deserving Navy pilot be bumped from the roster. Selection for Top Gun billets was a highly competitive process, and regardless of what Carson might have done to steal a ticket to the front of the line, he could count on repercussions when he arrived at the squadron.

If that wasn't enough, there was more. Tarbox's stipulation that Carson be temporarily reflagged as a Navy officer also required that he would deploy under a pseudonym rather than his own name. There was a pragmatic reason for the ruse; although Carson's feet would likely not touch dry land during his combat stint, it was still theoretically possible that he might land at an Air Force base—most likely Da Nang—if he sustained damage or experienced mechanical problems that would hamper a safe recovery aboard the carrier.

When and if he showed up at an Air Force base in Vietnam, all bets were off; certainly, the chances were good that if he arrived in a Navy jet then no one would pay him any undue attention, but it was equally likely that someone would recognize him and his flagged status would come to light. If that happened, he would probably be placed under arrest as he awaited the most expeditious means to return to the States for court martial.

As Tarbox and Wolcott had told him, the whole jaunt was planned under the notion that it was often easier to seek forgiveness than to ask permission; once Carson returned from Southeast Asia, his records would be restored to normal and his combat experience would be annotated accordingly. After all, it was a scheme cooked up between Wolcott and Tarbox; since Virgil was officially retired, there was little anyone could do to punish him, and Tarbox...well, save for perhaps the President himself, no one could lay a finger on the Ancient Mariner.

Although the flying came readily to Carson, the experience was still intensely challenging. The Navy was an entirely new and foreign culture to him. Prior to Pensacola, with the exception of a few impromptu "hassling" sessions in the air, most of his exposure to Naval aviators had occurred when they were off-duty. Navy pilots were widely renowned for their wild and crazy drunken antics, but Carson had discovered that even though they were swift to let their hair down in the bars, they were extremely disciplined in flight and shipboard operations. It didn't take him long to realize why. At sea aboard a carrier, launching and recovering aircraft were incredibly complex operations in the best of circumstances.

A casual observer might see only continuous chaos on a carrier's flight deck, but it was a tightly synchronized operation, with hundreds of interdependent actions occurring simultaneously. Space was always at a premium, and every square foot of empty deck spoken for. Every motion was calculated; no energy was wasted. Hazardous operations were everywhere, conducted in intimate proximity; planes were re-fueled only yards away from where ordnance was manually loaded onto wing shackles. *Everyone*—from the almighty Air Boss to the lowest ranking purple-shirted "grape" who topped off aircraft fuel tanks—relied on everyone else to be in their proper place and doing their job. Trust and discipline were the essential lubricants that allowed the cogs and gears to turn in order to make the big machine work.

For the aviators, perhaps one of the most grievous sins was to cause a fouled deck. A fouled deck could just as easily result from an overconfident aviator's simple blunder as it could from a nugget rookie's wobbly landing attempt. A fouled deck could not only derail a highly orchestrated strike operation, but if aircraft could not be brought aboard in a timely manner, it could cost lives as well. Consequently, Naval aviators were absolute sticklers for procedures.

Moreover, an aircraft carrier wasn't simply a big ship that sported a runway on its roof. It comprised many other things—ordnance depot, power plant, maintenance shops, commissary, fuel depot, dormitories, aircraft hangars—all rolled into a single self-contained ocean-going airbase. And even though thousands of men lived in tightly cramped quarters, so long as they were away from home, it was their refuge and sanctuary.

But even aboard floating sanctuaries of steel, safety and security could be fleeting notions. While the aviators left the carriers to fly high risk missions over North Vietnam, death was never far away for the men who remained aboard. Although the ships were theoretically out of harm's way, three horrible incidents had occurred aboard carriers in less than three years.

In 1966, a dropped flare aboard the Oriskany started a fire that killed forty-four sailors, injured 156 others, and severely damaged four aircraft. Next year, a freak calamity occurred aboard the Forrestal in which a Zuni rocket was apparently ignited by a stray electrical spark. The misfired rocket flew across the flight deck, set ablaze a parked A-4 Skyhawk and eventually caused a massive fire and several chain reaction explosions which killed 134 sailors, injured 161 others, and destroyed twenty-one aircraft. In January 1969, a Zuni rocket accidently detonated aboard the Enterprise, igniting a fire that killed twenty-eight men, injured 344 and destroyed fifteen aircraft.

Carson's thoughts were interrupted by the hoots and catcalls of the ensigns; they crowded around the television monitor to watch as one of their mates botched a landing and slowly boltered back into the air. Grinning, he shook his head. These guys *were* different, and they were just barely down the long road to being full-fledged Naval aviators. And if Badger's prowess was any indicator, his journey was also just starting as well.

If his circumstances were different and he had spent the past five years as just an anonymous Air Force fighter pilot, there was no question he would have flown in combat by now. But by virtue of his excellence, he had been set aside for more momentous—although secret—tasks. Now, if combat experience was his Holy Grail, he would have to venture forth with this strange new legion. It wasn't sufficient to be a superlative pilot, a solitary warrior; he had to adapt himself to this new culture and be accepted within their brotherhood.

As Badger had asserted, it was definitely not something that would be handed to him; it was his to *earn*. So now, his quest had become less about just enduring the grind to secure his ticket to glory, and more about his need to prove himself adequate in their eyes. After over a decade of flying, always at the pinnacle of his profession, Carson hoped that he would be found worthy.

4

THE PERILS OF AN INGROWN TOENAIL

Krepost Pre-Launch Processing Facility
Burya Test Complex, Kapustin Yar Cosmodrome
7:32 a.m., Thursday, September 7, 1972

This morning, Gogol and Vasilyev were granted a welcome break from training. As of yesterday afternoon, the freighter had been packed with its mandatory load of bulk supplies, and now they were bringing their "personal preference items" to fill out the remaining space available. The freighter would follow them into orbit two weeks after they occupied the *Krepost*, and provide supplies sufficient for them to remain aloft for the remainder of their five-week mission.

Presently, since they were working under entirely separate schedules, the two men had not seen each other in over four days. Gogol's time was committed to exhaustively detailed reviews of targeting data for the Egg, while Vasilyev's time was consumed with lectures and practical training associated with an experimental fuel cell that would be delivered on the first freighter.

They had previously loaded the personal preference items in the trunk and backseat of the *Zhiguli* sedan. After Vasilyev parked the sedan, a pair of enlisted soldiers unloaded the assorted goods into a pair of wagon-like carts and followed Gogol and Vasilyev into the assembly building. Raindrops spattered the dusty ground as a late summer shower approached. The air smelled like damp clay.

Vasilyev was happy to be out of the car and in the fresh air, because Gogol stank. His coveralls were rumpled and unwashed, as if he had pulled them out of the clothes hamper this morning and put them back on. And although Gogol had shaved, he clearly had neglected to bathe. Of course, Gogol was a man of many idiosyncrasies, so it was doubtful that anyone would question his personal hygiene, since it could likely be part of some pre-flight superstition. If it was some sort of pre-flight ritual, Vasilyev found it odd that he had not witnessed it before their previous missions. As they walked toward the brick building, he noticed that Gogol's pace was considerably slower than usual, and that he appeared reluctant to place too much weight on his left foot.

"Are you limping?" asked Vasilyev.

"I stubbed my damned toe," snarled Gogol. "And the less you say of it, the better, unless you want me to take a hammer to *your* foot, kitten."

They entered the assembly building to see that the workers were in the final stages of mating the *Krepost* to the Proton booster. It was difficult for Vasilyev to comprehend that it was only a matter of days before they would depart Earth and not return for over a month. A massive aerodynamic shroud, constructed primarily of fiberglass, was suspended from a nearby crane.

The *Krepost* would be launched with its nuclear warhead topmost. An escape rocket would be bolted to the fairing to complete the stack. In the event of a launch accident, such as the Proton booster spontaneously exploding on the pad, the solid-fueled escape rocket would fire to yank the Egg skyward, before parachutes deployed to allow a safe descent to earth. For all intents and purposes, the *Krepost* and Proton were considered effectively expendable in an emergency scenario, but for obvious reasons, the Egg had to be safeguarded by whatever means necessary. When Abdirov lobbied for the special Proton launching facility to be situated at Kapustin Yar, a significant portion of his argument centered on the dire setback that would befall the Soviet manned spaceflight program if the Egg was accidently cracked during a botched launch at Baikonur. For all of the arguments he had to justify before the General Staff of the High Command, that one was by far the easiest to win. Standing before them, mangled and horribly burned in the Nedelin catastrophe, he presented a very tangible example of someone who personally knew the terrible consequences of launch accidents.

They walked past a large assembly bay where their own Soyuz was being readied, and entered the clean room where the freighter waited. With its blunt nose pointed to the tall ceiling, the freighter was mounted on a massive support frame. The cargo-packing crew waited on a second level in the processing bay.

Gogol and Vasilyev ascended a ramp and reported to the chief packer, who was responsible for everything that would be loaded aboard the freighter. To a casual observer, the loading process might appear to be a haphazard operation in which boxes and bags of supplies and gear were crammed aboard the cargo module. In fact, it was a very sophisticated process, in which each item was meticulously assigned a position on a loading manifest, which took into account weight, volume and balance. Several heavy sacks, filled with various types of ballast, waited by a wall. Once the cosmonauts' personal preference items were squeezed aboard the freighter, the sacks would be temporarily loaded in order to simulate the bundles of perishable foodstuffs—fresh fruits, vegetables, and bread—that would be stashed aboard the freighter immediately before launch.

The chief packer, an overweight RSVN major with brown hair and sparse moustache, wore spotless white coveralls and bore a clipboard. "I can accommodate fifty kilograms," he declared.

"What? *Fifty* kilograms? That's all? You had led me to believe that we would have more room," groused Gogol.

"It's that new fuel cell," explained the chief packer. "When we discussed this a few weeks ago, we did not think it would be ready in time for your flight. It occupies a lot more volume than we had anticipated. I am truly sorry, Comrade Commander, but there's nothing that I can do about it."

"Fuel cell?" sniffed Gogol. He looked askance at Vasilyev; his sour expression made it abundantly clear that he blamed the shortcoming on the junior cosmonaut, since the fuel cell was notionally his responsibility.

The two cosmonauts walked over to the edge of the platform and peered downwards through the opening into the payload module. Partially obstructed by the pressure hatch that swung inwards into the dry compartment, the top of the fuel cell was barely visible with all of the bags and boxes packed around it. It was encased in a cylindrical metal shell, approximately the size of a standard two-hundred-liter fuel drum. The shell would be removed after the resupply freighter had been unloaded in orbit. Drawing from cryogenic tanks of oxygen and hydrogen in the cargo module's wet compartment, the fuel cell would produce a substantial portion of their electricity and drinking water once it was activated. When they received a new freighter, after unloading its supplies, they would physically dismount the fuel cell from the old freighter, mount it in the new freighter, and connect it to the new freighter's cryogenic tanks before jettisoning the depleted freighter.

Since Soviet engineers had virtually no success in creating a functional fuel cell on their own, they had taken their traditional detour to overcoming design shortfalls: they replicated an American design from stolen blueprints. The cylindrical unit that would partially power the *Krepost* station was almost an exact duplicate of the technology used in NASA's Apollo spacecraft.

"So that's it?" asked Gogol. To him, the internal workings of the fuel cell was almost the equivalent of some inexplicable alchemy, and he lent scarce stock to the prototype device functioning properly for the duration of their mission. "That's the fuel cell? I don't like experimental gadgets. I'm not fond of gadgets or experiments, period."

Gogol slowly made his way back to the table. He tugged a satchel from the cart and heaved it onto the table. The satchel was constructed of sturdy rubber-coated green-colored canvas, had two shoulder straps so that it could be worn as a backpack, and was slightly more than a half-meter long.

"First things first," said Gogol, sliding the bulging satchel across the table's shiny metal surface. "This bag contains my choices to supplement the survival gear on the Soyuz."

Frowning, the chief packer shook his head. "The Soyuz contingency kits are very substantial, Comrade Commander. Their contents have been carefully selected by survival experts who know their business. The kits are more than ample to protect you in any region of the world, so I see no need to augment them."

Gogol's gruff manner made it abundantly clear that he was not in the mood to provide a more detailed explanation. When he chose to display his wrath, his glare could probably melt case-hardened steel. "Survival *experts?*" he sneered defiantly. "I wonder how many of those damned survival experts have been compelled to spend three years stranded in the desert, left to fend for themselves. If they had, then perhaps they would understand why I feel the need to bring some of my own stuff."

"Okay, I cannot argue with your experience," said the chief packer, prodding the bag with his pencil. "Open it, please."

"Is that really necessary? It's just some odds and ends," said Gogol. His coarse demeanor swiftly dissipated, and a wry smile crossed his face. He reached out and gently touched the chief packer's hand. "Besides, aren't we old friends, Sergei? Is there really such a need to make a fuss over this?"

Obviously unnerved by the gesture, the chief packer yanked his hand away and replied, "Sorry, Comrade Commander, but I have my responsibilities. I cannot allow any hazardous items aboard."

Shaking his head, Gogol undid the clasps, opened the bag, and dumped its contents onto the table.

Vasilyev was engrossed by the contents of Gogol's satchel. Amongst a variety of other items, it contained a hatchet, a file, a short machete-like knife, several hanks of parachute shroud cord, a compass, a small Geiger counter, and a simple sextant. There was a limited collection of medical supplies, including four large plastic bottles of potassium iodide anti-radiation pills and several pre-loaded hypodermic syringes loaded with morphine.

The bag contained several maps. Gogol's choice of maps, most of which were of the variety issued to strategic bomber crews to assist with escape and evasion, was particularly interesting. The regions covered were Australia, South America, North America and Africa. Vasilyev realized that the maps were painstakingly trimmed, so that they depicted only areas where the latitudes would fall under the *Krepost's* orbital track, with all other superfluous regions carefully trimmed away. They were protected with plastic lamination, and various locations were marked with dots. The numbered dots were strategically situated in desolate expanses of wilderness. Vasilyev deduced that the dots denoted contingency landing sites, but none were locations that he had ever been briefed on, nor had they rehearsed for.

In addition to the official maps, there were also tourist maps for the same areas, as well as phrase books and translation guides for commonly spoken languages. There was also a battery-powered multi-band receiver, manufactured in Japan, that could also receive short wave frequencies. A small hand-cranked generator was taped to the back of the portable radio.

A Kalashnikov folding stock assault rifle was the *pièce de résistance* of Gogol's collection. Vasilyev was aware that Gogol had complained bitterly that his survival gear had lacked a decent weapon when he crash-landed in the Mongolian desert. Other cosmonauts had lent their voices to Gogol's lament, and a robust multi-barrel survival weapon was currently under development. Gogol had obviously selected the Kalashnikov as an interim substitute.

"A Kalashnikov?" asked the chief packer incredulously. "Hah! You must be kidding, right? I thought you both will have Makarovs at your disposal."

"Try popping an angry bear with a damned Makarov," sniffed Gogol. "He might eventually die of lead poisoning if you empty the entire clip into him, but will more likely die of severe indigestion first, after he makes a meal of you."

"Fair enough," said the chief packer. "I can see your point." He examined several boxes of ammunition and slid one box back to Gogol. "The regular rounds can go up. I can't allow these tracers, though."

Gogol replaced the items in the bag and refastened it. The chief packer snapped his fingers before handing the bag to an assistant to be weighed.

"Okay, next priority," said Gogol, placing a wooden box on the table. The box contained rubber hot water bottles; each one was sealed with a cork that was further held in place by a radiator clamp. Vasilyev was aware that the flexible containers contained a considerable volume of vodka and brandy. They would be wedged into any and all available voids and spaces between cargo containers, rendering the resupply freighter into a veritable flying liquor cabinet. Of course, in the weeks that passed before the freighter flew, the flexible flasks would almost certainly impart a rubbery aftertaste to the liquor they contained. But vodka was vodka, mused Vasilyev, even if it did taste like the gum sole of an old gym shoe.

Vasilyev was thankful that his superiors weren't stricken with the Americans' insistence on puritanical temperance in space. Besides, for most Soviet military personnel, sobriety was a vague concept at best, since the relentless pursuit and wanton consumption of alcohol were indelible aspects of Soviet military life. After all, history clearly showed that inebriated soldiers were *brave* soldiers, or at least more prone to acts of suicidal foolhardiness that might otherwise be construed as bravery. And intoxicated or not, the destructive capacity of the Soviet soldier was a force to behold. In an orgy of pillaging and rape, leaving scarcely more than scorched earth, broken bodies, and destruction in their wake, the vengeful Soviet Army had rolled off the Russian steppes to brutally pound the Nazis into abject submission.

In their never-ending quest for stout potables, Soviet soldiers left no stone unturned and no liquid—however foul—untried. They were notoriously prone to imbibe on anything even modestly inebriating, regardless of toxicity, to include antifreeze, metal polish, cleaning solvents, brake fluid and de-icing fluids. Even rocket fuel was not off limits. In the relatively short history of Soviet rocket designs, one of the most momentous achievements was the development of the R-2 in 1949, not just because it had nearly double the range of its predecessor, the R-1—which was essentially a direct copy of the Germans' A-4—but because its engineers were successful in fabricating a rocket motor that relied not on *ethyl* alcohol, but its highly toxic cousin—*methyl* alcohol—which even thirsty RSVN rocket troops would not consume. Well, more realistically speaking, they wouldn't consume methyl *twice*, since the first hangover was usually fatal.

But stocking *liquor* aboard a nuclear-armed space station? It seemed like a sure recipe for disaster, since common sense should dictate that nuclear weapons and alcohol don't mix. Packing the liquor in the resupply freighter was not an *officially* sanctioned activity, but a "wink and a nod" tradition overlooked by the mission-planning authorities. But there was a method to the planners' apparent madness. At this juncture, the relationship between the American and Soviet manned space programs was like the disparity between a sprinter and a marathon runner. While their American counterparts concentrated their efforts on short duration events like going to the moon, the Soviets focused their energies on developing equipment and procedures for the long duration missions that would be the mainstay of real space exploration. Consequently, even though the Americans would soon join the marathon race with their three-man Skylab, the Soviets were swiftly becoming the undisputed authorities on extended space flight. With every additional day that they toted up on the board, the Soviets were progressively more aware that a prolonged mission was anything but a pleasure cruise. So the mission planners did whatever they could to fend off boredom and complacency, to include plying the station's occupants with liquor, a relatively decent selection of foodstuffs and other distractions.

Besides, the Egg's arming process was extremely simple, so much so that Vasilyev was entirely confident that he and Gogol would successfully execute a deployment even if both of them were inebriated. And as peculiar as Gogol was, he had already established some clear-cut guidance about when and how they would imbibe. His rule was that they could only partake during the "blind" orbits, which were those revolutions where they would not pass over communications windows in the Soviet Union and also would not pass over the continental United States. Since it was so unlikely that they would deploy the Egg during a blind orbit, they were logical periods to relax and let off steam. The frequency and duration of blind orbits varied for any given twenty-four hour period, but there were often intervals where they might be blind for up to six or more hours, which was plenty of time to get snockered, with adequate time for recovery.

"And here's my extra chow," said Gogol, depositing another box on the table as he winked at the packer. Vasilyev noticed that despite Abdirov's assertions that Gogol would no longer be allowed to smoke in orbit, the box contained several packs of *Zolotoye Runo* cigarettes.

The chief packer submitted the box's contents to an inspection that was cursory at best. With a stern look on his face, like a schoolmaster preparing to scold an unruly youth, he picked up a pack of smokes. He held out the pack towards Gogol and said, "Comrade Commander, General Abdirov *specifically* ordered me not to permit any cigarettes to be packed aboard the freighter."

Gogol whistled quietly and replied, "I guess orders are orders, then."

Slipping the pack into the breast pocket of his coveralls, the chief packer smiled faintly, winked, and said, "Then you understand why I am compelled to confiscate this contraband."

Gogol flashed a metal grin, then turned towards Vasilyev and ordered, "Chuck your box up, too. Let's put this chore behind us."

Other than a couple of paperback novels and an extra toothbrush, Vasilyev's box contained nothing but foodstuffs. Remembering how Travkin had been compelled to sacrifice his meals so that Gogol could stuff his face, Vasilyev had intentionally restricted his culinary fare to items that he knew Gogol did not like. The sole exception was three jars of *Khrenovina* horse radish sauce; for whatever reason, the sense of taste was dulled by spaceflight, so both men were prone to dousing their meals with the pungent additive. Vasilyev hoped that he had packed enough for both.

As his own stuff was being weighed, Gogol dug through Vasilyev's box, auditing the contents. He tugged out a tin of pickled herring, scrutinized the label, and then nonchalantly tossed it into a nearby rubbish bin. He discarded several other items in the same manner. Vasilyev fumed, but was careful not to provoke Gogol's volatile temper.

Satisfied with his inventory, Gogol shoved the box toward the chief packer. "Weigh this stuff, and let's see the final tally."

Less than a minute elapsed before the chief loader pronounced, "You're three kilograms over."

"Cull out some more of your goodies," ordered Gogol. "*Now.*"

Vasilyev resisted the urge to swear out loud, but knew that it was futile to argue. He knew that he would suffer many worse indignities in the coming weeks, so he should learn now to hold his tongue.

Krepost Project Headquarters
11:15 a.m., Thursday, September 7, 1972

"So, there are no new developments concerning the second freighter, Rustam?" asked Yohzin. "There's still a potential that it might be delayed?"

"Correct," replied Abdirov, holding up a memorandum. "And that damned Korolev bureau still insists on playing both ends against the middle. All I'm getting from them is back channel stuff, where they are supposedly keeping me informed, but even though they are blatantly aware that the next freighter probably won't be consigned on schedule, they aren't making official notification to the High Command, so I stand to be left holding the bag when they fail."

Abdirov slowly pushed himself up and out of his chair and then turned to face the window behind his desk. As he looked at the massive Proton processing building approximately a kilometer distant, he spoke. "What's worse is that we are at the point of no return. If we're going to delay the launch until the next window, I have to make the decision today."

"Today?" asked Yohzin.

"*Da*. I must formalize the orders today to set everything in motion, or there's no telling when we might be able to launch, if ever."

Yohzin speculated on the potential outcomes. If Abdirov elected to postpone the *Krepost* launch and the Korolev bureau delivered the next freighter on schedule, despite their dire predictions, then he would be blamed for not acting in a decisive manner. If he chose to launch on schedule, and the second freighter was delayed, he would still be admonished for not acting with sufficient caution. Given either outcome, with his reputation indelibly tarnished, the *Krepost* effort would likely be snatched from his grasp. Most likely, it would be handed to the *Perimetr* bureaucracy, which was exactly what Abdirov feared, since he suspected that the *Perimetr* leadership and the Korolev bureau were conspiring against him.

The underlying problem was that postponing the launch wasn't like holding a train at the depot while awaiting an important boxcar, and then simply sending it on its way later. If it was a single rocket to launch a single spacecraft, the decision would be easy, but a delay of the *Krepost* launch would disrupt a carefully orchestrated chain of follow-on events. All of the subsequent crew and freighter launches, at least for the first two crew cycles, would have to be shifted, which was definitely not a casual endeavor. There were countless other variables that figured in as well. Rescue forces had to be deployed. Communications and tracking networks had to be synchronized. It was all an incredibly complicated process, and they were but one entity competing for a very scant allotment of unique resources required to support a launch.

"This is terribly frustrating," grumbled Abdirov, turning around before slowly lowering his broken body into his chair. "This project is on the verge of collapsing around our ears, and all because that damned Korolev bureau is holding me hostage with this freighter and their shoddy welding. I could lodge a formal complaint against those scoundrels, but what good could come of it?

"Not much, I suppose," said Yohzin.

"Well, damn them all," declared Abdirov, dipping a pen into an inkwell. "We are going to launch on schedule. Maybe this will light a fire under them, when they realize that this freighter debacle will eventually come to roost on them, and all of their frail excuses will be for naught." He signed several documents, slid them into a large envelope, and then affixed a wax seal.

Abdirov chuckled and then said, "It's ironic that the second freighter won't even be necessary, after all."

"Then you still intend to pursue your plan with the Egg?" asked Yohzin.

"I do, Gregor, but I haven't forgotten my commitment to spirit you out of here with your family. Perhaps saving you and Luba may be the only good thing that comes of all of this."

"And the timing, Rustam?"

"I'm going to wait until after Gogol and Vasilyev have received the first freighter. A few days after that, I will signal Gogol to drop the Egg."

"Are you sure that you can count on them?"

Abdirov paused as if to ponder the question, and then replied, "I am sure that Gogol will do exactly as he's instructed. I'm not nearly as confident about Vasilyev. He seems like one who might harbor second thoughts. In any event, I've made appropriate plans to ensure that my plan is executed."

Yohzin nodded. As peculiar as Gogol was, he was immensely reliable, almost to a fault, like a dog faithful to his master. Vasilyev, on the other hand, was a very capable officer but also extremely aloof. He was definitely one who might question orders, especially an order to extinguish most of human civilization.

"There's another matter I need to share with you, Gregor, and it's a situation that's less than pleasant," said Abdirov. "As if there were not enough balls for me to juggle, the director of the GRU Internal Security office paid me a visit this week."

"That idiot major?" asked Yohzin. "I can't conceive how that buffoon was able to find a job with the GRU."

"Precisely, but that's not the issue." Abdirov was momentarily quiet as he shifted his weight in his chair. "The GRU suspects that there's a security breach here at Kapustin Yar."

Yohzin's stomach immediately twisted into knots. "A breach?" he asked. "Ludicrous! With almost a battalion of GRU spooks slithering around, watching everyone's comings and goings, how could there possibly be a leak?"

Abdirov waved a hand as if to shush him. "This is not something that originated at the local office," he said. "The report came from GRU headquarters. Apparently, they just stole some information from the Americans that made reference to some information that they'd previously stolen from us, which referred to some information that we had earlier stolen from them, ad infinitum. Suffice it to say that they happened upon some scraps of choice information that could have only come from here, from Kapustin Yar, and from a very finite group of individuals."

Yohzin strained not to swallow. "About the *Krepost*?"

"*Nyet*. Actually, it's old stuff, about a missile concept that we've already discarded. It was something that was tested at your old bureau."

"Oh." Yohzin strained to control his breathing. This did not sound good.

"Frankly, the GRU major is somewhat of a blabbermouth, or he felt compelled to impress me, because he spoke about the situation at length, probably much more than he should have. Not very prudent on his part."

Abdirov continued. "The GRU has identified ten personnel who were potentially exposed to the information in question. Only four of them are currently assigned to Kapustin Yar. One is dead. Obviously, for very practical reasons, the five who have departed here are under the greatest suspicion, because they would certainly have greater freedom of movement to transmit information."

"If I may ask, Rustam, why are you telling me this?"

"Because, Gregor, you are one of the four that they have identified here at Kapustin Yar."

Yohzin swallowed. "But, sir, did you not say that they've identified ten individuals who were *potentially* exposed to the information? Is that to imply that some of them might not have *actually* been exposed?"

"Correct, but I think that you can be confident that the GRU will eventually divine precisely who *actually* had access."

"Obviously."

"I don't know if you're familiar with how the GRU works, Gregor, but let me explain a few things to you, so you'll know what to expect," said Abdirov. "You can anticipate that the GRU will place you under even greater scrutiny than usual, even though they will probably focus their investigation on the five men who have left here, at least initially."

"Sir, do you think they'll haul them in for interrogation?"

"Certainly," answered Abdirov. "But only after they're surveilled them for a while. I think that it goes without saying that those men are due for some tremendous misery.

"In due time, if the GRU hasn't ferreted out the leak, then it's highly likely that they will whisk you away to Moscow to be interrogated. I don't know if you've experienced that before, but I have, and I can assure you that it will be an extremely unpleasant interlude. On a positive note, it's not the end of the world, but you won't be conscious of that fact until the GRU has pitched you out onto the street.

"Of course, what's truly ironic here is that the man who died might have been the guilty party, which leaves the surviving nine set to suffer mightily for his indiscretion. I suppose that doesn't lend you any solace, though."

"True," croaked Yohzin.

"So, tell me, brother, are you familiar with an American missile called the Sprint?" asked Abdirov, consulting a note card.

"I am," replied Yohzin. "The Sprint is a hypersonic, solid-fueled anti-ballistic missile with a nuclear warhead. It works in conjunction with a phased array radar system on the ground, which provides guidance to intercept reentry vehicles."

"Correct. To be specific, the compromised information in question concerns radar frequencies for the Sprint prototype."

"Sprint radar frequencies?"

"*Da.*"

Yohzin smiled. "I clearly remember how that information was handled, and I can assure you that I never even saw those frequencies," he explained. "All the information about the Sprint was secured in a special notebook. We were required to sign an access log to use the notebook, and it could only be viewed in a special area within the archives. Since radar information fell outside my area of expertise, I never looked at the notebook because I had no reason to. The GRU can check the access log to verify my claim, plus they can immediately identify those individuals who were granted access to the notebook."

Looking toward the floor, Yohzin felt a great sense of relief, as if a tremendous weight had been hoisted from his shoulders. What he told Abdirov was the absolute and unadulterated truth. Restraining an urge to laugh outright, he was almost giddy with delight; the fearsome GRU was not looking for *him*. He felt strange, though, since it was now abundantly obvious that someone else had been secretly passing information to the Americans. And then he also felt trepidation, since it was quite feasible that he could be inadvertently ensnarled in the GRU's dragnet, despite his meticulous efforts to cover his tracks.

He looked up to see Abdirov staring at him. The disfigured general's one-eyed gaze was like a powerful x-ray beam penetrating deep into his soul. As his scarred eyelid twitched, the corners of Abdirov's distorted lips turned up momentarily, as if he was blissfully content with the evidence that his trusted associate could not be found culpable, and then his smile faded just as quickly as it had appeared. He obviously realized that someone who was wily enough to purloin an intensely guarded secret code from a *Perimetr* engineer's notebook would find little difficulty in snatching a few radar frequencies from a much less secure source.

Abdirov spoke. "I think that it goes without saying that I need you here until we have the *Krepost* in orbit and the crew aboard. The timing is such that the GRU should still be working on their five priority suspects even as we launch. With any luck, they will have this unfortunate mess resolved long before it becomes too much of a nuisance for us, but as I said earlier, you should expect to be placed under greater scrutiny than usual. You should act accordingly, little brother, so that you don't arouse undue suspicions."

"I understand, Rustam."

"Gregor, I am confident that you are not the man that the GRU is searching for. I think that I can keep them at bay and steer them toward that access log you described. If need be, if worse comes to worse, I can pull some strings and have you placed under special duty here, under sort of a house arrest, at least until we get the *Krepost* upstairs. That way, the GRU can maintain a constant vigil on you while you work. I know that would be extremely awkward, but there is much work yet to be done, so I want you to be prepared to endure some hardships until this little storm subsides. Fair enough with you, Gregor Mikhailovich?"

"*Da*, Rustam. Plenty fair."

Residential Complex # 4
Znamensk (Kapustin Yar-1), Astrakhan Oblast, USSR
7:12 p.m., Thursday, September 7, 1972

Yohzin arrived home almost an hour earlier than he normally did. Excited, he was winded as he quickly climbed the stairs. Magnus bounded ahead of him with his pink tongue flailing from the side of his mouth.

Arriving at the third-floor landing, Yohzin paused to catch his breath and wait for his thumping heart to settle. His heart pounded in his chest not only because of exertion, but also out of sheer excitement. As terrible as it was, Abdirov's news about the leak was truly momentous. But there were important items to report as well, and only a limited amount of space available to summarize the developments.

As his sons studied their lessons in their room and Luba busied herself preparing dinner in the kitchen, Yohzin closed the door to his study, quietly locked it, and then sat down to write his report. He donned his reading glasses, slid open the bottom drawer of his oak desk, felt for a concealed latch, and removed a small box from a hidden compartment.

The box contained several objects, including a tiny cipher book, as well as several small sheets of graph paper specially treated to burn spontaneously and completely. There were also several strips of an almost translucent paper like onion skin, each approximately the width of a fingernail and about sixty centimeters—two English "feet"—in length.

He unscrewed the lid to a small plastic vial and extracted a dark brown pellet. To the untrained eye, the tiny pellet would resemble a medicine capsule, but a closer examination would reveal that it was not made of gelatin that would swiftly dissolve in gastric acid, but of a special acid-resistant plastic intended to remain intact during a transit of the digestive tract. First, after carefully composing his thoughts and distilling them down into the simplest phrasing, he wrote his weekly report on the graph paper, so that each individual letter occupied a single block. Since there was so much information to convey, brevity was crucial. In terse language, he conveyed that the *Krepost* was set for launch on Wednesday next week, with the Soyuz following the next day. Since his American handler—"Smith"—clearly knew that he was at Kapustin Yar, he felt no need to specify the launch site. He briefly described the potential delay with the second freighter and mentioned that it might cause the *Krepost* crew to return to Earth earlier than planned.

He reported the potential security breach that Abdirov had described; with that, he indicated that he would be under more intensive GRU scrutiny, but felt confident that he could continue reporting, at least until the noose grew too tight. The last portion of his message was the longest; he submitted a preliminary request for his family to be extracted. He didn't specify a time and date, but did identify a location—the remote cabin occupied by Luba's parents—by precisely plotted latitude and longitude coordinates. If nothing else, the Americans should be able to ascertain that his tenure at Kapustin Yar was drawing to an end, so they could formulate the necessary plans to be executed on a contingency basis. Not wanting to alarm the Americans, he did not report Abdirov's intent to drop the Egg.

Then, consulting the cipher book for the day's transcription code, he entered a second letter in each of the adjacent blocks. Satisfied with the transcription, he tore the day's page from the cipher book. Carefully writing in the tiniest of script, he painstakingly transcribed his encrypted report to one of the flimsy strips of onion paper. It was a much longer dispatch than usual, so it required writing on both sides of the strip. Once done, and after he had triple-checked it for any errors or other discrepancies, he lit a match and ignited the graph paper and today's cipher page. The two pieces of chemically treated paper disappeared in a flash.

He lit a cigarette from the still-burning match before carefully wrapping the strip of fragile paper around a sewing needle. He then gradually wound it progressively tighter until the snugly furled spool would fit neatly into the plastic capsule. Using the needle as a guide, he poked the spool into the capsule and then assembled the two halves. Holding the capsule in his left hand, he stashed the cipher book and his other secret writing materials in the drawer's hidden compartment.

Yohzin rolled the capsule between his palms as he contemplated his actions. He was amazed that he had maintained his deceitful routine for years, but there had never been the slightest indication that his activities had been detected or his behavior suspected. He contemplated his family's forthcoming departure. Ideally, as Smith had explained, an extraction agent would make physical contact with them before escorting them to a safe house, where they would remain for several days—or even longer—until it was safe to spirit them out of the country. From the way Smith had described it, the process would be like passing through a magic tunnel into an utterly new and prosperous life. They would be provided with *all* of their needs—including an automobile, a home, and brand new identities—in America. As far as the Soviet Union was concerned, Yohzin and his family would just simply cease to exist.

Yohzin wondered how Luba and the boys would react to the revelation that he was a traitor. Even though he had long been placing them all at enormous potential peril, they knew absolutely nothing of his clandestine activities. In fact, he literally passed his secret messages under Luba's nose. While he was certainly driven by mostly selfish motives, he hoped that they would eventually understand that he had risked everything for their benefit, so that they could live a better life outside of this repressive society.

Strolling out of his study, he consulted his wristwatch. He glanced into his sons' room to see the teenagers hunched over a ponderous stack of textbooks. Concealing the plastic capsule in his left palm, he walked into the kitchen and sniffed the air. He smiled; Luba was cooking lamb dumplings—*chebureki*—one of his favorite meals.

Magnus sat beside his dish, patiently waiting for his master's command. Glancing up to ensure that his wife was facing away, Yohzin dropped the opaque pellet into the dog's bowl and ordered, "*Essen.*"

Magnus responded obediently. The small kitchen was filled with chomping and crunching noises as the Alsatian swiftly devoured its dry kibble—and the capsule—in short order. In moments, the stainless steel bowl was scoured clean.

Yohzin examined the bowl, making sure that it was absolutely empty, patted the dog on the head and quietly said, "*Gut. Braver hund.*"

"You and your dog," said Luba, grinning as she turned away from the stove. "I don't know which one of you is crazier."

Yohzin walked into the kitchen and placed his hands on his wife's narrow hips. He brushed a smudge of flour from her nose before kissing her gently on the left cheek. "Luba, dearest, you know that I love you more than anything else in this world, don't you?"

Smiling, she asked, "Oh, Gregor, where is this going? Let me guess: You want another child, right?"

"I do, but we can wait. Luba, I need you to do something, but you cannot tell anyone. Not your friends and not even the boys," he said quietly.

"What is it, Gregor?"

"You know that they test some dangerous things here, don't you?" confided Yohzin, looking toward a window.

She nodded. "I've heard rumors that they sometimes launch rockets with atomic bombs on them. Is that really true?"

"Luba, you know that I can't be very specific."

"I know. So you are concerned that we are in even more danger now than we usually are?"

He nodded, and said, "You know that suitcase that we are required to keep packed for emergencies?" He looked toward an open closet in the living room; the valise, made of lacquered pasteboard, was visible at the bottom of the closet. It contained some basic things—an extra set of clothes for them and the boys, a few cans of food, some basic toiletries, as well as a few other items—two books and a deck of cards—to keep them entertained for a short interlude. Like the other residents in the apartment complex, they were mandated to always have the suitcase ready in case of a rapid evacuation.

"*Da.* I checked it just last week, Gregor. The women's auxiliary captain came by with some iron rations to stick in there. Anyway, are you concerned about anything specific? Would you like for me to open the bag so you can examine it?"

He cleared his throat and replied, "No. I am just very concerned that if we have to evacuate, then we may not be allowed to return here."

"Because of contamination?" she asked.

He nodded solemnly. "Assuming that we may not be able to come back, I want you to put some extra items in the suitcase: photographs, immunization records, the boys' report cards from school. Try not to be too obvious about it, but take down those family portraits in our bedroom, especially the ones of your parents and mine, and put them in there as well. I just want the boys to see where we came from, just in case we have to suddenly depart from here."

"Okay."

"And one more thing," he added. "I know it might be a little aggravating, but whenever we go on a trip, I want us to bring that suitcase with us, just in case."

"Just in case of *what?*" she asked.

"In case an accident happens here while we're travelling, and we are not able to return."

"Oh. That makes sense. You had said that Rustam might grant you a furlough so that we can visit my parents. If we go, should we bring it with us then?"

He nodded. "*Da*. Good idea. Better to be safe than sorry. And Luba..."

"Yes, dear?"

"Remember that you can speak of this to no one. We don't want to cause a panic."

Flight Crew Isolation Quarters
Burya Test Complex, Kapustin Yar Cosmodrome
11:15 p.m., Sunday, September 10, 1972

Four weeks prior to an impending mission, the flight crew and some critical support personnel were quarantined—in a manner of speaking—in a small cluster of secluded cottages. Vasilyev wasn't fond of his musty little cottage, but at least it was a quiet asylum; it was virtually impossible to get any decent rest in the dormitories where bachelor officers were housed. There, every night was filled with raucous parties and boisterous escapades. Of course, it was likely somewhat quieter than usual in the dorms tonight, since most of the residents had packed into the dayroom after midnight to watch the Soviet basketball team beat the vaunted Americans at the Munich Olympics.

Probably not by happenstance, the isolation cottages were maintained by a trio of spectacularly attractive maids, who just happened to reside in their own dwelling in the center of the cluster. In a bout of loneliness, Vasilyev did avail himself of a maid's company one night. He still felt terribly guilty the next morning, even though it had been over a year since his wife— Irina—was killed along with his daughters in a car crash.

His day had been long and excruciating; in the hectic pre-launch schedule, every single minute was spoken for. Unfortunately, today was largely devoted to tedious lectures concerning the internal wiring of the experimental fuel cell. Tomorrow, he and Gogol were slated to spend several hours in a launch procedures simulator. Right now, gazing at his cot by the wall, he just looked forward to putting his head on the pillow and falling asleep.

He climbed out of his coveralls, folded them neatly, and placed them in a chair. He brushed his teeth, washed his face, switched off the lamp, and eagerly slid into bed.

He had no sooner fallen asleep before he was jarred awake by pounding at the door.

Gogol's voice. "Open up!"

Holding his breath, Vasilyev was frantic. *What to do?* He could pretend like he wasn't here, not answer the door, wait until the commotion passed, and hope that Gogol would move on. With any luck, Gogol might assume that he had joined in on the drunken carousing over at the bachelor quarters.

"Open up *now!*" bellowed Gogol.

An idea flashed in his mind. He pushed the coarse wool blanket aside, jumped to his feet, quickly wrapped himself with a sheet, and quietly padded across the room. Grinning, he cracked open the door, put a finger to his lips, and winked at Gogol.

"So, you're entertaining one of those damned trollops?" blared Gogol. "So what! Let me in!" He jammed his broad shoulder against the heavy wooden door and shoved it open. As he staggered inside, Gogol was clearly in severe pain. He was no longer merely limping, but actually hobbling.

"I've depleted my fuel supplies," he declared, holding out an empty *Minsk Kristall* bottle. "I've come for a resupply." A half-smoked cigarette dangled from his lips. He was woefully disheveled, and although it seemed implausible, he was still clad in the same rumpled coveralls that he had worn yesterday.

Vasilyev retrieved an unopened bottle from the nightstand and handed it to Gogol. When the cosmonaut demonstrated that he lacked the dexterity to open it, Vasilyev twisted off the cap. Gogol threw the cigarette into the sink and took a long pull from the bottle.

Vasilyev sincerely hoped that the loathsome monster would be content to take the bottle and leave, so he could anesthetize himself with strong drink in the privacy of his own cottage, but that was not to be. Gogol belched loudly and then flopped onto Vasilyev's narrow bed, evidently intent on lingering.

"I thought...you had a woman in here," observed Gogol. "Or was that just a figment of your imagination? Anyway, bring yourself a cup, kitten. We who are doomed to die together will drink together."

"I don't think that it's a good idea, sir," replied Vasilyev. "You know that we have a very early day, and..."

"I said that we will drink *together*," declared Gogol. "And I am your commander, Comrade, so consider yourself commanded. Tonight, we drink, and we will accept our fates when the sun rises." A string of saliva dribbled from the corner of his mouth.

Vasilyev fetched the plastic tumbler that he used to brush his teeth. He reluctantly held it out, and Gogol filled it.

"Your toe still hurts?" asked Vasilyev. "You're limping badly for just a stubbed toe." He sipped from the tumbler, hoping that he could nurse the alcohol until Gogol eventually went on his way.

"I didn't stub my toe, idiot," muttered Gogol. "It's just an ingrown toenail, but now it's become infected and it's killing me." He dug two large white codeine pills from his chest pocket, crunched them between his stainless steel teeth, and chased them with a gulp of vodka. He repeated the process with two large yellow pills.

Gogol was as drunk as Vasilyev had ever seen him. Not only was his commander drunk, but he appeared to be in a drug-induced daze as well. *But that didn't mean that he couldn't still be dangerous*, thought Vasilyev. He glanced around the small room for potential weapons and made a contingency plan to bash Gogol with the empty vodka bottle if he got too frisky.

"Here," said Gogol, sticking his foot out as he leaned back on the bed. "Make yourself useful. Help me get this damned boot off."

As Vasilyev tugged at the boot, Gogol remained silent but still writhed in agony. His face was contorted in pain. Struggling to maintain his composure, he gritted his teeth as tears streamed from his eyes. Previously, Vasilyev wasn't even sure that Gogol knew the meaning of pain and suffering, now it was clearly evident that he was as mortal as anyone else.

As much as Vasilyev struggled, the boot just wouldn't budge. Suddenly, a few things made sense; Gogol had probably worn the same grungy coveralls for the past two days, or possibly longer, simply because he couldn't pull them off over the boot.

"The boot won't come off," stated Vasilyev. He found his pocketknife in the bedside table. "I'll have to cut it off."

"Do what you have to do," grunted Gogol. Turning the bottle up, he gulped down a copious measure of liquid fortitude.

Vasilyev sawed at the stiff leather with the serrated edge of his blade. Satisfied that he had slashed away enough of the boot, he gripped it tightly and yanked until it finally gave way.

Wincing, Gogol unleashed a torrent of expletives as the boot flew off and rebounded off the opposite wall.

The room was immediately filled with a reeking stench much like rotting meat. Gogol's swollen foot was encased by a cotton sock that had once been white, but was now tinged with yellowish-green pus and blood.

"That's infected!" declared Vasilyev, gagging at the powerful stench. Aghast, he slurped down the rest of his vodka. Gasping for breath, he immediately felt woozy.

"You have such a flair for the obvious, kitten," replied Gogol, slowly peeling off his sock as he quickly regained his normal persona. He dropped the sodden sock on the floor, guzzled from the bottle, and examined the foot. "*Da*, it's pretty damned ripe, isn't it?"

"Have the doctors seen it?" asked Vasilyev. Gogol's foot was badly discolored and swollen. He was obviously deluding himself as to the severity of the infection.

"I'm not showing it to the doctors before it heals up, because they'll ground me. And mark my words, Vasilyev, I'll clobber you if I catch you saying anything behind my back."

Grimacing as he wiggled his big toe, Gogol grunted. He gingerly squeezed the digit, causing thick green pus to ooze from the edges of the nail. "This is infuriating," he said. "I've soaked it in salts and peroxide and have even tried turpentine, but nothing helps."

"Then why don't you go to the flight surgeon? Certainly, he could do something." Vasilyev could not believe that he had let it fester this long without seeking medical help.

"*Nyet*. One of the orderlies snuck me some antibiotics yesterday, and those will likely fix it in due time. After all, it's only a damned toenail that's ingrown and infected. It's not a problem," he asserted. "If need be, I'll borrow some tin snips at the sheet metal shop and just lop it off."

Familiar with the cosmonaut's incredible exploits, particularly his three-year trek across Mongolia, Vasilyev had to wonder if Gogol was serious about amputating his own toe. "You would really just hack it off?" he asked.

Gogol's calm reply dispelled any doubt. "I would not hesitate. I swore to sacrifice my life in the defense of the Motherland, so a toe is nothing to weep over. I would take an axe to the whole foot, if necessary."

Vasilyev heard a faint tapping at the door, followed by a feminine voice softly speaking. "Are you all right, Pavel Dmitriyevich? I heard shouting, and it frightened me."

"I'm fine, Natalya," replied Vasilyev. "Go back to your cottage. Everything is okay."

"But I'm lonely, *lubov moya*. Can I come in? *Puzhalsta?* Don't you want me to spend the night again?"

"Not tonight," answered Vasilyev. "I'm very tired, and I have an early morning."

"But I'll be good. I'll just keep you company, I promise. You shouldn't sleep alone."

"Leave us be!" bellowed Gogol, clearly losing his patience. "Can't you take a hint, woman? If *mily* Pavel wanted *your* company, he would have taken it. Now, be on your way. Find another bed to crawl into."

The maid left without further comment; Vasilyev heard the faint crunching of gravel on the walkway outside and listened as the sound gradually faded away. If there was a bright side to the strange events of this evening, he probably needn't worry about Natalya darkening his door again.

"It seems like you've been a busy boy," said Gogol. As he spoke, blood and pus dribbled from his bare foot onto the scuffed linoleum floor.

"Just once, Comrade Commander, I promise. I was very lonely and needed a distraction."

"Trust me, Pavel, I know all about loneliness, probably more than you could ever fathom. I'm not judging you."

"*Spasiba*, Comrade Commander."

Almost an hour passed as they shared the bottle and talked about the pre-flight preparations to ensue in the coming days. As drowsy as he was, Vasilyev did his best to keep pace with the conversation. Remembering that Gogol could be dangerous, he also kept an eye on the empty vodka bottle on the floor by his feet.

Vasilyev had seldom seen Gogol nearly so talkative and assumed that the commander's relaxed state had to be the product of mixing vodka with powerful painkillers.

With Gogol as inebriated as he had ever seen him, Vasilyev thought that he could seize upon this opportune moment to fill in some gaps in the puzzle. "Comrade Commander," he said meekly. "Would it offend you if I asked about the materials that you stowed on the resupply freighter?"

"*Nyet*," grunted Gogol. "What do you want to know?"

"Those items that you stowed on the freighter to augment our survival kits…"

"That's nothing to be curious about," answered Gogol. "I don't think that our standard survival kits are adequate, so I'm bringing the items that we will really need when we return to Earth. Weren't you listening when I explained that to the chief packer?"

"I was, but don't you trust the search and rescue people to find us?"

Gogol chuckled. "Not only do I not trust them to find us, I don't expect that they will be looking at all, especially since they will likely already be dead by that time."

"Dead? Do you really think that it's even remotely possible that we will drop the Egg?"

Gogol was silent for a moment and then spoke with surprising candor. "Not only do I think it's *possible* that we will drop the Egg during our stint, I am absolutely *sure* of it, so we must plan accordingly."

"How can you be so certain?" asked Vasilyev.

"Because I am!" declared Gogol, glowering. "And nothing more need be said."

Vasilyev nodded in silent affirmation and gingerly held out his plastic tumbler.

"Is there anything else you need to know, kitten?" asked Gogol, refilling the tumbler.

"The maps," said Vasilyev. "I saw the marks. Are those recovery sites?"

"*Da,*" answered Gogol. He raised the bottle, a signal for Vasilyev to drink.

Vasilyev swallowed a mouthful of vodka. His face felt warm, and the rustic interior of the cottage started to gradually rotate. Gathering his bearings, he looked at the floor and made sure that the depleted vodka bottle was still within reach. Confident that he could still defend himself, he asked, "But the locations on your charts are not the recovery sites that we were assigned, correct?"

"Correct," replied Gogol. "I chose them myself. I selected places that are remote and uninhabited. We can forage and hunt for food, no matter where we come down.

"I don't suspect that there will be much left of North America. Most of the major cities in the United States will be gone. The interior should be fairly safe, except for places where there are military bases or ICBM fields. Canada should be unscathed, and most of Alaska should be safe as well, but with our inclination, they will be out of reach. The Southern hemisphere will probably be our best refuge. Most of Australia should be okay. South America should be safe. I would prefer to stay clear of Africa, since we obviously aren't going to blend in very well there. I'm sure that we will have to fend for ourselves for some time, but eventually we will have to make contact with people. I would prefer to land where English is spoken, since we both speak a little of it from our training for international flight.

"I packed the multi-band radio to determine how nearby populations might be faring, and also to determine which areas are safe and which ones might be contaminated. Obviously, we'll need to pay close attention to the winds, since they can carry fall-out."

Vasilyev was surprised and also very impressed; despite his drunkenness, Gogol's thoughts were evidently very lucid; he had obviously reflected on these matters for a very long time. "I'm very impressed," he said. "You've certainly applied some thought to this."

"I have," replied Gogol. "I have thought about it for both of us, Gregor." He scooted close to Vasilyev and placed his arm around his shoulders. Vasilyev flinched at the unwelcome gesture. As if to calm him, Gogol caressed his arm.

"I know that you're not comfortable with me," said Gogol. "And I'm sure that you don't approve of my behavior, but I think your reluctance will melt after we've spent more time together."

Vasilyev shuddered at the ominous implications. His stomach turned as he envisioned the two of them nestled together in the intimate quarters of the *Krepost. Five weeks? And then, perhaps, roaming a shattered planet for the rest of our lives?*

Gogol kissed his cheek and tousled his hair. "I may never win your heart, Pavel, but I think that you'll come to appreciate me in due time," he said. "If nothing else, once we return to earth, if you want to live, you'll learn to do whatever I say, without question."

Numbed by the codeine and alcohol, Gogol's speech grew progressively more slurred. Only barely coherent now, he seemed only moments away from lapsing into a stupor.

"I've enjoyed our little chat, but I'm very tired, kitten." With that, Gogol slumped over and promptly fell asleep.

Krepost Project Headquarters
2:15 p.m., Monday, September 11, 1972

In the decades that he had known Abdirov, Yohzin had never witnessed his friend so irate. The general was livid after reading an official report from the *Krepost* program's chief flight surgeon.

Abdirov crumpled the report and pitched it against the wall. "Damn it! Damn it! Damn it!" he ranted, pounding his fist on his desktop. "An *ingrown toenail!* How can something so simple cause such an awful predicament? How could Gogol have let this damned thing fester for so long without anyone knowing?"

"It's merely a setback, Rustam," offered Yohzin. He heard loud klaxons blaring and looked past Abdirov, through a gap in the curtains, to watch the hectic activity associated with moving the Proton rocket and its *Krepost* payload from the processing facility to the launch site. Transported on a specially built flatbed railcar, rolling slightly faster than a snail's pace, the massive booster had just barely emerged from the processing building.

"A setback, Gregor? *Nyet*. Not hardly. I have committed to this launch," said Abdirov frantically. His voice was filled with unadulterated anguish. "Everything is set, and there's no turning back. I am such an idiot. I wagered *everything* on Gogol, and he has screwed us royally."

"But, Rustam, perhaps Gogol can fly later, and then…"

"There is *no* later. Those *Perimetr* demons are already on the verge of snatching this entire mess from me. It's merely a matter of time before they perfect their new interlock. And if the second freighter is not delivered on time and we are forced to abandon the station on orbit, then it's all over as well."

Yohzin was relieved; it was very likely that this new development would almost certainly derail Abdirov's scheme to deploy the Egg, particularly if the second freighter could not be launched on schedule. If nothing else, the delay might allow him sufficient time to persuade Abdirov to just accept the ongoing stalemate between East and West.

He was still confounded by Abdirov's seeming inability to comprehend that he—or anyone else, except perhaps that psychopath Gogol—did not necessarily share in his unbridled enthusiasm for the destruction of civilization. Sometimes, he wondered if a portion of Abdirov's soul was destroyed in the same fires that scorched his body.

Abdirov's secretary quietly tapped on the door and stuck her head in. "Comrade General, Majors Vasilyev and Travkin are here to see you," she said. "They are in the anteroom."

"Wait five minutes and send them in," replied Abdirov.

"Would you like me to serve tea, Comrade General?" asked the secretary.

"*Nyet*," snapped Abdirov. "We will have no time for tea or cookies."

Abdirov swallowed a white tablet, followed it with a sip of water, and then closed his eye as if to take a brief nap. In short order, the outward signs of his anger seemed to dissipate. He seemed calm, if not somber.

Vasilyev and Travkin marched in and formally reported. Standing before Abdirov's broad desk, their postures were so stiff that they resembled lead soldiers or mannequins in a Moscow department store.

"At ease, gentlemen," said Abdirov. "Vasilyev, were you aware that your commander had an ingrown toenail that was infected?"

"*Da*, Comrade General. I was. I am also aware that he was trying to treat it himself."

"And you did not think it might be prudent to inform someone, perhaps the flight surgeon, of Gogol's condition?"

Vasilyev swallowed deeply and then replied, "Comrade General, I felt confident that he was managing the situation, and that his toe would be adequately healed in time for flight."

"You were mistaken, Major. Gogol is presently occupying a bed in the infirmary, and will be incapacitated for at least the next two weeks. His foot is badly infected, and there's a very strong possibility that the surgeons might be compelled to amputate it in order to spare his life."

Travkin gasped audibly.

"I take it that this is all a surprise to you, Major Travkin?"

"It is, sir."

"Then so long as we're letting cats slip the bag, here's another tidbit of news," said Abdirov. "Both launches will proceed as planned, on schedule. Travkin, you will accompany Major Vasilyev in the Soyuz. I have directed the launch preparation crew to pull Gogol's couch out of the Descent Module and replace it with yours."

"Comrade Major Vasilyev," said Abdirov, pivoting in his chair.

"Comrade General!" barked Vasilyev in reply.

"You will command the *Krepost* mission," said Abdirov. "You'll receive the formal orders later today."

"*Da*, Comrade General."

"Not that you have an option, Vasilyev, but do you feel confident that you and Travkin can be ready to launch on time?"

"We will be ready, Comrade General," replied Vasilyev. "But the freighter will have to be cracked, so Travkin's items can be substituted for Gogol's."

Contemplating the situation, Abdirov closed his eye and tapped his few remaining fingers on his desktop. "*Nyet*," he replied, opening his eye. "We're not going to fool with that. Right now, your sole priority is to ensure that Travkin is proficient on launch and rendezvous procedures, so we can get you two off the ground on schedule. Understood?"

"*Da*, Comrade General," replied Vasilyev. "But you should be aware that Travkin has understudied us on the…"

"No matter," snapped Abdirov. "You'll lock into the simulators, Major, and be ready. We're not going to fritter away any valuable time swapping goodie bags on the freighter. Travkin can live with whatever Gogol cached. Understood?"

"*Da*, Comrade General."

"Good," replied Abdirov. "You gentlemen have much to do, and none of it can be accomplished here, so be on your way."

5

ON ORBIT

Krepost Station, On Orbit
12:02 p.m. GMT, Sunday, September 17, 1972
GET (Ground Elapsed Time): 3 Days 11 Hours 15 Minutes, REV # 55

Vasilyev had been in orbit for three days. Staring out through a porthole, he was captivated by the sublime and tranquil beauty of the heavens. Stargazing was one of his favorite pastimes, whether here on this majestic perch or back on earth. On summer nights, he and his wife used to bicycle to a nearby pasture, spread out a blanket, and lie under the sky for hours on end. But Irina was gone now, and he would never share the stars with her again.

As much as Vasilyev enjoyed looking outwards, Travkin preferred to look down at the earth. The fact that they could share the limited portholes without bickering was a testament to their compatibility. Like human embodiments of yin and yang, the two men were effectively polar opposites, but they were also the best of friends. They were almost ideal companions, compatible in every regard, perfectly suited to live and work together for their five-week excursion. Their personalities meshed perfectly. Vasilyev was a self-admitted slob, entirely content to dwell in clutter and disarray, while Travkin was a compulsive neatnik who contributed his extra moments to tidying up the station, fixing broken gadgets, and fastidiously organizing their domicile.

Vasilyev lightly pushed off the wall and floated over to the opposite porthole to glance down at the earth. Just a few minutes from orbital dusk, they were passing over the Kazakh Republic, just north of the Caspian Sea. He glanced at the shiny pilot's chronometer on his wrist: his eight-hour watch was drawing to a close. It wasn't as if they paid strict attention to the clock and lived their lives by the chiming of bells. Travkin had been puttering around for almost six hours now. Their actual assigned watches overlapped by two hours.

"You're on, Petr," announced Vasilyev. "I reluctantly relinquish the controls to you, if but briefly. No big changes: the stars still twinkle above us, and the good earth beckons below. All is sound and shipshape, my slothful comrade, so the bell is rung and the watch is officially yours. "Floating up from the docking hub, Travkin yawned broadly, zipped up his light blue coveralls, and then replied, "I officially acknowledge that I have the watch, my most officious commander. So, Pavel Dmitriyevich, are you up for another marathon round of *durak*?"

Vasilyev pondered the thought of playing cards for the next several hours, groaned, and shook his head.

"Cribbage perhaps? Or maybe chess?" queried Travkin. "Unless you've grown weary of constant defeat..."

"*Nyet*," answered Vasilyev. "No games today. We have a targeting update in ten minutes. I'll spell you on that one and give you a break."

"Thanks. My hand is still cramped from yesterday's."

Although their responsibilities were great, and the consequences of their actions potentially greater, their watch chores weren't particularly demanding. As they passed over their communications windows in the Soviet Union, they monitored the radios for any updates or instructions. During daylight passes, they were expected to spend at least part of the time peering down through powerful binoculars, watching for missile contrails, mushroom clouds, or any other telltale evidence that the civilized world had suddenly pitched into wide-scale thermonuclear war.

The station was also outfitted with a substantial battery of detection systems to warn of missile launches, including a sophisticated electro-optical sensor array. The electro-optical sensors were intended to detect sudden flashes, but the cosmonauts had found that they were almost too effective at doing so. The detectors could be accidently triggered by lightning flashes, such as those common to the massive thunderstorms passing through the same areas of the American Midwest where ICBMs were maintained in underground silos. Consequently, Vasilyev and Travkin were compelled to occasionally tweak the detectors to inhibit their sensitivity. Additionally, Control routinely kept them posted on weather conditions at the locations where American ICBMs were known to be stationed.

They regularly rehearsed deploying the Egg, usually during drills initiated by their mission controllers back on Earth. Additionally, at least once during every twenty-four-hour watch cycle, they were expected to conduct at least one emergency drill, to rehearse their procedures for such contingencies as an immediate evacuation, an electrical failure, a micrometeorite strike or onboard fire. But since their main tasks were merely to watch and wait, they had a tremendous abundance of leisure time. Vasilyev was constantly amazed at the amount of idle time he had, particularly since he rarely slept more than six hours at a stretch. And six hours was plenty, because if he slept any more than that, then he would dream, and his dreams invariably dwelled on the past that he wanted to forget.

10:07 a.m. GMT, Monday, September 18, 1972
GET: 4 Days 9 Hours 22 Minutes, REV # 70

Hoping to snatch a catnap, Vasilyev had barely gone to sleep when he was jolted awake by the emergency alarm, a wailing klaxon not unlike the alarm used on submarines. He punched the red plunger button on the bulkhead, silencing the earsplitting noise, grabbed the cloth satchel that held his emergency breathing mask and other gear, and immediately soared "downstairs" toward the control module.

"Petr! What's up?" he yelled, skimming effortlessly through the air.

"Action message!" replied Travkin, as yet unseen at the far end of the station. "Immediate deployment!"

"On my way!" shouted Vasilyev, listening to the action message amplified over a loudspeaker in the control module: *Krepost! Krepost!* Action...Action...Action...Enemy bomber attack in progress...Deploy device on first available contingency target...Deploy device on first available contingency target. Authorization code: Eight-Zero-Two-Two-One...Authorization code: Eight-Zero-Two-Two-One."

As he had been conditioned to do in countless drills, Vasilyev immediately memorized the all-important deployment authorization code. He was aware that Travkin would already be entering the sequence into the Egg's targeting computer.

He arrived in the control module just as Travkin punched in the last digit of the authorization code into the Egg's targeting computer. Hovering over the Egg's control console, he studied the read-out display as he plugged in the jack to his communications headset. He corroborated the numbers, keyed his microphone and concisely stated, "Control, this is *Krepost.* I verify receipt of deployment authorization code Eight-Zero-Two-Two-One to deploy device on first available contingency target."

"I confirm your receipt of authorization code Eight-Zero-Two-Two-One," replied Control. "Execute deployment."

"Key check," said Vasilyev, displaying the arming key attached to a small chain around his left wrist. Travkin held out his left hand to solemnly display a matching key. Vasilyev nodded and said, "Keys are verified. Open the code locker."

Travkin opened the safe and extracted the targeting book that Vasilyev had updated just slightly more than three hours ago. He traced his finger down the target list, double-checked the mission clock, and announced, "Comrade Commander, the first available contingency target is Seattle, Washington: Latitude 47 Degrees 36 minutes North, Longitude 122 Degrees 20 Minutes West."

Travkin's manner was usually lighthearted and jovial, but he was strictly business right now. As well he should be; both cosmonauts stood to earn a mission bonus of ten thousand rubles if they precisely executed all of their tasks in a timely manner. Although this could be an actual Egg deployment, where the stakes were at their highest, it could be one of the many scored reliability exercises they would undergo during their mission. So, for very tangible reasons, even the simplest error was unacceptable; their incentive payoff was contingent on receiving nothing but perfect scores.

Vasilyev examined the target list. "I verify that the optimal target is Seattle, Washington: Latitude 47 Degrees 36 minutes North, Longitude 122 Degrees 20 Minutes West. Enter the information."

Without comment, Travkin dutifully punched in the coordinates into the Egg's targeting computer.

"I verify the target coordinates," stated Vasilyev, looking at the numerical readout over Travkin's right shoulder as he compared the displayed digits to the target list. "Lock the fix."

"I am locking the fix," replied Travkin, swiveling the red key that locked the coordinates into the computer. "Platform aligning. Stand by for deployment data estimate."

"Standing by."

While he waited for the computer to complete its calculations, Travkin tucked the targeting notebook into its cloth pouch and replaced it in the code safe. Two minutes later, a green light pulsed on the computer display. He checked it, and announced, "Comrade Commander, platform is aligned. Twenty-two minutes to braking rockets."

Twenty-two minutes? Vasilyev whistled shrilly and then ordered, "I verify twenty-two minutes to braking rockets. Man your interlock station and stand by for deployment." With that, he lightly kicked off the lower bulkhead and drifted upwards to his own interlock station approximately four meters away. Once the target fix was locked in the computer, both men had to simultaneously turn their arming keys. The final step to complete the sequence was for Vasilyev to press the Arm button.

After their keys were turned and the Arm button pushed, they would linger at their stations just long enough to receive verification that it was a live drop or a reliability exercise. If it was not just another reliability drill, explosive bolts would fire to jettison the Egg.

After activating a time-delayed explosive charge to destroy the *Krepost*, Vasilyev and Travkin would scramble to the Soyuz docked at the other end of the station, close the hatches, assume their positions in the cramped cockpit, power up the electronics, and undock. Afterwards, they could linger in orbit for up to five days before descending to a designated recovery site. Of course, if the Egg was actually dropped, then it was a foregone conclusion that they would eventually return to a world vastly different than the one they left.

"Insert arming key. Rotate arming key on my mark," declared Vasilyev.

"Arming key inserted. Standing by for your mark," replied Travkin.

"Five, four, three, two, one, mark." He gritted his teeth as he swiveled his arming key and then pushed the Arm button to initiate the deployment sequence. *Was this the end of the world? Was this the nuclear showdown that he so dreaded?*

"Deployment sequence commenced," confirmed Travkin.

Holding his breath, Vasilyev stared at the verification light. Several seconds passed before it blinked yellow three times, signifying that the drill was merely a reliability exercise. He breathed a sigh of relief, checked his chronometer, smiled, and said, "Stand down from the drill. Once again, Petr, we have instigated Armageddon in record time."

Aerospace Support Project Headquarters
8:12 a.m., Monday, September 18, 1972

Ourecky trudged into the Project's headquarters, presented his credentials to the guard at the security desk, and then slowly slogged up the four flights of stairs to the Flight Crew Office. As part of his morning ritual, he rinsed out his white porcelain coffee mug in the sink and poured a cup from the percolator. Turning around, he was surprised to see Carson seated at his desk.

"Back already?" asked Ourecky, stirring a teaspoon of sugar into his coffee. "I thought you were supposed to be in Pensacola for at least another week."

"Virgil called me back last night," replied Carson, barely looking up from his newspaper. The compact pilot wore tan chinos, a powder blue Ban-Lon knit shirt, and Adidas athletic shoes.

"So, could you tell me again why they sent you down there?"

Carson folded the paper, set it aside and replied, "To get qualified to land on aircraft carriers. The Ancient Mariner has a wild scheme about using them as contingency landing sites. You know that, Scott."

"Still smells very fishy to me," observed Ourecky.

"*Everything* that Tarbox says is at least a little bit fishy. That's new to you?"

Ourecky nodded as he struggled to clear his throat. He sat down and examined the newspaper's front page. The headlines were dominated by stories concerning a massive wheat export deal with the Soviet Union, a clash between Israeli and Syrian forces, and President Nixon's campaign to curtail the flow of heroin into the United States.

"Landing on carriers? I just feel like there's a lot more to the story," said Ourecky.

"Maybe. So, how was therapy?" asked Carson, changing the subject.

"Miserable. It's a lot like getting your lungs scoured with a steel wire brush," grumbled Ourecky. "And how was *your* morning?"

"Clearly, much more pleasant than yours. Hey, Scott, don't get too comfortable yet. I bumped into Virgil before I came up. They want us downstairs at ten hundred. Tarbox arrived this morning, too, so I'm guessing that something big is going on."

"Hmmm...Tarbox? I didn't think he was supposed to visit again until next week." Ourecky blew lightly on the surface of his steaming coffee, cleared his throat, took a tentative sip, and mulled over the situation. More often than not, when he paid a visit to the Project's facilities, Admiral Tarbox usually arrived with an entourage of Pentagon bigwigs in tow. "Do you think it's *another* VIP visit? You know, for an organization that's not supposed to exist, we sure are getting more than our share of visiting dignitaries lately."

"Nope," replied Carson. "I don't think it's another VIP boondoggle. I'm pretty sure it's an operational situation, maybe even another mission."

Ourecky grimaced. "This does sound ominous. Do you really think we're going up again?"

"Looks like it," answered Carson, folding his newspaper. "Something is definitely in the wind. I swung by Gunter's shop after I saw Virg, and he had all hands on deck, running trajectories and intercepts on some mystery target. I asked Gunter about it, but he wouldn't share."

Ourecky nodded; he was more than familiar with the German engineer's tendency toward stubborn reticence. He coughed repeatedly and then blew his nose. "I'm still on the mend," he observed. "Surely they can't expect to send me back up in this condition, no matter what's upstairs."

"Maybe," answered Carson. "But if this thing is as big as I suspect it is, what choice do they have?"

Ourecky sat down at his desk, pondered Carson's rhetorical question, and replied, "Yeah, Drew, I suppose you're right: What choice do they have?"

9:55 a.m.

Taking his place in the conference room, Ourecky studied his surroundings as he waited for the meeting to commence. Much had happened in the brief interlude following their last mission. As Admiral Tarbox gradually gained influence, the Project was swiftly evolving into a different place, with a decidedly different atmosphere.

Looking around, Ourecky realized that this very room was a tangible reminder of Tarbox's broad sway. Once an austere briefing area, with spare trappings reminiscent of a battlefield command post, the space had been recently outfitted as a lavishly appointed, extravagant conference room. No expense was spared in the renovation. It was certainly the most well-appointed meeting space that Ourecky had ever been in, and probably rivaled similar facilities in the headquarters of major corporations. An adjacent room had been repurposed as an audio-visual projection room replete with a state-of-the-art 16-millimeter movie projector, a 35mm slide projector, and an elaborate sound system.

One wall of the conference room was devoted to official photographs of the Air Force pilots assigned to Blue Gemini, as well as the aviators who participated in the Navy's MOL program. Most struck gallant poses in their space suits. The subjects of the ornately framed portraits included "Big Head" Howard and "Squeaky" Riddle, as well as Ed Russo. Commander Chris Cowin's picture also hung on the wall, but the Navy astronaut's smiling visage hardly resembled the bloated, discolored face that Ourecky had witnessed in orbit. Of all those who had been assigned to the Project or the MOL, whether they had flown or not, the sole missing man was Tim Agnew; there was apparently adequate space for dead heroes, but not for the faint of heart who weren't anxious to join their ranks.

A larger wall was adorned by a rogue's gallery of Soviet satellites targeted by Blue Gemini. The centerpiece was the crisp image of the brass data plate from the first mission Carson and Ourecky had flown. In retrospect, Ourecky regretted taking *that* photograph, since his present circumstances would be immensely different if that mission had been a dismal failure rather than a serendipitous success. After the fatal launch accident on the first mission, which killed Howard and Riddle, the Project likely would have been quickly curtailed if he and Carson had returned to Earth as Tew had directed. He would have likely left the Air Force as soon as his initial contract expired and would already have his doctorate from MIT. Instead, even when confronted with a catastrophic failure of their Gemini-I's main batteries, he and Carson had pushed their luck, against Tew's orders, but had delivered the goods.

He glanced at his Timex watch and saw that he was almost overdue for one of his prescription medications. He fumbled with a plastic medicine bottle and spilled most of the pills. His clumsiness had nothing to do with his current medical condition. Most of the fingers of his right hand were literally numb from almost constantly shaking hands. In the past two weeks, a steady stream of dignitaries—generals, admirals, and elected officials "read on" to the highly classified Project—flowed through the space. They made the pilgrimage to receive a terse informational briefing—heavily slanted towards the new joint program—and to heap kudos on the heroes who had made it all happen.

Carson helped him collect the last of the pills just as Tew and Wolcott swaggered through the door. Ourecky was mystified by Tew's demeanor. In recent months, the general's health had been on a rapid decline. He had devolved into a fragile wisp of the man Ourecky had once known, but this morning he seemed vibrant and animated. Obviously sharing some sort of secret agenda, he and the grizzled ex-cowboy grinned and acted almost like a pair of fraternity brothers scheming up new hazing rituals for freshman rush.

Minutes later, Tarbox strolled in, trailed by Gunter Heydrich and Blue Gemini's intelligence officer, Colonel Seibert. In contrast to the jubilant generals, the admiral seemed like his normal acidic self, glowering like Ebenezer Scrooge compelled to dole porridge to unwashed foundlings in a Victorian orphanage. Frowning, he nodded toward Wolcott.

"Ted," drawled Wolcott, gesturing toward Seibert. "Would you be so kind as to share your latest scoop with Admiral Tarbox and our valiant heroes?"

"Gentlemen, we have very preliminary evidence that the Soviets successfully orbited a nuclear weapons platform five days ago," disclosed Seibert tersely. "A Proton was launched with a large payload into a stable circular orbit at roughly fifty-one degrees inclination. This object was initially cataloged as a discarded booster stage, until we received an intelligence report from a well-placed source that indicated it was, in fact, a weapons platform."

Exuberant, Tew acted like a fidgeting child eager to rip the shiny paper from his neatly wrapped Christmas presents. If nothing else, thought Ourecky, the monumental piece of news—if it was true—certainly vindicated Tew's oft-stated assertions about the Soviet's intent to station nuclear weapons in space. If the intelligence was valid, the general's long-standing obsession had finally come to fruition.

"Correct me if I'm wrong, Colonel," said Tarbox skeptically. "But is this not the same intelligence information that you disavowed just two weeks ago? Didn't a source indicate that the Soviets were preparing to launch a Proton, and you claimed the source was just spewing malarkey?"

Chagrined, Seibert stated, "Sir, our initial analysis was incorrect. As it turns out, the Soviets were preparing to launch another Proton, just as our source claimed, only it was launched from Kapustin Yar, rather than Baikonur. Moreover, the same source provided intelligence of *another* object, possibly manned, originating from Kapustin Yar. They apparently launched an R-7 on the following day and there are clear indications that its payload was a manned spacecraft, probably a Soyuz."

"*Hmmph*," snorted the wizened admiral, shaking his head. "I'm plenty familiar with Kapustin Yar. The Soviets do *not* launch manned flights from there."

"Perhaps, sir," said Seibert, "but NORAD verified that the second object was also injected into orbit at fifty-one degrees inclination, just as the source claimed, and we have ample reason to believe that it rendez-voused and docked with the first object, so we strongly feel that our source is indeed telling the truth. We contend that this is a manned nuclear weapons platform."

"So, this is the *Krepost* you warned us about?" asked Tarbox.

"Apparently, it is the *Krepost*, Admiral," replied Seibert.

Ourecky whistled quietly.

"Do you have something to contribute, cowpoke?" asked Wolcott.

Ourecky shook his head and quietly said, "I'm just taking all this in, Virgil. It's a lot to absorb."

"So, when do we go?" asked Carson, stifling a yawn. He seemed almost bored.

"It's hard to say," answered Tew. "Maybe soon, maybe in a few weeks, maybe in a few months, but maybe never. Obviously, we can't send up Ourecky anytime soon, at least until he heals up."

"Speakin' of which," drawled Wolcott. "How's your lung therapy coming, pardner?"

"Good, sir," replied Ourecky, stifling a cough. "I should have a clean bill of health within a month, possibly even sooner."

"That's good," interjected Tew. "But that's not our only concern. I've considered it at length, and I'm also not willing to potentially jeopardize the lives of you boys by sending you up against a manned target. That's just too dangerous. We will have to be patient and bide our time until we are confident that there is not a crew aboard. But unfortunately, we don't know when or even if that will be the case, so we can't afford to just sit on our hands and wait."

"All of this is well and good," snapped Tarbox. "But even if this *Krepost* is up there, as you say, can we reach it? We certainly can't attain fifty-one degrees inclination launching out of Vandenberg. Your Johnston Island site is wrecked beyond repair. Is this situation even worthy of serious discussion?"

"It is, Admiral," interjected Gunter Heydrich. "We've analyzed the situation, and we're confident that we can safely achieve fifty-one degrees out of Cape Kennedy. Our tentative plan is that we will launch at night, under the pretext of a limited notice contingency test. Our launch personnel are already headed to Florida to assess available launch sites. When and if an opening presents itself, assuming that Ourecky is physically capable of flying at that juncture, we intend to be ready to exploit the opportunity."

Ourecky looked at the polished walnut surface of the new conference table. *Our tentative plan? Launch personnel are already headed to Cape Kennedy? We intend to be ready to exploit the opportunity?* None of this sounded very tentative to him. It seemed very obvious that he and Carson were going up again, more likely sooner than later.

"Perhaps it won't be necessary to send up Carson and Ourecky again," noted Tarbox. "My most proficient crew has been here for two weeks already, working in your simulators. I am confident that they will be ready to go in short order."

"Duly noted," said Tew. He coughed, then turned towards Heydrich and asked, "What's your assessment, Gunter?"

"I concur with the Admiral that they are a very solid crew," answered Heydrich, referring to his notes. "When they arrived, they were already extremely proficient in retrograde and paraglider flight procedures. Honestly, since they were already familiar with the Gemini-B cockpit and systems, all we had to do was acquaint them with the Gemini-I layout. Obviously, launch and ascent abort procedures are significantly different between the two platforms, but they are quickly adapting to our methods."

"Good," replied Tew. "And the intercept procedures?"

Frowning, Heydrich answered, "They are making excellent headway, General, but they are still weeks away from achieving the degree of proficiency where we might consider them operationally ready."

"Weeks?" asked Tarbox.

"*Ja*," answered Heydrich. "And that is a generous estimate on my part. It might be months before they are anywhere close to the same level as Carson and Ourecky, if that's even humanly possible."

Ourecky smiled slightly, and then frowned as he realized that while he and Carson meshed so well that their expertise might not ever be replicated, their shared proficiency might also be their downfall. At this point, he welcomed the notion of being replaced by another crew.

"So, truthfully, they may not be ready anytime in the relatively near future, correct?" asked Tew.

Heydrich nodded solemnly.

"I think that Gunter's bein' just a tad bit harsh on these Navy boys," observed Wolcott. "I'm sure that it ain't goin' to take months to get them ready to fly."

"Perhaps, Virgil, but assuming that we must be prepared to execute on short notice, Carson and Ourecky will have top priority on the simulators," stated Tew. "Gunter, make any adjustments necessary to run two shifts, if need be, so we can accommodate our Navy brethren."

"That's a problem," said Heydrich. "Since we have lost so many people in the transition, I barely have enough personnel to staff one shift, much less two."

Tew turned toward Tarbox and asked, "Can you bring in some of your people from California to reinforce Gunter's crew?"

Nodding, Tarbox replied, "I will."

"Good," said Tew. "I'll see if we can dredge up some emergency funds to hire back some of Gunter's staff, at least on a temporary basis. And just for the sake of clarity, Gunter will be our certifying authority for your crew. When and if he blesses off on them, I will consider assigning them to the flight, but Gunter will make the call."

"So be it," said Tarbox grudgingly.

Ourecky knew that Tarbox was anxious to push Tew aside. The meddling opportunist clearly wanted to gain control of Blue Gemini sooner than later, but this new *Krepost* revelation had created an entirely new wrinkle in the transition plan. Now, with Soviet nuclear weapons actually orbiting overhead, the very reason the Project had been authorized, Tew didn't seem inclined to surrender the reins to Tarbox or anyone else until the job was done.

"Let's move on," declared Tew, lightly slapping the polished surface of the table. "Refresh my memory about this *Krepost*. Have the Soviets made any significant changes since you briefed us about it in March?"

"They've made a few minor changes that we are aware of," answered Seibert. "But the core design has remained essentially the same." The intelligence officer flicked a switch. The lights dimmed, and a projection screen spooled down from a slot in the ceiling. Seibert flicked another switch to turn on a slide projector.

Adjusting the knot of his regimental tie, Seibert said, "As we've seen in the past, the *Krepost* is primarily a combination of components from other Soviet spacecraft. The overall design and assembly are apparently orchestrated by a military directorate, probably part of the Strategic Rocket Forces, but the various components are fabricated by at least two of their premier aerospace design bureaus.

Seibert gestured at a diagram projected on the screen. "The main building block is derived from the *Almaz* military space station design, a product of the Chelomei design bureau. In addition to the main block, the Chelomei bureau also contributes the Proton boosters used to launch the *Krepost.*

"The Korolev bureau supplies Soyuz spacecraft to deliver crews to the *Krepost* and return them to Earth. They also are manufacturing a modified version of the Soyuz to serve as a resupply ferry for the station."

Tracing his finger along the rear-projected image of the space station, Seibert continued. "The nuclear warhead is located in the aft end of the *Krepost.* It's contained in its own reentry vehicle, which is roughly cone-shaped. As best as we can determine, our initial estimates of the yield were grossly understated. It is a much larger warhead than we had previously believed, perhaps yielding somewhere in the neighborhood of fifty megatons."

Wolcott whistled and stated, "That's a danged city-killer."

"The most significant changes have been to their docking hub," said Seibert. "The docking hub is located at the far forward end of the station. The original plan had three docking ports set at one-hundred-twenty-degree angles, like the spokes on a wagon wheel. There are still three docking ports, but two are perpendicular to the long axis of the hub, and the third one is in the nose of the hub.

"This new docking hub is structurally reinforced and aligned with the center line of the station, so we suspect that they will eventually use a modified version of the Soyuz as a propulsion module to nudge the Krepost into a higher orbit, if need be, to mitigate orbital decay. As with the original design, the docking hub also contains an inflatable airlock, as well as a separate module that houses the automatic cannon and its targeting radar."

"So their gun has not gone away?" asked Tew.

"Not in the least, General," replied Seibert. "The Soviets are apparently convinced that we have an operational satellite interceptor…"

"Which we do," interjected Wolcott, glancing toward Carson and Ourecky.

"Correct, sir," said Seibert. "In any event, they are tremendously concerned that the *Krepost* could be vulnerable to attack, so not only has the automatic cannon remained a key part of the design, I anticipate that later versions will probably be equipped with even more substantial defensive systems."

And the news just keeps getting better and better, thought Ourecky. As he strained to focus on the details of Seibert's presentation, which was primarily a regurgitation of technical information that he had already seen, he found himself pondering the likelihood of surviving this next mission. *At this juncture*, he mused, *the odds just didn't look too promising.*

Tew glanced at his watch, and then asked, "Our last stack is still in San Diego, correct? At the HAF?"

"It is, boss," answered Wolcott.

"Then it's on the far side of the continent from our prospective launch site," noted Tew. "We can move it by rail, correct?"

"Yup," answered Wolcott. "But that probably ain't a practical option. We would have to crack the encapsulation and move the stages and spacecraft on separate railcars. On the other hand, if we go by water, we can keep the stack encapsulated, pack it back on the LST, and move it to the Cape."

"And that would entail making a transit through the Panama Canal?" asked Tew.

Wolcott nodded. "Yup. But I still think it's the most viable option, boss."

"Then move it," ordered Tew. "Load it on the LST and get them underway as expeditiously as possible."

"As you wish," replied Wolcott. "We'll have it at the Cape in two shakes."

Tew shifted his gaze to Heydrich and said, "Gunter, pass on to your team in Florida to focus their energies on Launch Complex 41. If it's available, then we can maintain the stack in their assembly building until it's time to go."

Launch Complex 41 was situated at the north end of Cape Kennedy Air Force Station, approximately two miles southeast of NASA's Pad 39, where the Apollo lunar explorers had left the earth on their Saturn V rockets. The complex had been built in the sixties as a self-contained Integrate-Transfer-Launch—ITL—facility, expressly for military launches, including manned launches for the now defunct Dyna-Soar and Air Force MOL. Since its designers had anticipated an aggressive launch schedule, the complex was configured like a factory assembly line, to aggressively streamline the process of matching payloads to launch vehicles before expeditiously firing the combinations into space.

"Will do," answered Heydrich. "I'll check on its status, and I'll have an answer for you by the end of the day."

"I know that we likely have weeks or even months to prepare, gentlemen," said Tew. "But I want us to be ready in *days*, just to err on the side of caution."

6

FREIGHTER

Internal Security Office
Glavnoye Razvedyvatel'noye Upravleniye (GRU)
Kapustin Yar Cosmodrome, Astrakhan Oblast, USSR
9:42 p.m., Tuesday, September 19, 1972

GRU Colonel Felix Federov had a new assignment, one that he didn't particularly relish but a task that he was compelled to execute to the best of his abilities. After all, his entire career was at stake. Incognito in dark mufti and a straw fedora pulled low over his eyes, he quietly entered the anteroom of the GRU's Internal Security Office at Kapustin Yar. An overweight sergeant sat behind a pinewood desk, engrossed in a novel.

"Where is your boss?" asked Federov, leaning over the desk.

"The major is presently indisposed," smirked the sergeant, barely looking up from his book. "He is interrogating a subject and is not to be disturbed."

After driving twelve hours from the Aquarium, Federov had absolutely no patience for the insolent or ignorant. He reeled back his right hand as he stepped around the desk and then swatted the left side of the sergeant's pudgy face. The small room was filled with the resounding sound of the impact, like a thunderclap of a summer storm swiftly sweeping across the steppes. Knocked clear out of his chair, the sergeant tumbled into a heap on the floor. His reading material—*The Idiot*, by Dostoyevsky—plopped down beside him. Growling, clenching his fists, the sucker-punched sergeant rolled to his side and started to push himself up, obviously anxious to strike the next blow.

Federov glanced at the title of the sergeant's book and mused over how appropriate it was, given the circumstances. Then he removed his fedora, revealing his unmistakable mane of curly red hair.

"The *Crippler!*" gasped the sergeant, rubbing his jaw as he slowly climbed to his feet. "Uh, uh, I mean Colonel Federov. I am woefully sorry, sir, but we expected you to arrive next week. Honestly, I did not recognize you. Your reputation precedes you. How can I be of service?"

"Fetch your commander, moron, before I lose what little remains of my patience," growled Federov.

Tugging up his trousers and clumsily fastening his suspenders, the commander of the station emerged from his office and demanded, "Just *what* the hell is going on out here?! Who *are* you?"

Federov looked past the disheveled major to observe a partially naked blonde girl, probably in her early teens, cowering in the office behind him. Angered by the sordid scene, he stepped toward the major and unleashed a mighty uppercut into the officer's chin. The major crumpled to the floor as the girl screamed.

Turning to the loathsome sergeant, Federov said, "When and if that buffoon wakes up, tell him to pack his belongings. He is relieved. He will report to the Aquarium tomorrow, where he will face charges."

Federov strolled into the major's office, sat down at the desk, and told the crying girl, "Get dressed. The sergeant will take you home."

Trembling, clutching her clothes to her chest, she stared at the wooden floor and whimpered.

"You have no need to fear, *zhuk*," added Federov. "If the sergeant touches you, I will personally chop his hands off. Just be forewarned: I never want to see you here again."

Federov yanked open the drawer of a pinewood filing cabinet. He sifted through several folders before he located the surveillance records he was looking for. He dropped the first folder on the desk, flipped it open, examined the index, and thumbed through its pages.

He had spent the past year at the Aquarium—the GRU headquarters at Khodinka Airfield in Moscow—as an administrator overseeing background checks on Army personnel being considered for transfer to the GRU. The marginal assignment—effectively a punishment tour—had been levied upon him following his last posting as the GRU Station Chief in Washington, DC.

While in Washington, he had been caught up in an asinine chain of events initiated by his predecessor, but he was held to account when an operation went woefully sour. The GRU had expended thousands of US dollars in an effort to exploit an American airman at Wright-Patterson Air Force Base, Ohio. The airman supposedly possessed inside knowledge of the US Air Force's Project Blue Book and other activities related to the study of UFOs, and allegedly could recruit Air Force personnel working in a secret warehouse at Wright-Patterson where captured alien spacecraft were studied. The whole endeavor turned out to be a gigantic fiasco. The airman was an unreliable drunk who was later murdered by gangsters because he didn't pay his gambling debts. Unable to adequately justify the expenditures, Federov was summoned back to the Aquarium to endure clerical work and humiliation.

Now, he had a chance to redeem himself at Kapustin Yar. This was a momentous task, exactly suited to his abilities, and if he was successful, he would be rewarded with a prize assignment. If he unraveled the tangled yarn at Kapustin Yar, his next posting would be as the Director of the Bureau of Special Cooperation, an office that oversaw the exploitation of American POWs and equipment captured in North Vietnam. Although he would be physically stationed at the Aquarium, he would supervise the *spetsgruppa* operating from Hanoi. The *spetsgruppa* examined downed US aircraft and monitored—and in some cases, supervised—POW interrogations, particularly in those instances where the prisoners had some form of "special knowledge," such as training in nuclear weapons or other strategic systems, as a result of previous assignments.

Besides routinely briefing the big bosses at the Aquarium, the Special Cooperation posting also entailed frequent travel and temporary duty in North Vietnam. It was an absolutely perfect assignment: a golden opportunity to hobnob with the bosses, impress them with his abilities, as well as to participate in the field work that he so enjoyed. *And who knows?* He might even have an opportunity to kill an American or two. Grinning, Federov cracked his thick knuckles as he considered the possibility.

But why was he here? This station was a counterespionage operation, which monitored the activities of scientists, engineers, and military personnel stationed at the cosmodrome. Although Federov had cut his teeth as a ruthlessly efficient commando leader in the GRU's *spetsnaz*—Special Purpose Troops—he was also well steeped in spy tradecraft and clandestine activities. He was particularly adept at stalking spies, sniffing out subterfuge, and ferreting out subversives. He was absolutely tenacious at his tasks, a relentless stickler for details who left no stone unturned.

For several months, the GRU had been aware that there was a substantial leak at Kapustin Yar. The initial clues had come indirectly from a source in the United States. The source—an American engineering technician—was involved in testing the new "Sprint" surface-to-air missile. He revealed that the Sprint's designers were privy to extremely detailed information concerning the characteristics of Soviet medium-range ICBMs that weren't yet even fielded.

The GRU's Department of Archives and Operational Research—the "Encyclopedia"—conducted exhaustive research and analysis of the suspect data, and identified ten possible culprits. Assigned to a testing range that evaluated medium-range ICBMs, all ten had potential access to the applicable information. At this juncture, one was dead and five others had departed Kapustin Yar for other assignments. GRU counter-intelligence operatives had expended several weeks surveilling those five, had recently subjected all of them to intensive interrogations, and had finally concluded that the men were not involved in any form of espionage activities. Of course, the five men had paid a high price to prove their innocence; all were still incarcerated pending completion of the investigation, and two were very close to succumbing to physical injuries incurred during their brutal inquisitions.

Logically, the GRU's attention shifted to the four still assigned to Kapustin Yar, even though it was a mystery how anyone could possibly sneak the information out of the cosmodrome, since the facility was so secluded and security was virtually airtight.

The local Internal Security Office had mounted an elaborate surveillance operation, deploying a veritable legion of snoops to stalk the remaining four suspects. They went so far as to film the comings and goings of the suspected turncoats. Additionally, a special radio interception team was brought in from Moscow to constantly monitor the airwaves for clandestine transmissions. The security office had been at it for weeks, but the incompetent bunglers were never able to figure out how the messages were being conveyed. These oafs had delivered excuses, and little else, so now it was Federov's turn to catch the spy.

He cocked his head and listened to the commotion in the outer office. He smiled; he was absolutely confident that he would be placed on report before the day was over. After all, he had only been here a few minutes and had already physically assaulted two subordinates. But his superiors certainly knew that cracking heads was his forte, and if that's what it took to motivate these malingerers to efficiently execute their duties in a timely manner, then Federov would knock as many skulls as necessary. And if he delivered the goods, which he surely would, the world would be his oyster.

Simulator Facility, Aerospace Support Project
10:35 a.m., Tuesday, September 26, 1972

After being effectively dormant for several months, the Project's Simulator Facility was again operating at full swing. Carson and Ourecky occupied the simulators during the day, and their Navy back-up crew was given priority at night. It was definitely no time to dawdle; the Blue Gemini engineers were engaged in a frenzied round-the-clock initiative to devise options to attack the sinister *Krepost* while simultaneously safeguarding the Gemini-I astronauts from danger.

As quickly as the engineers refined new procedures, Carson and Ourecky tested them in the Box—the procedures simulator—where they were now, evaluating a new "Cold Nose" intercept protocol. The protocol was so named because they would fly the intercept profile with their nose-mounted acquisition radar switched off—hence the cold nose—and instead rely primarily on radar data and instructions sent up from the ground. The rationale behind the tactic was that the mission planners were concerned the Soviets had likely developed radar detection systems similar to those being installed in modern American aircraft, and there was always the possibility that such a detector—even aboard an unmanned station—could trigger some catastrophic event if the Soviet spacecraft was illuminated by radar energy.

Besides minimizing the crew's reliance on the Gemini-I's powerful radar, the engineers had quickly fabricated a radar detection system as an interim defensive measure. The hastily built multi-band sensor, not unlike the "Fuzzbuster" radar detectors that were now becoming increasingly popular with interstate truck drivers and other habitual speeders, would be mounted in the blunt nose of the Gemini-I spacecraft. Another team of radar specialists were also at work concocting a small pod that borrowed from radar-jamming technology presently deployed in Vietnam. The pod would be carried within the Gemini-I's adapter and launched after the radar detection system was tripped. It was intended to spoof the *Krepost's* target acquisition radar as the Gemini-I initiated evasive maneuvers.

Besides the radar detectors, Carson and Ourecky had another tool at their disposal. A "ferret box," approximately the size and dimensions of a small suitcase, had been installed in the crowded cabin. Borrowing from airborne signals intelligence gear currently used in Vietnam, the ferret box scanned radio frequencies to detect and classify transmissions.

Presently, the Gemini-I's primary means of attack was the multi-function Disruptor, which had proven effective on seven of nine previous intercept missions against Soviet satellites. The remaining two intercepts failed not as a result of the Disruptor's shortcomings, but rather because the crew—Jackson and Sigler, in both instances—failed to close with the targets.

To complement the Disruptor, the engineers were toying with different options, including a stand-off "torpedo." The description was somewhat of a misnomer; the device, a sphere roughly four feet in diameter, more resembled a mine than a torpedo. As presently envisioned, it would be carried in a cradle in the Gemini-I's adapter section, to be ejected and then guided—by remote control augmented by a television camera—to demolish the *Krepost* at a safe distance. If need be, if the other attack options were not practical or had already failed, Carson and Ourecky would also be prepared to conduct an extravehicular activity—"EVA"—spacewalk to interdict the *Krepost*. But although Ourecky would wear a bulky EVA suit to orbit, and an ELSS—Extravehicular Life Support System—chest pack and thirty-foot umbilical would be stored in the already crowded cockpit, their EVA options were extremely limited. Some rudimentary tools were stowed in the adapter section, so that in a pinch, Ourecky could leave the Gemini-I to physically damage critical components of the *Krepost*, like the docking mechanisms and communications antennas.

Regardless of how they attacked the monster, the engineers' primary concern was to devise some means to defeat or at least reduce the effectiveness of the 23-mm radar-guided automatic cannon mounted in the space station's prow. At this juncture, short of a radical evasive maneuver which would also terminate the mission, there was no practical way to dodge the gun's threat. The entire mission hinged on them making a stealthy approach. The risks were too high otherwise; if they were detected on the final approach, their only recourse was to abandon the pass and save their skins.

For the time being, as the hardware was being perfected, a key aspect of their defensive scheme was to make certain that the Gemini-I's four hundred-pound aft thrusters were ready to fire at a moment's notice. Ordinarily, the thrusters required at least a few minutes of preparations to ensure that they were ready to burn, because in a weightless state, their fuel and oxidizer naturally tended to form useless blobs at the center of their storage tanks. Consequently, prior to a "normal" scheduled burn, the combustible liquids were physically shifted to a ready state by using helium to inflate Teflon bladders that lined the spherical tanks. In this situation, even as Carson carefully manipulated the controls to fly the final approach of the intercept, he had to also regularly monitor and tweak the helium pressure settings.

"Fuel and ox pressure are nominal," stated Carson, checking the tank gauges. A well-chewed toothpick protruded from his lips.

"Copy good pressure on OAMS fuel and oxidizer," replied Ourecky. "No tone heard."

Carson shifted his eyes away from the instrument panel to the optical reticle. The target was approximately two thousand meters distant. "Closure rate is still good."

"I have tone on X-Band," announced Ourecky, listening to the warning buzzer in his earphones.

"*Maneuvering,*" replied Carson, immediately shoving the maneuver controller forward to fire the four aft thrusters. The burn would shoot them forward initially, towards their target, but then their increased velocity would also radically propel them into a slightly higher orbit. "Firing aft thrusters thirty seconds. Burn is uniform, all four thrusters burning."

"Copy uniform burn," replied Ourecky.

"*Good* burn," commented Carson, watching the clock as he held the maneuver controller. "Very even…Ten seconds…Fifteen seconds…

Twenty seconds...Shut down in six, five, four, three, two, one. Mark. Throttling down. All stop."

Ourecky coughed repeatedly as he listened to Gunter Heydrich's heavily accented voice over the intercom: "Good job, gentlemen. I think you have the timing down just about as close as it can be. I just hope that it's enough. Did you have any problem hearing the tone, Scott?"

"No," replied Ourecky. "It's plenty loud."

"Good," answered Heydrich. "We'll be a few minutes setting up for the next run, so it's an opportune time for a break. Do you want out?"

Ourecky looked towards Carson and shook his head.

"Gunter, we're fine in here," said Carson. "Unless it's a problem, we'll just keep rolling through until lunchtime."

"Sure thing," replied Heydrich. "We should be ready in another ten minutes at most. Go ahead and reset your controls, please."

Ourecky removed his headset and massaged his ears.

Removing the foil from a stick of Juicy Fruit, Carson said, "So, Scott. I haven't even seen Bea since we got back in August. How's she doing? How's Andy?"

"Uh...they're fine," replied Ourecky. "For a little guy, he's a handful. He sure keeps us on our toes." He swallowed nervously; he was too embarrassed to tell Carson about his marital difficulties, so the pilot had no way of knowing that he and Bea were currently living apart.

Carson popped the gum in his mouth and chewed. "I have an idea. We should be out of here at a relatively normal hour, so why don't I take you and Bea out to dinner so we can catch up? Maybe we can grab some schnitzel at that German place downtown?"

"I don't know, Drew. We would have to hire a babysitter, and that's not going to be easy on a Monday night, plus Bea is on a new sched-ule and is also taking care of a friend who's really sick. She's over there just about every night." Resetting the switches on his instrument panel, Ourecky considered the ease at which he was willing to bend the truth about his circumstances. Of course, he had plenty of practice; he had been doing the same thing with Bea for almost as long as he had known her.

"Sick friend?" asked Carson, consulting a reference card as he reset his own controls. "Anyone I know?"

"Yeah," answered Ourecky. Yanking a tissue from a small box, he blew his nose before adding, "Do you remember Jill? She was Bea's maid of honor at our wedding."

"Jill? Oh, sure...I remember her. So she's sick? Is it serious?"

"It's *very* serious. Terminal cancer."

"*Oh*. Sorry. Maybe we can try some other time."

"Maybe. I'll let Bea know that you asked about her." Ourecky decided to change the subject, before he inadvertently revealed anything really significant. "So, Drew, what do you think will happen after this last mission?"

"Honestly? There's really no way of knowing. With Tarbox and the Navy coming on board, it looks like this monster is expanding instead of contracting."

"I suspect that you're right," said Ourecky, flexing his fingers to alleviate cramps in his hands.

"How about you, Scott?" asked Carson. "Do you really think that we'll ever be free of this?"

"I think so," replied Ourecky. "Well, I *hope* so. General Tew has assured me that he's going to take care of us after this last mission is over."

"Well, no matter what he's promised us, once Mark Tew retires, he's not going to have very much influence over our fate. I suspect that Virgil will conjure up some way to rope us back in, if given the opportunity. I just think that we're deceiving ourselves if we believe that we'll ever be able to move on."

"Maybe. So, what do *you* want, Drew?"

"I'm not sure. I'm ready for a change, but I'm not sure that there's anything else out there for me. I guess you know that Tew promised me command of a fighter squadron in Germany, once we fly this mission, but he's sticking with the party line and won't let me go to Vietnam. That's what I *really* want."

"Is that really so important?" asked Ourecky.

"Yeah. I believe so; otherwise I'll spend the next ten years competing against guys with combat time on their records."

Interrupting their conversation, Heydrich's voice boomed over the intercom: "We're ready to re-start. Are you gentlemen ready in there?"

"Let 'er rip, Gunter," answered Carson, adjusting the microphone on his headset. "We're up."

Krepost Station, On Orbit
6:10 a.m. GMT, Thursday, September 28, 1972
GET (Ground Elapsed Time): 14 Days 5 Hours 25 Minutes, REV # 227

The big day had arrived, and Vasilyev did not want to leave anything to chance. After spooning up the last of his breakfast—reconstituted *owsianka* porridge that he sweetened with honey—he tore the plastic wrapping from a vitamin-enriched wafer. Newly developed, the concentrated foodstuff was intended as a nutritional supplement when the cosmonauts were otherwise too busy to consume a regular meal. Sampling the wafer, he grimaced; the dense white hardtack had the consistency and taste of a paving stone.

He shifted slightly to gaze out through the porthole and saw that they were currently passing over the Solomon Sea, just east of Papua-New Guinea, in daylight. He floated up into the cozy galley area, where Travkin was preparing his own breakfast.

"The supply boat is due in less than six hours," he stated, nudging Travkin's shoulder. The two men looked at each other and grinned. "I hate to be a stickler, but I want to rehearse our docking procedures again."

"Excellent idea, Pavel," replied Travkin, crumpling the wrapper of his own half-finished serving of oatmeal before stuffing it in the trash receptacle. "I've had my fill of this damned mush. I'm ready for some real chow."

Vasilyev inserted the water dispenser's metal spout into a beverage envelope and then squirted the package full. Kneading the plastic envelope, he saw that the liquid wasn't sufficiently warm enough to dissolve the evaporated tea crystals, much less make a decent serving of hot tea. "Damn it!" he snapped. "Is the heating element on the fritz? This is going to be a long tour if I can't have my tea."

Travkin shook his head and answered, "It's not the heating element. The rheostat is shorted out, so we can't regulate the water temperature." He tapped his finger on the finicky water dispenser, which emitted a faint buzzing noise.

"So we're stuck with lukewarm water from here on?"

"Hopefully not. I'm fairly certain I can scrounge another rheostat out of our Soyuz. It's in the ascent panel, which we don't need any more. You'll be able to brew your precious tea, Pavel. I promise."

"*Spasiba*. You're my hero, Petr. I'll see to it that you're decorated accordingly: Illustrious Mender of the Tea Machine." Vasilyev floated back into the control area and drew his knees to his chest as he gazed through a porthole and scratched his itching toes. As was his practice on all missions, from his first stint in orbit two years ago to this one, he wore his "lucky socks." The woolen footwear, lovingly knitted by Vasilyev's late wife, were once dark brown, thick and cozy; now, the socks were faded and well worn. Threadbare in several spots, where pink patches of bare skin peeked through, the ragged socks barely kept his feet warm, particularly when he ventured into the poorly heated docking hub or the dormant Soyuz. He was finally beginning to accept that this would probably be the last mission for his treasured socks, and it was highly likely that they would fall off his feet long before this flight was over.

"Hey! Maybe the cargo packers stashed some new booties for you," noted Travkin, reaching out to pick an errant piece of yarn out of the air. "Those are really starting to reek. If you want, I would be delighted to loan..."

Silent, Vasilyev glared at him.

"Hey," muttered Travkin, floating close and clutching his companion tightly. "I didn't mean it, Pavel. I was just kidding. I know that you miss Irina. I'm really sorry, I am."

13:03 p.m. GMT
GET: 14 Days 12 Hours 18 Minutes, REV # 232

The two cosmonauts quietly waited in the docking hub at the forward end of the *Krepost*.

Besides the three docking ports—one of which was currently occupied by the Soyuz crew vehicle—the module also housed the controls for the station's rendezvous and self-protection systems. Anxious, Vasilyev checked his watch as Travkin nervously picked at a bothersome hangnail.

Breaking the awkward silence, the proximity alert radar chirped. The low-powered sensor, which emitted a modulated pulse roughly every minute, was intended to warn of approaching space vehicles, particularly those suspected to be hostile. Although they apparently had no substantive proof, certain members of the Soviet intelligence apparatus were convinced that the American military had secretly developed an unmanned system to surreptitiously intercept and destroy satellites. This notion had caused the design bureau's engineers considerable consternation as they planned the *Krepost*. To counter the vague threat, they armed the station with a Nudelman 23-mm automatic cannon.

Of course, using the automatic cannon—"Sparky"—to fend off a satellite interceptor was contingent on having sufficient notice to take action. But there was also the distinct possibility that the Americans might attack with no notice, using nuclear weapons to blast the *Krepost*—and other Soviet satellites—to smithereens. After all, there were ample rumors that the Americans had already fielded an operational system to do just that, and might likely use the capability in conjunction with a first strike thermonuclear attack. So, he and Travkin might be obliterated without warning or even the slightest hope of defending themselves.

"*Target,*" announced Travkin quietly. He occupied the weapons console that controlled the target acquisition and tracking radars associated with Sparky, as well as the automatic cannon itself. He flipped up the safety switch, which caused the acquisition display to be illuminated in the aiming reticle as the weapon was slaved to the radar.

On the opposite side of the docking hub, Vasilyev monitored the controls for the *IGLA*—"Needle"—automated rendezvous and docking system. The *IGLA* system had guided the Soyuz freighter to them, and—without the slightest of human intervention or interaction—the amazing contraption could actually dock the unmanned cargo vehicle to the *Krepost*. Theoretically, at least, *IGLA* could handle the terminal flight and docking, but since it was not absolutely trusted yet, Vasilyev would take control of the Soyuz freighter for the last phases of the critical operation.

"Five thousand meters and closing," stated Vasilyev, switching on a television camera mounted on the docking hub. "Right in the bubble." Suddenly, he heard a pervasive whirring sound and realized that Travkin had activated the automatic cannon's stabilization gyros and was now manually sighting Sparky on the approaching freighter. "What the hell are you doing, Petr?"

"Practicing," replied Travkin, squinting through the telescopic sight as he pretended to trigger bursts from the cannon. "This is the *perfect* opportunity to practice engaging a target. After all, I want to be prepared in case one of those American killer satellites pays us a visit. We cannot afford to be lax."

"Damn it, Petr! Knock it off!" growled Vasilyev, with a mock air of sternness. "That's our chow for the next three weeks, and if you blast it out of the sky with your goofing around, I will *personally* hold you responsible."

"Ha! Big talk, Pavel," sniffed Travkin. "I imagine that your tone will be a bit muted when we're wasting away from starvation."

Vasilyev chuckled. "I've contemplated that, and I'm not planning to starve," he vowed. "I've already formulated my contingency plan. You're a big healthy specimen, Petr, so after I bash your noggin with the adjustable spanner from the tool kit, then I'll just spend the remainder of the mission gazing at the stars and stuffing my belly with *cosmonaut tartare.*"

"You damned glutton! I hope you choke!"

14:18 p.m. GMT
GET: 14 Days 13 Hours 33 Minutes, REV # 233

As the sun slipped up over the horizon and Vasilyev anxiously watched the freighter's approach on the television monitor, he thought longingly of the boisterous party they would have later today, as they stocked the station's pantry. In contrast to the bland fare they had been subsisting on since launch, a feast awaited them. At this point, he longed for anything with *flavor*. And if things went well, he would have just that before the day was over: the freighter's capacious dry compartment was jam-packed with food, and most of the victuals were quite delectable. As he was fond of telling his friends, he usually ate better in space than he did on Earth.

Manually assuming control of the freighter, Vasilyev adroitly manipulated a joystick to make minor steering corrections as the Soyuz flew the last few meters to the *Krepost*. With hundreds of hours in the simulator, rarely not making the connection on the first attempt, he was the undisputed master of the crucial docking process.

After the freighter's supplies were depleted, in roughly three weeks, they would stuff it with trash and other waste, and eject it. The mission planners had so much confidence in Vasilyev's docking prowess that he would also dock the next freighter, just before he and Travkin left orbit, which would sustain their relief crew. That way, even before they departed Earth, the next crew would be assured of having adequate supplies to complete the first half of their six-week mission.

Staring at the television monitor, Vasilyev lightly bumped the joystick to make a minor correction, observed the green light that indicated the ferry's docking probe had entered the matching receptacle on the docking collar, and then pushed a button to finalize the mating sequence. He listened to the muted clicks and clanks of the locking claws swinging into their corresponding latches. "Capture," he announced, confident that the connection was solidified.

"You did it!" exclaimed Travkin boisterously, embracing Vasilyev in a bear hug and pounding his back. "On the first go, no less. You're *my* hero, Pavel!"

15:43 p.m. GMT
GET: 14 Days 14 Hours 58 Minutes, REV # 233

It took the two cosmonauts almost an hour to dismount and stow the docking mechanism. After finishing that task, they connected a fitting to equalize the pressure between the *Krepost* and the freighter. Finally, the famished pair swung open the freighter's hatch to reveal their treasure. Several light green bags floated out of the dry compartment. The mesh bags contained perishable "goodies," like fresh fruit and bread, packed just hours before the freighter was launched. Vasilyev dug down and found the light blue bags that contained his "personal preference" items. He immediately snatched one of the bundles and rummaged through it. Grinning, he quickly peeled the metal lid from a tin of smoked horse mackerel.

Accustomed to the almost sterile atmosphere of the fledgling space station, Travkin almost retched at the pungent new odor. "Ugh," he noted, pinching his nose. "That's *so* revolting. You just had to eat the stinkiest thing you could find."

With his chin smeared with the packing oil, Vasilyev picked out the chunks of mackerel with his fingers. Quickly devouring all of the fish, he carefully licked the interior of the flat tin to capture all of the oily residue. "Scrumptious!" he pronounced with gusto, wiping his greasy face with a disposable towel. "At long last, something that doesn't taste like cardboard."

"Pavel, you disgust me," said Travkin, turning up the docking hub's ventilation fans in the hope of dissipating some of the fishy stench. "You remind me of a monkey feeding at the zoo." Carefully sorting through a bag of fresh fruit and vegetables, he found a ripe lemon and sliced it in half. He smeared some of the juice on the wall, to function as an impromptu air freshener, and sucked the remainder from the sour fruit.

"Maybe," answered Vasilyev. "But speaking of despicable monkeys, you should be thankful that you're sharing this bucket with me, rather than our old friend Gogol."

Travkin literally shuddered at the sound of Gogol's name. "You're right," he said in a soft voice, barely audible above the buzz of the fans. "I am truly fortunate to be flying with you, but you should also remind yourself that I was not slated to fly with him in the first place, so you should consider yourself the lucky one."

"That's very true. At least now I know that I have the option of grounding myself by cutting off an appendage or two, should I ever be ordered to fly with him again."

Travkin sucked on the lemon, nodded, and then observed, "Desperate times require desperate measures."

"Unfortunately."

"What's this?" asked Travkin, letting go of the lemon as he pulled a light green satchel from the freighter. Resembling a pockmarked yellow planet, the depleted citrus fruit floated free in the access tunnel. "It feels like there's some sort of hardware in here. Spare parts for the fuel cell, maybe?"

Vasilyev immediately recognized the bag. "That's some junk that Gogol insisted on packing in the freighter to supplement the survival gear on the Descent Module. I had to give up most of my good chow so that queer bastard could shoot his toy collection into orbit."

Travkin opened the satchel, looked inside, and pulled out a bottle of anti-radiation tablets. "Wow! What could have possibly possessed Gogol to pack all of this stuff?"

"At first, I just thought it was because of what happened to him in Mongolia, but he got drunk one night and insisted that we would drop the Egg while we were up here, so he had made elaborate plans to survive afterwards." Vasilyev tugged out a plastic-bound notebook from Gogol's satchel. He opened the notebook and studied its pages; it contained intricate instructions—timing, spacecraft orientation, engine burn times—for reentry to the various alternate sites that Gogol had identified. He was actually very impressed with the meticulous level of detail; Gogol had clearly exerted a tremendous amount of thought and energy toward creating the instructions.

Vasilyev heard a metallic clacking noise and looked up to see Travkin racking back the bolt on the Kalashnikov.

"Excellent!" declared Travkin, pulling the trigger to dry-fire the weapon. "Now we have something substantial to repel boarders. We are ready for all contingencies."

"Quit screwing around with that damned gun before I'm forced to rap my knuckles on your thick skull," declared Vasilyev. "Put everything back in the bag. I want it and all of Gogol's other junk, including his liquor and cigarettes, stowed in his compartment." To afford the cosmonauts at least a modicum of privacy while they rested, the *Krepost's* designers had provided three narrow 'sleeping compartments' just aft of the docking hub. Neither Vasilyev nor Travkin used the compartments for their intended purposes. Rather than jam themselves into the claustrophobic spaces, both preferred to sleep in the docking hub, where they could float free in relatively quiet peace. Moreover, they could immediately man the station's defensive systems if the proximity alarms sounded. Since available space was at a premium, the superfluous sleeping compartments now served as overflow storage lockers.

Returning to the docking hub, Travkin asked, "How about Gogol's spare uniforms and underwear?"

Vasilyev retrieved the yellow bags that contained the extra clothing, verified that they were emblazoned with Gogol's name, and gently tossed them toward Travkin. "Stick them in his compartment as well. Since this stuff won't fit either of us, we'll just have to share mine."

"*Spasiba*," said Travkin. "You're very kind, Pavel. Only a true gentleman would offer clean skivvies in such dire circumstances."

"I'm sorry," replied Vasilyev. "Did I somehow imply that you would be getting clean ones?"

The process of unloading and stowing the cargo would consume hours. Even though they would periodically take breaks to sample their new bounty, it was going to be an extremely long and arduous day. Once they stored all of the supplies in the galley's aluminum cupboards, their next chore was the time-consuming and complicated process of activating the fuel cell carried aloft in the Soyuz freighter.

After attaching various couplings and wiring the fuel cell to the station's power supply, their next job was to begin the laborious process of shifting propellants from tanks in the freighter's "wet" compartment to storage tanks aboard the *Krepost*. The transfer would take place over the course of the next several days; since the station's designers had not furnished the station with an automatic pump, the two men would have to manually crank a centrifugal pump—normally for two to three hours at a stretch, until their shoulder muscles burned from exertion—to siphon the propellants.

Once the fuel was brought aboard the *Krepost*, there was little to do but stand their watches, play cards, read books, play chess, sleep, eat, watch the stars, study the Earth, tend the ship, and wait for the dreaded call that would pitch the world into nuclear oblivion.

Residential Complex # 4
Znamensk (Kapustin Yar-1), Astrakhan Oblast, USSR
8:35 p.m., Friday, September 29, 1972

With Luba attending her monthly women's committee meeting, Yohzin walked Magnus by himself. A harsh wind blew in from the west, so very few people ventured out this evening. Given a choice, Yohzin would have preferred to remain indoors, but his schedule dictated that he write his messages on Thursdays and "deposit" them—courtesy of Magnus— on Friday evenings. Yohzin had no clue what happened after the encapsulated message was placed in the dead drop. He didn't have the slightest inkling of who picked it up or how the information was transmitted— physically or electronically—from Kapustin Yar to the Americans.

It had been a relatively quiet week, so there was little to report. Vasilyev and Travkin had successfully received their first freighter, which meant that the crew could remain in orbit, tending to their nuclear warhead, for another two weeks. Abdirov had revealed nothing new concerning the delivery status of the second freighter, so it appeared that the mission would proceed apace.

As he strolled along his usual circuit, pausing briefly to allow Magnus to relieve himself, Yohzin felt distinctly uncomfortable. Since Abdirov had confided about the leak, he was much more conscious of the notion that he was probably under more intensive scrutiny. For whatever reason, even with that knowledge, he felt even less at ease. He paced on, not knowing the source of his anxiety, and then he realized that this was the very first time that he had ever taken his evening constitutional when he did *not* detect someone observing him. *Are the counter-espionage goons taking the night off?* Surely, this could not be so; as blatantly incompetent as they were, they were never lax, but always on watch.

He wasn't so naïve as to think that all of the GRU counter-espionage spooks had suddenly departed on holiday. Decades of living in this closed society had taught him that seemingly unwarranted paranoia was nothing to be laughed at; just when it seemed that you were not being watched, it was a virtual certainty that you were.

Leaves rustled as a cold breeze gusted down the open collar of his coat. He shivered as a chill passed over him, and remembered the adage about someone stepping over his grave. He smiled before fastening his collar button, but the chill remained.

Krepost Station, On Orbit
20:15 p.m. GMT, Friday, September 29, 1972
GET (Ground Elapsed Time): 15 Days 19 Hours 30 Minutes
REV # 253

"*Check*," stated Travkin authoritatively. "Ready to forfeit?"

Stroking the greasy stubble on his chin, Vasilyev sipped cognac before shaking his head. Briefly holding the liquor in his mouth to savor its taste, he considered his decimated array of chess pieces. As he pondered his options, he looked up from the magnetic chessboard, sniffed the air, and commented on an unusual new odor: "Is that electrical? I don't think I've smelled that one before."

"I don't see how you could *possibly* detect any unusual fumes, considering the sheer volume of noxious gases that you see fit to constantly contribute to our limited atmosphere," smirked Travkin, twirling a black bishop in the air with his finger. "Seriously, Pavel, you're even worse than Gogol, if that's even possible."

"Hush!" said Vasilyev, sniffing the air.

Grinning, Travkin said, "So is this just another gambit to throw off my concentration? Why don't you just capitulate, Pavel, and admit that you have once again been defeated by my vastly superior intellect?"

"I said *hush*, Petr," replied Vasilyev curtly. "I'm serious: I *do* smell something odd. It's faint, but it's *there*."

Like a gourmet chef sampling the aroma of a freshly cooked entree, Travkin used his palm to lightly whisk air toward his bulbous nose. "You're right," he confirmed, nodding his head solemnly. "It smells kind of like scorched cloth, like some wiring insulation is overheating."

"*Fire protocol*," ordered Vasilyev calmly.

Both men opened nearby satchels containing their emergency gear and swiftly donned flame-resistant gloves, hoods, and breathing masks. Vasilyev despised the cumbersome rubber mask, particularly the plastic lenses that obscured his vision. Grabbing a fire extinguisher from a bulkhead bracket, he headed toward the control area at the aft end of the station. According to their protocol, they would first conduct a cursory visual check of the station's compartments to spot any obvious flame or smoke.

Moments later, he heard Travkin's voice, heavily muffled by the rubber mask, at the far end of the station. "Pavel, I've checked all of the compartments. No smoke or flame."

"Okay!" he yelled. "I have the same situation down here. Doff masks for sniff test." During the next phase of the fire emergency protocol, they would retrace their routes though the station, sniffing for any indication of fire. He tugged off his rubber ventilator and then switched off the ventilation fans, since it would be much easier to detect the source of the odor if the air was not circulating. Although he wasn't particularly worried about an open fire in the cabin, he was concerned about the potential of a small smoldering fire, perhaps concealed behind an equipment panel, which could eventually overwhelm the air scrubbers with smoke. If the air exchange scrubbers were compromised, he and Travkin could be smothered.

After working their way through the station, the two cosmonauts met in the docking hub, to execute the next phase, which entailed a detailed check of the two docked Soyuz spacecraft. Vasilyev didn't trust the cantankerous fuel cell and its attendant cryogenic tanks, particularly after the mishap on America's Apollo 13 lunar flight. Unlike similar equipment on the Apollo spacecraft, which was positioned outside the pressure vessel occupied by the astronauts, their fuel cell was physically *within* the docked Soyuz freighter, so a similar explosion could have very sudden and very catastrophic consequences.

More than concerned about the potential hazard inherent in the fuel cell, he decided to depart from established protocol. Pointing to the abbreviated tunnel that led to the freighter, he stated, "While you're checking the furnace in the basement, I'll get the lifeboat ready to go, just in case we have to abandon ship."

"Abandon ship? But how about the Egg?" blurted Travkin. Their orders were clear, if not painfully stringent: if need be, they would perish at their stations to protect the weapon. They were authorized to depart the *Krepost* in only two circumstances: either after they had deployed the Egg, or after they had been relieved by another crew.

"Damn the Egg! We sure can't deploy it if we're dead, can we? So what's the point of staying on board if the fuel cell decides to rupture?"

Travkin nodded. As his crewmate disappeared through the access tunnel into the freighter, Vasilyev made sure that their own Soyuz was ready for immediate departure. He checked the hatch and docking port fittings to ensure that they were free of obstructions, should a rapid escape become necessary. Even the most miniscule piece of debris could preclude the hatch from properly sealing, and the results could be disastrous. After verifying that their lifeboat was functional, he returned to the docking hub.

Moments later, Travkin emerged from the access tunnel. His ventilator mask trailed behind him from its corrugated rubber hose. "No smoke or flame," he reported, stuffing the mask back into its cloth container. "But you were right about the sour smell, Pavel. I found a spot where the wiring harness had been scuffed, and the wires were warm to the touch. I flagged it, and I'll make a point to keep an eye on it."

"Good plan," replied Vasilyev, removing his own mask before scratching his nose. He used a small sponge to wipe out his perspiration before stowing the mask in his emergency satchel. "Anyway, this should suffice for our daily emergency drill."

The two men returned to the snug galley, where Vasilyev took a quick sip of water from the dispenser. "I appreciate your loyalty, Petr," he said, putting away their magnetic chess set. "But I'll be damned if I'm staying aboard to babysit the Egg if that damned fuel cell blows. We don't have to tell Control, but I think we need to adjust our emergency protocols to beat a hasty retreat, if need be. Are you with me?"

"*Always*, Pavel, always," replied Travkin. "So, are we going to play cards or are you going to waste the day gawking at the stars again?"

7

STUMBLING BLOCKS

Krepost **Project Headquarters**
10:30 a.m., Wednesday, October 4, 1972

"Well, this is it," declared Abdirov somberly, holding up a formal memorandum. His voice was raspy and hoarse, like he had been screaming all morning. "Our goose is now officially cooked. Our stalwart comrades in the Korolev bureau have finally seen fit to officially announce that the next freighter will not be delivered on schedule."

"But it will still be delivered?" asked Yohzin.

"Eventually."

"Did they present any prospective dates, Rustam?" asked Yohzin. "What would be the best case?"

"No earlier than the end of October," answered Abdirov. "That's ten days after our cosmonauts will have exhausted their supplies."

"And the worst case?"

"It might take another month, so theoretically the freighter might not be delivered until the end of November."

Shaking his head, Yohzin replied, "That *is* bad news."

"It's all over, little brother," mumbled Abdirov, slumping in his chair. He was highly distraught; almost incoherent, he sounded like he was on the verge of babbling, as if he had completely lost his equilibrium. He looked haggard; even with the mass of pink scar tissue, his face was deathly sallow.

Yohzin found himself assuming an unusual role, serving as a counselor to the well-grounded man who normally offered Yohzin his counsel. He sought to soothe Abdirov's anguish. Hoping to coax his friend back from the precipice, he said quietly, "I really don't think it's over, Rustam."

"But it is. Can't you see that, Gregor? It was a monumental struggle just to get this project approved. It was approved with the understanding that once the weapon was armed, it would be constantly manned at all times. Now, unless some miracle happens, it looks like we'll have at least a two-week lapse where there will be no one up there. The weapon will be unattended and the *Krepost* will be vulnerable."

"So, Rustam, maybe you'll be fired. That's really not so bad."

"I would be fortunate if being fired was all that I had to be concerned with," replied Abdirov. "I don't think I'll be so lucky."

"What do you think they'll do?"

The spindly general tilted his scarred face upward, held his hand above his head and made a gesture like he was being hanged. "The bosses will demand a scapegoat to sacrifice," he lamented in a quavering voice. "So I'm the most logical candidate to dangle from a noose."

"Rustam, the freighter setback was entirely outside your span of control," argued Yohzin. "I strongly believe that you will maintain control over the *Krepost* if you go to Moscow with a solid plan to salvage the mission. The Korolev bureau failed you. It is simple as that, and the General Staff of the High Command will surely recognize the Korolev bureau's shortcoming as well. If you present the High Command with a viable plan, then you will surely keep sway over the *Krepost*."

"That's all well and good, but that notion is predicated on too many variables that we cannot control," replied Abdirov. "For example, what if the Americans swat the *Krepost* out of orbit while it is unoccupied and vulnerable?"

"So what if they do?" replied Yohzin, raising his hands as if to surrender. "What can we do but hope that they don't? Likewise, what if the station is destroyed by orbital debris or perforated by micrometeorites? We have no control over those, either. Rustam, why don't we focus on those matters that we *can* control and disregard those that are beyond our grasp? Why gnash our teeth and fret over things that can't be helped?"

"Perhaps you're right, little brother, but without the next freighter, we will still be compelled to abandon the *Krepost,* at least for a while. Those damned *Perimetr* scoundrels are anxious to wrest this thing away from me, and that's all the excuse they would need."

Yohzin shook his head. "I suspect that you're right that the *Perimetr* people yearn to control any and all nuclear weapons in orbit, and they want to eventually transition to an unmanned system like the *Skorpion.* But you should remember that I have spent a considerable amount of time hobnobbing with their engineers, Rustam, and I think that they have come to realize that an unmanned orbital bombardment system is far too vulnerable to the Americans.

"I assure you that the *Perimetr* leadership is scared to death of this satellite interceptor system that the Americans allegedly possess. Until we develop some very sophisticated countermeasures against it, then it would be foolhardy to send up an unmanned system. That's why I don't think they will be in a tremendous rush to transition from the *Krepost* to the *Skorpion.* And so long as *Kreposts* are overhead, you'll keep your hand in the game, so you should not be so despondent now."

"Perhaps," said Abdirov. "But if we are compelled to abandon the *Krepost,* even for a brief timeframe, then I will not have any recourse but to appear before the High Command and beg for forgiveness."

"Sincerely, Rustam, I don't think that you will be compelled to beg or plead for anything. I think it's simply a matter of appearing before the High Command to present your case. Merely state the facts as they are and then present a plan to move forward. As I said, it's the Korolev bureau that failed you, not the other way around."

Abdirov nodded. "Maybe you're right."

"I'm sure that I am, Rustam. The greatest technical hurdle in our path is that we do not have an effective plan for this contingency, since we had never anticipated leaving the station unmanned. On a positive note, since we have roughly fifteen days of consumables on board, we have more than enough time to develop an orderly plan to power down the *Krepost* to minimum operational requirements, so it can be reoccupied as soon as we are confident that the next freighter is ready."

"*Nyet*, Gregor. Not until the next *two* freighters are ready," countered Abdirov, evidently regaining his confidence. "And formal assurance that the third and fourth freighters will be delivered on schedule. I will not allow the Korolev bureau to hold me hostage again."

Residential Complex # 4
Znamensk (Kapustin Yar-1), Astrakhan Oblast, USSR
8:34 p.m., Thursday, October 5, 1972

Yohzin's heart thumped in his chest as he composed his message on the special graph paper. As he had done countless times before, he encrypted it and then transcribed it onto the special paper strip. He let the ink dry before tightly rolling the strip and then inserted it into the capsule. In a few minutes, the capsule would go into Magnus's kibble bowl, and roughly twenty-four hours later, on command, Magnus would deposit the capsule into the dead drop.

Sighing, he contemplated the latest events as he collected and cached the implements of his secret writing kit. As much as he respected and admired Abdirov, Yohzin was anxious for this nonsense to end. He had arrived at the conclusion that the *Krepost* was just too damned dangerous to leave in orbit. Even if Abdirov failed to make good on his scheme to drop the Egg, it was just a matter of time before an accident occurred and a catastrophe ensued.

He provided the Americans with the detailed ephemeris that described the *Krepost's* orbital track, and made it abundantly clear that the station would be abandoned, at least temporarily, beginning on

October 19. Short of drawing up a diagram, he could not have made it any simpler. From his perspective, assuming that the Americans did in fact possess the means to knock the *Krepost* out of the sky, he was providing them with more than enough information and sufficient lead time to execute. If they failed to do so, then they should be ready to accept the consequences.

Krepost Project Headquarters
9:45 a.m., Saturday, October 7, 1972

Yohzin sat with Abdirov at a large walnut table at the rear of the *Krepost* mission control facility, reviewing the tentative plans to power down the station before Vasilyev and Travkin returned to Earth.

Working nonstop, both shifts of *Krepost* engineers had dedicated the past forty-eight hours to hammering out a plan, but it was far from complete. Presently, except for a handful who were decisively engaged with monitoring the mission, they were huddled around a pair of tables cluttered with large schematics and diagrams of critical systems. The objective was to gradually and systematically switch off non-essential systems so that the station's batteries, augmented by the solar panels, would keep the critical systems functioning. Above all else, the plan had to ensure that the vacated *Krepost* could be quickly re-occupied once the freighters were available to sustain operations; to this end, the *IGLA* automatic rendezvous system, the docking mechanisms and stabilization gyros had to remain powered on. Swilling lukewarm tea and chain-smoking cigarettes, the nearly exhausted engineers patiently labored toward their goal.

Since the *Krepost* was not intended to be shut down, developing an orderly plan to do so was purely a process of trial and error. As they drafted a new iteration of the plan, the engineers tested it, step by step, on a fully functional systems mock-up housed inside the Proton processing facility. Once they arrived at a workable and reliable solution, they would brief it to Abdirov. If he approved of their scheme, then he and Yohzin would fly to Moscow on Monday morning to present it to the General Staff of the High Command.

The mission control facility was divided into two equal parts. Normally, the twenty *Krepost* flight controllers occupied three rows of consoles on the right side of the large room. Six *Perimetr* controllers occupied a row of consoles on the left side of the facility. A waist-high wooden partition, like one found in a courtroom, divided the two spaces. Aluminum-bladed ceiling fans slowly stirred the stale air, barely churning the hanging pall of smoke. The large room was marginally lit by a series of incandescent light bulbs, dangling on electrical cords.

A large global map display, featuring a mechanical tracking device that followed the parabolic traces of the *Krepost's* revolutions, took up most of the front wall on the *Perimetr* side. Beside it, a regularly changing screen showed the next three potential targets, as well as pertinent data—city name, population, latitude, longitude, elevation and theoretical time of detonation—associated with each target.

The half-dozen *Perimetr* controllers were rarely gainfully employed, except during contact windows in which they received and analyzed diagnostic data from the Egg. Otherwise, they pretended to be busy, read books, and played chess. Today, their major source of amusement appeared to be observing the hectic activity on the other side of the divider.

Yohzin casually waved at Bogrov, the *Perimetr* engineer who had unknowingly surrendered the magic code to unlock the interlock; Bogrov grinned and waved back. Yohzin watched the *Krepost* flight director approaching. The flight director was a full colonel in the RSVN. Dark-haired, tall, thin and well-groomed, he was an exceptionally competent officer who regularly exhibited calm and effective leadership in crisis situations.

Standing before Abdirov, the flight director cleared his throat and said, "The flight surgeon respectfully requests a word with you, Comrade General."

"The flight surgeon?" growled Abdirov. "And pray tell, what the hell would he want to chat about? Perhaps he has he grounded someone else? Gogol would be up there right now, if the doctors had not fallen down on the job."

"It's not about Gogol, Comrade General," explained the flight director. "Sincerely, sir, I think it's something that you'll really want to hear."

Abdirov gestured for the flight surgeon to approach. "Make this quick," he said.

Dressed in his doctor's whites, the bespectacled flight surgeon saluted and then said, "Comrade General, I have consulted with my nutritional scientists, and we have devised some viable alternatives to abandoning the station when the supplies run out."

"Go ahead."

The flight surgeon placed a meticulously drawn graph before Abdirov. "Comrade General, you should be aware that we assume the fuel cell will continue to properly operate at its current capacity, so water and electricity will not appreciably factor into our calculations. Consequently, the availability of food is our principal concern."

"Agreed," said Abdirov. "Go on."

"At the present rate of consumption, Vasilyev and Travkin will exhaust their food in eleven days."

"Doctor, don't waste my time with things I already know. I am very aware of when Vasilyev and Travkin will run out of groceries, which is why we are planning their return on the nineteenth."

"*Da*, Comrade General. Now, to continue, if you ordered them to subsist on half-rations, they could theoretically extend their stay to twenty-two days, but we don't recommend that option."

Looking over Abdirov's shoulder, Yohzin noted that twenty-two days would carry them almost to the end of October, which was the Korolev bureau's best-case scenario for delivery of the next freighter. But even if they delivered the freighter earlier, it would still have to be loaded with commodities, and could only be launched if an optimal window was available to facilitate a rendezvous with the *Krepost*. He remembered that the launch parameters for the freighters were much more stringent than the manned Soyuz vehicles, because the crewed spacecraft had considerably more latitude to make orbital adjustments to ensure a successful rendezvous.

"We don't have empirical data for extended duration missions," explained the flight surgeon. "But we must assume that surviving on severely reduced rations would exact quite a toll on their bodies in weightlessness. We are aware that cosmonauts suffer considerable loss of bone and muscle mass during prolonged flights. Frankly, your men might be too debilitated to survive reentry."

"Agreed," replied Abdirov. "But that's a risk that I might be willing to accept, if it might buy us some more time."

"There's yet another option, Comrade General," said the flight surgeon. "One that might yield even more time, possibly enough to keep the *Krepost* occupied until the next freighter can fly."

"That does sound intriguing. You have my ear, or what's left of it, so go ahead."

The flight surgeon leaned over Abdirov's desk and traced his finger along a faint dashed line on the graph. "Here's what we can theoretically achieve if we send one of your men home tomorrow. We can prolong the mission to at least thirty days, assuming that your cosmonaut subsists on reduced rations. We think three-quarters rations would be sufficient, but suffice it to say, he would still be considerably weakened by the decreased caloric intake."

Abdirov nodded his head. "But it still sounds like a very viable concept," he said.

"I'm not sure that I would agree, Comrade General," interjected Yohzin. He was trying diligently not to display any outward signs of panic. He had just told the Americans that the *Krepost* would be unoccupied after the nineteenth, and ripe for attack, and now it appeared that Abdirov was seriously considering a hastily conceived concept to keep one cosmonaut in orbit. Magnus had "transmitted" the message last night, so it was likely well on its way to the Americans, and there was no way to recall it.

"You don't agree, Gregor?" asked Abdirov.

"The cosmonaut left in orbit would have no means to escape if there was an emergency aboard the *Krepost*," explained Yohzin. "Moreover, as the flight surgeon implied, he might not even survive reentry later."

"Correct," replied Abdirov, obviously growing impatient with Yohzin's discord. "And realistically, we have no solid evidence to indicate that he could indefinitely remain alive in weightless conditions, not to mention the potential exposure to other environmental factors like cosmic rays and the like. Regardless, Gregor, you're missing the point."

"I suppose I am," said Yohzin quietly.

"Well, then, let me enlighten you," said Abdirov. "Those cosmonauts overhead are Soviet officers. Just like you and me and every other officer who has sworn the oath to defend the Motherland, they have volunteered to sacrifice their lives, if necessary, to accomplish their assigned task. Certainly, there can be no more valid situation where such a sacrifice would be warranted. Would you not agree, Gregor?"

"I suppose so, Comrade General."

"There's more, Comrade General," blurted the flight surgeon. "We can likely extend the mission by a few more days, if you're willing to assume some additional risks."

"What risks?"

"The Soyuz Descent Module contains a NAZ-3 survival kit that includes vacuum-packed rations sufficient to sustain three men for three days. Theoretically, that would grant your man roughly twelve more days in orbit, provided that we transfer all of the emergency rations from the Soyuz to the *Krepost* and adhere to the three-quarters consumption rate."

"I consider that an acceptable risk," declared Abdirov.

"And there's yet another variable to the equation," said the flight surgeon. "It's an unconventional approach, but we are aware that at least five liters of potent alcohol in various forms went up on the freighter."

"*Five* liters of alcohol?" asked Abdirov. "I wasn't aware that we were running a bar in orbit."

"Gogol," noted Yohzin. "It's his stuff up there."

Abdirov nodded, and then gruffly said, "Continue."

"There's roughly two thousand calories in a liter of vodka," explained the flight surgeon. While we certainly don't recommend that your man drink to excess, a drink or two every day would add to his overall caloric intake, and that would allow him to stretch his food that much more. Assuming a basal metabolism rate of fifteen hundred calories a day, that's almost seven more days."

"Noted," said Abdirov. "Right now, I'm not sure that I'm willing to travel that particular route, but we can certainly keep it in reserve if things get sketchy later. Anything else?"

"Just one item, Comrade General. Even if your man is obligated to go to significantly reduced rations, once the freighter arrives, he should have ample opportunity to eat well and recover at least some of his strength before he returns to Earth."

"Excellent. Ingenious plan, Doctor. *Spasiba.*" Abdirov turned towards the flight director and ordered, "Do it. On the next communications window, send word to Vasilyev to immediately start preparing Travkin for return. Rally the controllers to commence work on the specifics."

"As you wish, Comrade General," replied the flight director. "And the shutdown procedure? What are your orders concerning putting the *Krepost* into hibernation?"

"Cease work immediately, but ensure that your engineers keep a good record of their work to this point, just in case we are compelled to revisit this issue in the future."

"But there's still an issue," interjected Yohzin. "The *Perimetr* inter-lock can only be disabled if two cosmonauts are aboard to man their key stations. I doubt that the *Perimetr* leadership will concede to this new concept."

"They will have *no* choice, Gregor," snapped Abdirov, as if he were admonishing a recalcitrant child. "My orders were to keep the *Krepost* occupied at all times, and that is significantly more important than their damned interlock. Since I am obligated to implement such an unusual solution to the dilemma that the Korolev bureau has created, then the *Perimetr* goons can adjust their procedures as well. If nothing else, they could issue us an Independent Action Code. That would certainly resolve the interlock problem until we can get someone else up there. They need to adapt to the circumstances, not the other way around."

Krepost **Station, On Orbit**
20:32 p.m. GMT, Saturday, October 7, 1972
GET (Ground Elapsed Time): 23 Days 19 Hours 47 Minutes
REV # 381

Vasilyev was dreaming of Irina when he was roughly jarred out of slumber. The dream was so vivid that it seemed absolutely real.

Prodding his shoulder as he switched on the light, Travkin whispered, "Rouse, rouse, Pavel. Time to wake up."

Woozy, Vasilyev struggled to resist consciousness so he could return to his wife and her filmy nightgown. Then he remembered that his wife was dead, but he was in orbit and very much alive. "Damn it!" he hissed.

"Wake up, Pavel."

"This had *better* be good," growled Vasilyev, rubbing the scraggly growth of beard on his chin. Squinting, he glanced at the luminous face of a small German-made alarm clock wired to a nearby pipe. "You're supposed to be standing watch for another three hours. Could you not let me sleep, Petr Mikhailovich? Is it too much to ask that I get a few hours of rest?"

"Control is transmitting on the guard channel," explained Travkin curtly.

"Guard channel? *Shit*," exclaimed Vasilyev. His head spun with the dire possibilities. The guard channel was strictly reserved for emergency messages. His right hand flew instinctively to his left wrist, feeling for the thin strand of chain that anchored his all-important brass arming key. "Launch alert? Action Message?"

Travkin shook his head. "*Nyet*. Apparently, the specter of nuclear annihilation will just have to wait. They want to speak directly to *you*, Pavel."

"About what?"

"They didn't say."

Vasilyev scurried out of the docking hub, kicked off a nearby bulkhead and swooped toward the radio console at the other end of the station. He clamped the headset over his ears, adjusted the volume dial, listened for few minutes, and then jotted down the new instructions that Control read up. He acknowledged receipt of the message before switching the radio back to the regular channel setting.

"*So?*" demanded Travkin. "What is it, Pavel? Is there a change to our mission?"

"*Da.* Our mission plan has been revised, comrade, but it's excellent news. *One* of us is going home today," announced Vasilyev, pulling off the headset and kneading his temples.

"*One* of us?" asked Travkin incredulously. "How can that possibly be?"

"The plant producing the Soyuz freighters is woefully behind schedule," explained Vasilyev, referring to his notes. "They won't have one ready for launch any sooner than the end of this month."

"But our supplies will run out by the 19th," bemoaned Travkin.

"Correct, numbskull. And that's why *you* are going home today. Our bosses think that sending one man home and keeping one on station will stretch the consumables until the next cargo boat can come up. One less mouth to feed, you know."

"Interesting logic," mused Travkin.

"Obviously, they don't want to leave the Egg unattended."

"So, what happens if they can't get a relief crew back up here in time?" asked Travkin.

"No matter. The priority is on the freighter. They'll finish that first and send it up, and if need be, I'll stay up until I'm relieved or until I run out of supplies."

"Well, if I'm supposed to go home today, did they send up the reentry profile?" asked Travkin, referring to the data that would be manually entered into the computer of the Soyuz descent module. "If you have it, I'll go ahead and plug it in."

"Here," replied Vasilyev, handing him an index card with the reentry instructions. He stuck a second index card into the breast pocket of his coveralls.

"What was that?" asked Travkin. "What else did they say?"

"Nothing that you need be concerned with, shirker. Move quickly now, and get your stuff together."

00:13 a.m. GMT
Sunday, October 8, 1972
GET: 23 Days 23 Hours 28 Minutes, REV # 383

Normally, the process of preparing for reentry would occupy them for days, but the abbreviated schedule mandated that procedure be accomplished in a matter of hours. As a result, the two cosmonauts worked at a frenzied pace to restore the Soyuz from its three-week hibernation. As Travkin painstakingly followed checklists to restore power and verify that the essential systems were ready, Vasilyev worked in the docking hub to ensure that the connection would be cleanly broken when the appointed time came.

Despite the hectic pace, although they didn't speak of it, both men were beginning to understand the potentially dire implications of Travkin's hasty return. Vasilyev certainly understood the rationale behind Control's decision that he would remain in orbit to continue the mission. After all, he was the mission commander and should always be the last to depart the ship, regardless of the severity of the circumstances. Moreover, given his unrivaled mastery of the docking process, it only made sense that he should stay aloft, given that the protracted mission was so dependent on receiving the freighter when it was ready. And as a lesser note, probably not an item even considered by Control, Travkin had a wife and family to return to, and Vasilyev was probably closer to his now than if he returned to Earth.

But as much as he understood and appreciated the necessity of his stay in orbit, Vasilyev also realized that he would be far from home with scarce chance of rescue if anything went wrong. For a vessel made mostly of titanium, one of the strongest metals known to man, the *Krepost* was actually a very fragile home. It was an extraordinarily complicated machine, and although its major systems had performed well to date, there was always the possibility that something might break. And if there was a solitary logical reason that Travkin should remain rather than him, that was it; Vasilyev was extremely conscious that Travkin possessed a much more intimate knowledge of the station and its components, and was far more accomplished at diagnosing problems and fixing them.

He heard Travkin's muffled voice echoing through the access tunnel. "Power's good, Pavel. The reentry solution is locked in the computer, and all systems look adequate. I'm activating the warm-up heaters for the thrusters now. Everything will be just fine, I know it. They'll dispatch a freighter up here in no time, and then a relief crew will follow soon after. In no time, you'll be headed back to..."

"An empty apartment? An empty bed?"

Travkin swallowed. "I'm very sorry. I'm such an idiot."

Hugging Travkin's shoulders, Vasilyev smiled and said, "Buck up, cosmonaut. You're right: everything will be just fine. This is just a farewell, not a damned funeral."

"You're right, Pavel. I apologize."

"Look, Petr, there's only a few more minutes before I have to send you off. I don't know if you've considered it, but your weight and balance is really skewed. Your Soyuz is going to be out of kilter if I'm not in the other couch. We need to load some ballast to center up your boat, or you can kiss *dosvedanya* to your dreams of triumphantly parachuting into Red Square."

Swimming through the air, the two men hurried to transfer additional weight—Travkin's personal gear, books they had both read, and a couple of bulky trash bags—from the *Krepost* into the Soyuz. After positioning the bags in place and tying them down with strips of tape, Vasilyev declared, "We're very close, but I still think you're going to be short a few kilograms."

Hovering in the docking hub, the two men looked at each other, mentally inventorying what spare items could possibly be left in the *Krepost*. "Obviously, I can't take any of your food," said Travkin.

"And I won't let you sneak out of orbit with my damned liquor. Or my caviar or truffles."

"Then what else is there?" asked Travkin.

Suddenly, it dawned on both men what non-essential material was still left on the station. Simultaneously, they grinned as Vasilyev noted, "We left the human waste container in the lavatory."

Travkin hustled back into the station to retrieve a fairly large rubber-coated sack that in turn contained smaller carefully sealed bags that held three weeks' worth of their collected feces. Pushing the bag before him, he swooped down the narrow access tunnel into the Soyuz. He lashed the lumpy bundle into Vasilyev's reentry couch and wiped a sheen of sweat from his brow. Cinching a restraint strap, he said, "Well, Pavel, this definitely confirms an assertion I have been making for several years."

"And that would be?"

"Don't be offended, my esteemed commander, but I've always argued that you could be replaced by a bag of shit. This proves it."

The two men returned to the docking hub, where Vasilyev assisted his companion in donning his bulky orange SK-1 space suit.

"Well, you're all dressed up for your journey," said Vasilyev, glancing at his wristwatch. "Time to go, cousin." He tousled Travkin's greasy hair, enveloped him in a bear hug, and gave him a big smooch on both cheeks.

"*Dosvedanya,* Pavel," muttered Travkin, wiping a tear from his eye.

"*Dosvedanya,* yourself, rockhead," replied Vasilyev, nudging his companion down the access tunnel. "No time to dawdle. Give your ticket to the conductor and hurry to your seat. I have *real* work to do and no time to waste on a lackadaisical slacker like you."

Peering through the access tunnel, he watched as Travkin sealed the Soyuz's nose hatch. He checked his watch again; they were running uncomfortably close to the timeline. In order for Travkin to hit his mark for retro fire, the Soyuz needed to undock immediately. He knew that Travkin should have already retreated to the bullet-shaped Descent Module, closed the hatch that connected it to the front-most Orbital Module, and was probably now connecting his umbilicals before strapping into his padded reentry couch.

Vasilyev swung the docking collar's hatch closed, wiped a tear from his own eye, and listened as the dogging latches clicked into place. He twisted the valve that depressurized the access tunnel, and then threw the lever that disengaged the docking mechanism. Now, he had to accept that his companion was gone. He made his way back to the control module. Peering through a porthole, watching the Soyuz fade in the distance, Vasilyev felt lonelier than he had ever felt in his entire life. His was now the ultimate in solitude.

He reflected on his earlier wake-up call, the unexpected radio message from Control. Besides the abrupt change in mission, the transmission contained two details that he had not shared with Travkin. First, the Soyuz freighter production delay was considerably worse than he had let on. He should expect to be up for considerably longer than just a month, so he would be compelled to stretch his supplies. He resolved himself to stringently ration his water and food, to consume just barely enough to sustain life.

Second, Control had revealed that there was ample reason to believe that the Americans were very much aware of the *Krepost* and its mission, and so it was highly likely that they would deploy a satellite interceptor to destroy it. He would have to remain on almost constant vigil to protect the station from destruction. Now he suspected that Control had another motivation in leaving him here. Rather than protecting the *Krepost* from assault, they were probably more interested that he remain aloft to actually bear witness to an attack, especially if he could describe the details. He had heard rumors that the Americans had apparently successfully knocked several critical satellites from orbit, but no one had yet figured out how they were doing it. The only logical theory was that the Americans possessed an extremely sophisticated robotic satellite interceptor, and now he was concerned that he was being deliberately offered as bait.

Krepost Project Headquarters
7:35 a.m., Sunday, October 8, 1972

Abdirov was pleased with himself. Acting on the flight surgeon's insightful suggestions, he had devised a plan to snatch victory from defeat. Best of all, since he was still certain that the Korolev bureau was striving to undercut him, to force him into an untenable position, he was now able to manipulate their treachery to his own ends.

The *Krepost* project's dedicated intelligence officer quietly entered Abdirov's office and posted squarely before his desk. "Lieutenant Colonel Malenkov, reporting as ordered, Comrade General," he declared.

Abdirov studied Malenkov. Square-jawed and handsome, with receding blond hair and a brushy moustache, he was the nephew of Georgy Malenkov, one of Lenin's confidantes. He was a young officer on the rise in the RSVN, and a man that Abdirov had come to trust for his intellect and reliability.

"Have a seat, Nikolai Danilovich," said Abdirov. He waved his hand over a personnel dossier lying open before him. "Your records indicate that you have considerable experience in preparing training exercises for high-level strategic staffs."

"That is correct, Comrade General," replied Malenkov, casually brushing red dust from his trousers as he sat down.

"I want to enlist your assistance in developing an exercise," said Abdirov, closing the dossier. "Are you acquainted with the current developments with the *Krepost* mission?"

"*Da*, Comrade General," replied Malenkov. "I am aware that the next freighter will not be delivered on schedule, which will likely necessitate that the *Krepost* be abandoned, at least until the freighter production can keep pace with the missions."

"You're correct about the freighters, Nikolai, but there's an entirely new wrinkle. I have authorized a plan to extend the mission until the next freighter is available. Major Travkin will return to Earth, and Major Vasilyev will remain in orbit. With only one man aboard the *Krepost*, we can make the most effective use of food supplies and other consumables."

Malenkov nodded. "Excellent plan, Comrade General, but I am not sure how I might be of assistance to you."

"It's simple," answered Abdirov. "I have decided that this is a perfect opportunity for an exercise, a sort of strategic simulation. This situation creates an almost perfect petri dish to evaluate a cosmonaut's decision-making capacity when placed under intense stress."

"I completely agree, Comrade General, but don't you think that being in orbit by himself would be stressful enough?" asked Malenkov.

"Certainly, that's true," replied Abdirov. "The circumstances grant us an incredibly unique opportunity that we could not ever hope to duplicate on Earth. Just as you indicated, Nikolai, Vasilyev is already under tremendous stress. When we conduct exercises on Earth, we try our utmost to induce high stress levels, but such stress is still an artificiality. Short of occasionally killing exercise participants, there's little that we can do to perpetually keep them on their toes."

"I suppose that the High Command would frown on that," said Malenkov.

"But you agree that we cannot readily replicate such a degree of realism on Earth?"

"Absolutely, Comrade General."

"And there are other factors we can exploit," said Abdirov. "Besides the obvious physical isolation, Vasilyev is significantly isolated in other ways. Since the *Krepost* lacks any means to receive radio or television broadcasts, Vasilyev is effectively deaf to the news of the world, except for the information that we pipe up to him on the intelligence channel."

"Very true, Comrade General," replied Malenkov. "It is a truly unique situation."

"I'm glad that we agree," said Abdirov. "And here's what I want. With your experience, I want you to create an overarching exercise scenario and the same sort of mock-up reports that we use for strategic war games. I want the reports to describe a series of incidents leading to a gradual breakdown in international relations, followed by the enemy's gradual escalation for a major attack."

Malenkov's mouth dropped open. Regaining his composure, he asked, "You want to deceive Vasilyev into believing that the world is at the brink of thermonuclear war?"

"I do. As I said, we could not possibly generate such a realistic training environment on Earth, so I want to take advantage of it while the opportunity presents itself."

"Begging your pardon, Comrade General, but this all seems very extreme."

"Nikolai Danilovich, stationing nuclear warheads in orbit is extreme," said Abdirov in a fatherly tone. "But we are compelled to do so, because the Americans have fielded weapons that we cannot readily counter. And if putting nuclear weapons in orbit is an extreme option, we certainly must be confident of the men who are tasked to deploy them, and that's why I consider this exercise to be so essential. Do you not agree? Does that not make sense to you?"

"*Da.* I concur, Comrade General," answered Malenkov. "It does make sense, and I can clearly see why you are so eager to exploit this unique opportunity."

"Look, there's another reason I want to pursue this exercise scenario," said Abdirov. "Vasilyev will be up there by himself, possibly for several weeks, without any direct human interaction. I want him to remain vigilant. I am very concerned that if we don't provide him with adequate motivation to remain alert, then he might readily be lulled into complacency."

"That makes perfect sense, Comrade General, but I am compelled to ask: What if he becomes too wrapped up in the mock scenario that we feed him? What if he becomes unhinged and decides to deploy the Egg of his own accord?"

Abdirov smiled, and then replied, "That's the best part. He can't. The *Perimetr* interlock prevents him from taking any sort of unilateral action. Regardless of how much he might want to drop the Egg, he cannot do so without an Independent Action Code, and we certainly aren't ever going to issue one to him. Moreover, if we sense that he is losing his equilibrium, then we just admit that it was an exercise."

"That's quite a relief, Comrade General," said Malenkov.

"Besides you, who else sees the intelligence reports that are sent up on Channel Three?"

"Just the strategic intelligence analysts who prepare them," answered Malenkov. "Two men on either shift. They summarize the daily intelligence reports and specify the reconnaissance targets for each pass over enemy territory. I review and approve the summaries. Once I sign off on them, then the intelligence analysts record them on a high speed recorder, encrypt them, and then they are burst-transmitted on Channel Three during the next available communications pass."

"How long are the voice recordings?" asked Abdirov.

"No more than fifteen minutes, Comrade General," replied Malenkov. "We are confined to that duration, because the communications window is limited. The burst transmission equipment reduces the fifteen minute voice recordings down to three minutes."

"And fifteen minutes is adequate?"

"Actually, sir, it's usually more than adequate. With very few exceptions, the reconnaissance targets were designated long before the mission launched. They are all identified by code number, so we have a brevity system that specifies when each target will come into view. As an example, San Diego Naval Base is Target N-26, so all we have to say is 'N-26, 16:38,' unless there are special instructions."

"Special instructions?"

"*Da*, Comrade General. For example, 'N-26, 16:38, verify departure of aircraft carrier battle group' to confirm intelligence that we have received from other sources."

"That makes perfect sense," replied Abdirov. "So, would it be an immense burden if I asked you to record the intelligence messages yourself, so that only you are aware of what is transmitted?"

"Not a burden at all, Comrade General. I am just glad that I can be of service to you."

Abdirov smiled. "I think that it goes without saying, that you are not to share this with anyone," he said.

"Of course, Comrade General."

"Good. Draft up a preliminary scenario, and bring it round this afternoon."

8

RUMORS OF WAR

Aerospace Support Project Headquarters
8:00 a.m., Tuesday, October 10, 1972

E ven though Mark Tew had been rejuvenated by the initial news
that nuclear-armed *Krepost* was in orbit, the ensuing three weeks
had turned into a laborious slog of paperwork and meetings. His
fragile health had deteriorated yet again; just this past Friday, his cardiologist had made it abundantly clear that Tew's unrelenting workload
was depleting his dwindling days at an ever exponential rate. But try
as he might to obey his physician's wishes, Tew's schedule just couldn't
accommodate regular rest and a healthy diet.

As he waited for Seibert to hand out some papers, he slipped two pills
into his mouth and sipped from a glass of lukewarm water. Struggling to
conceal a persistent tremor in his hands, he donned his reading glasses
and adjusted them on his nose. He picked up the report that the intelligence officer placed before him, but after seeing that its pages fluttered
in his unsteady hands, he laid the document flat on the table to digest
it. He looked up to see if Wolcott or Heydrich might have noticed his

discomfort, but the two men were already focused on their own copies of the report.

The tersely written intelligence summary amplified a cursory report they had received last week, indicating that the Soviets might not be able to sustain continuous manning for their nuclear-armed *Krepost*. The previous report, drawn from information provided by the highly placed source at Kapustin Yar, alleged that the assembly of Soyuz cargo ferries was significantly delayed; as a consequence, the *Krepost's* current crew would be obligated to return to earth if the station was not replenished in time.

As he skimmed through some charts and tables, Tew saw that this new summary not only corroborated the initial report, but also identified the specific facility responsible for the production delay. Moreover, it pinpointed the precise cause: a critical weld in the Soyuz cargo ferry's pressure hull had repeatedly failed a stress test stipulated in design specifications. The thoroughness of the summary was a testament to the responsiveness, broad scope, and immense reach of the American intelligence machine. Since learning of the *Krepost* facility at Kapustin Yar, US intelligence agencies had focused on the Soviet cosmodrome with laser-like intensity. All available intelligence resources—electronic listening posts to monitor rocket launch communications and telemetry, surveillance satellites and analysis centers—were brought to bear. Obviously, the effort relied primarily on technical means that could only view Kapustin Yar from a distance; the secretive test facility was otherwise immune to penetration by HUMINT—human intelligence—assets. As a result, although they were collecting a wealth of information from the program's periphery—like the facilities that produced the Soyuz modules—there was still much that they did not know.

Flipping over the red-bordered document, Tew asked, "Do they have any other alternatives to resupply the *Krepost*?"

"Not really, General," replied Seibert. "No viable options, anyway. As best we can tell from reading some cable intercepts, they're grasping at straws. They apparently have at least one Soyuz crew vehicle ready to fly. Theoretically, as a stopgap measure, they could modify it to fly as a cargo vehicle. But even if they were successful, there wouldn't be much free space available to stuff with consumables. It might buy them a little time, maybe a week or two at most."

Tew nodded. "Assuming that they fail to replenish on schedule, when will they have to abandon the *Krepost*?"

Seibert opened a manila folder to reveal a calendar. He pointed at a red-circled date and stated, "According to our source, their drop-dead date is the nineteenth. He stated that if they don't receive a cargo ferry by then, the mission planners will order the crew to come home. Our supplemental intelligence supports this; their recovery forces are already ramping up for a mission on the nineteenth."

"Okay. Assuming their crew comes home then, how long will the *Krepost* be vulnerable?" asked Tew.

Shrugging his shoulders, Seibert replied, "Honestly, General, I'm not sure. I have no idea of how long it might be vacant, but I also feel that we might not see a similar lapse ever again."

"So, Ted, do you feel that this report is reliable?" asked Tew, tapping a finger on the downturned document.

"Yes, sir, we've surmised that this intelligence is accurate," stated Seibert. "*Very* accurate, so I strongly recommend that we decisively act on this opportunity."

Tew swiveled his chair to face Heydrich. "Gunter, do we have adequate lead time to act on this?"

Heydrich nodded. "We've certified the profile," he stated. "Barring any changes or unforeseen consequences, we can launch from Kennedy on or slightly after the nineteenth. Based on our calculations, we should be able to effect an intercept in twenty-four hours or less."

"But, Gunter," interjected Wolcott. "How about all those new gadgets that our eggheads have been workin' on? Will any of those gimmicks be ready in time?"

"From an equipment standpoint, no. We've validated the radar detection system, but the jamming pod and the torpedo won't be ready for several weeks. Although we won't have the new gear, we have cultivated new procedures that should enable the crew to fly the entire intercept without energizing their radar. Of course, they will be relying on a lot of information fed up from the ground."

Heydrich continued. "In any event, other than the new procedures, our crew will be flying essentially the same profile that we've employed successfully on seven other missions. They won't have the new equipment, but they will still have the Disruptor, and that's a piece of technology that's proved itself time and again. In summary, if I believed the *Krepost* might be empty again, I would recommend that we wait for our technology to catch up, but I agree with Colonel Seibert. I don't think that we'll see this opportunity again, so I strongly recommend that we seize it."

"So that's it?" asked Tew. He consulted the tiny metal calendar clipped to his leather watchband. "We send up Carson and Ourecky in nine days?"

"Assuming your approval, General," replied Heydrich.

"Done," said Tew. "Granted, we'll have to run our plan past Kittredge and his group at the Pentagon, and they'll have to seek approval from the NCA, but you have my blessing to proceed."

Wolcott cleared his throat loudly and rapped his fingernails on the table. "Mark, I'm obligated to remind you that Leon is in California," he stated. "He can jump on his plane and be here tomorrow. Don't you reckon we at least owe it to him to wait one more day so he can be a party to this discussion? You seem absolutely insistent on flying Carson and Ourecky on this mission, and I think we need to at least consider flyin' his crew."

"*No*," replied Tew emphatically. "This is my decision, not his."

"Mark, before we do something rash that we might later regret, can't we at least take a minute to hash this around some more? Are you absolutely *sure* that you want to fly Carson and Ourecky?"

"I am," answered Tew. "How about you, Gunter? Just for the sake of discussion, how does the Navy crew look right now?"

"Honestly?" answered Heydrich. "They've made great headway with the intercept procedures. They're picking them up much faster than I had anticipated. I would say that they are at least on par with Carson and Ourecky before they flew their first mission."

"See what I mean, boss?" said Wolcott. "Even Gunter says that the Navy boys are on par with our best hands. Ain't that something you should take into consideration?"

"Virgil, I said that they were on par with Carson and Ourecky before their first flight," said Heydrich, obviously flustered. "But they probably will never reach their present level of competency."

"But they could probably still handle this mission, right?" asked Wolcott. "After all, this revised procedure is even new to Carson and Ourecky, and the way you've explained it, virtually everything will be fed up from the ground. If that's the case, the Navy boys shouldn't have any problem wranglin' this steer. This is an excellent opportunity for them to earn their spurs."

"I don't agree," said Heydrich, shaking his head.

"Why don't you talk to Carson, Mark?" asked Wolcott. "Personally, I ain't comfortable with Ourecky going up now. I don't think he's healed up sufficiently, despite what the docs say. Anyway, those two are all but joined at the hip, so why don't you ask Carson what he thinks?"

Tew felt veins pulsing in his forehead. Struggling to contain his anger, he closed his eyes as he let out his breath slowly. It infuriated him that Wolcott was so anxious to ingratiate himself to Tarbox. He sensed his temples pulsing, then felt a sharp twinge in his chest, and then felt his fingers grow cold and numb. His circulation returned to normal in a matter of seconds, but it was still a frightening episode.

The decision was his—and *only* his—to make. Of course, his recommendation would have to be ultimately approved at the highest levels in Washington, but they had not disagreed on any of his other recommendations concerning the previous eleven flights. Clearly, they trusted his judgment.

As much as he disliked the notion of putting Carson and Ourecky back into harm's way, he was absolutely convinced that the two men were the only crew even remotely capable of successfully executing this mission. And *this* mission was the very reason that the Project had been approved in the first place, so given this unique opportunity to finally swat a Soviet nuke from orbit, how could he *possibly* send anyone but his varsity team?

As someone who had commanded in combat, he certainly understood the terrible human costs that might be incurred. After all, he had personally ordered Bea's father to his death in Korea; as a result, Ourecky would never meet his father-in-law and Ourecky's son would never meet his grandfather. But as much as he felt compelled to protect Ourecky from harm, because the engineer had already acted far above and beyond the call of duty, Tew also understood the clear necessity to destroy the *Krepost* at all costs.

"Mark?" implored Wolcott. "Mark? Are you okay, buddy? I didn't mean to upset you. I just wanted you to weigh out this situation some more. Just this once, I think we can afford to lend Carson and Ourecky a breather."

Opening his eyes, Tew groaned audibly. "Virgil, I've ordered plenty of men to execute missions when I knew full well that the odds were stacked against them. In every instance, I was very aware that they probably wouldn't come back, but the harsh reality was that the mission was more important than their lives. Do you think I was ever comfortable sending *those* men to their doom?"

"No, but..."

"*But* nothing, Virgil. This damned *Krepost* is a terrible thing, just hanging over our heads, and we *must* destroy it. That's our mission. That's *always* been our mission, even though we've ventured off on a sidetrack. Should we send Carson and Ourecky? Since those two gentlemen have consistently intercepted every target we've sent them up against, they're obviously the only logical crew choice. Certainly, the Navy crew is extremely proficient and probably could be ready in time, but Carson and Ourecky are ready *now*."

"But, Mark, I think..."

Frustrated, Tew shot his erstwhile deputy an angry glare. "Virgil, there's absolutely no doubt in my mind: we are compelled to send our *best*, regardless of the risks, to ensure that this task is done. To me, this mission is less about destroying that station, and more about sending the Soviets a clear message that we *can* destroy anything they send up there, and if it's a threat to our people, we *will* destroy it."

"*But, Mark...*"

"*Enough*, Virgil!" howled Tew. His heart pounded in his chest. "Enough!"

Krepost Station, On Orbit
3:13 a.m. GMT, Wednesday, October 11, 1972
GET: 27 Days 2 Hours 28 Minutes, REV # 433

Vasilyev awoke from a fitful sleep and stretched. Feeling the onset of a migraine headache, he massaged his throbbing temples. After briefly contemplating the day's requirements, he immediately commenced his daily chores.

His self-appointed "day" was divided into a sleeping period, followed by an eight-hour watch. Although he had allocated four hours to every rest period, it was extremely rare that he actually dozed for more than a couple of hours at a time. It had been a few days since Travkin had returned to Earth. Since his companion's abrupt departure, he had settled into the simple routine that centered around the need for constant vigilance. Except for the infrequent occasions when he retreated to the galley to grab a Spartan meal, Vasilyev divided his time into shuttling between the distant points of his little world: the docking hub and the control area. When he did snatch an hour or so of fitful rest, he elected to remain in the docking hub, close by the controls for Sparky. His motivation was that if the proximity alarm's radar was triggered by the approach of a hostile space vehicle, he wanted as much warning as possible, so he could immediately activate the automatic cannon to engage the target at the greatest possible distance.

During the *Krepost's* passes over the United States, he diligently posted himself at the Earth-facing porthole, carefully scanning for any indications that the Americans might be launching a salvo of ICBMs in a surprise attack. Those were the only instances where he saw anything but the dismal gray interior of the station; as much as he enjoyed looking at the stars, he had abandoned his favorite pastime altogether. Since he routinely deprived himself to stretch his food reserves, hunger pangs constantly gripped his abdomen. He dreamed of the massive feasts that he would devour when he made it home.

He floated down the access tunnel that led from the docking hub to the freighter's dry compartment. His first task was to make a cursory check of the fuel cell. Examining the pressure gauges atop the cell, he saw that it appeared to be in good condition, but when he placed his bare palm on the cylindrical flank of the fuel cell, he thought that it felt slightly warm to the touch, more so than usual. He was a bit concerned that the coolant loop might be clogged or otherwise obstructed. To keep the device at a relatively constant temperature, a refrigerant liquid cycled through a loop that meandered through its inner workings. Exiting the fuel cell, the coolant loop's pipe circulated through radiator-like heat exchangers located on the exterior of the freighter's wet compartment.

Although there was scarcely little that he could do to diagnose a problem with the fuel cell, and even less to fix it, he was still apprehensive about a potential failure. He had recently noticed that the water produced by the cell was gassier than usual, and had a distinctly sour taste. He was also experiencing a lot of bloating and mild stomach discomfort, which he attributed to the water. Unfortunately, although he was aggressively rationing his food intake to stretch out the pickings in the pantry, the fuel cell was producing water faster than he could drink it. Initially after Travkin had departed, Vasilyev drank as much water as he could hold, thinking that it would fill his belly and alleviate some of the hunger, but doing so accomplished little more than making his stomach ache.

Next, he made his way down to the galley to methodically parcel out his food ration for the day. His daily fare fit neatly into the thigh pocket of his coveralls, which made it very convenient to grab a meal on the fly. He used a nozzle to squirt measured increments of hot water into a plastic pouch to reconstitute tea crystals, and then did the same with a similar container of dehydrated chicken broth. As he squeezed the pouches, he examined the rheostat dial on the water heater and remembered how Travkin had repaired it by scrounging a part from the Soyuz. As he stashed the pouch of chicken broth in his pocket, he thought of his friend and how much he missed him. Travkin was always an excellent companion, someone who could make even the worst of circumstances somewhat more bearable.

As he slowly savored the tea, his sole luxury for the day, Vasilyev continued with his daily routine. He looked out the porthole to orient himself and saw that he was passing over Sumatra. He drifted into the control area to check the diagnostic readings for the Egg. Everything was in parameters, so the Egg appeared to be in good order, patiently hibernating until it awoke to obliterate the unknowing inhabitants of an ill-fated city.

Conscious that he had over an hour before the next contact window, he listened to the radios, switching from one channel to the next, hearing nothing but static. He glimpsed a blinking blue light that indicated that the Channel Three receiver had automatically recorded the daily intelligence update while he was sleeping. Vasilyev rewound the recorder, toggled its speed to "normal voice" and donned a headset to listen.

He was immediately alarmed at the items contained in the report. It sounded as if the world was suddenly descending into chaos. American commandos had flown into North Vietnam on helicopters to raid an air defense training camp just outside of Hanoi; in doing so, they had captured several Soviet advisors and were now presenting them as evidence of direct Soviet involvement in the conflict.

One particularly disconcerting item described a major mishap which occurred yesterday in the North Atlantic, during the course of a scheduled NATO exercise. While operating in close proximity, an American missile submarine collided with a Soviet attack submarine. In the immediate aftermath of the incident, the Soviet sub plunged to the bottom with all hands aboard, but the American vessel was able to briefly limp to the surface to disembark most of its crew to a British frigate. An American Navy officer claimed that the Soviet sub had shadowed the

American sub for several hours before intentionally ramming it as it tried to turn away and break contact. The Soviet account declared that the American sub had been the aggressor. The maritime incident had devolved into a diplomatic confrontation, with both parties going before the United Nations Security Council to request strict sanctions against the other.

A long list of locations to monitor followed the intelligence updates. In almost all cases, he was being directed to spot evidence that American nuclear forces were being placed on heightened alert. Vasilyev found the list of surveillance targets very frustrating, because he realized that most of the sites were going to be in darkness when the *Krepost* passed overhead.

It was all very aggravating. According to the succinctly worded intelligence reports, the world was quickly unraveling, so theoretically he was in the best possible vantage point to witness it all. But try as he might, diligently staring down from the loftiest of perches, he could see no evidence of the strife and turmoil that was unfolding beneath him.

9

FUEL CELL

Krepost **Station, On Orbit**
5:50 a.m. GMT, Saturday, October 14, 1972
GET (Ground Elapsed Time): 30 Days 5 Hours 5 Minutes, REV # 483

Jotting down notes, Vasilyev listened to the intelligence update that he had recently received via the automatic recording system on Channel Three. The news was growing progressively worse. Soviet and American nuclear forces were at their highest alert status. American anti-submarine planes had damaged a suspected Soviet submarine just a few miles off the Outer Banks of North Carolina. The crippled sub was forced to surface, and its crew taken into custody by the US Coast Guard. A US Navy reconnaissance plane flying from Alaska had blundered into Soviet airspace, or so the Americans claimed, and was engaged by Soviet interceptors. The crew bailed out over the Bering Strait, with two of the ten men lost at sea. The other members of the United Nations Security Council were taking extraordinary measures to resolve the growing confrontation before the pot boiled completely over. As he listened to the reports, Vasilyev was absolutely sure that the world was teetering on the verge of all-out total war.

He was perplexed. Despite the ominous developments described on Channel Three, the communications traffic on the other radio channels was calm. Channel Two, the frequency reserved for *Perimetr* communications, was strangely silent. On Channel One, the Control frequency, the mission controllers seemed to act as though business was entirely normal.

There was not the least tone of alarm in their voices; they could have been housewives casually gabbing over the clothesline, sharing tidbits of gossip. Oddly, they seemed to be even more chipper than usual. Perhaps they were under strict orders to keep him as calm as possible. He also suspected that the mission controllers were not granted access to the intelligence updates, in order to ensure that they maintained focus on purely operational matters concerning the *Krepost*.

Rubbing his aching stomach, he floated up to the docking hub to check the *Krepost's* defense systems. He switched on Sparky's independent stabilization gyros to test the gun's sights. Instead of the usual whirring sound he had grown accustomed to, he heard a clicking noise, followed by a series of loud pops, and then a persistent hum. The docking hub's lights and fans sputtered for a moment and then resumed their normal functions.

Vasilyev was certain that it was an electrical problem of some kind and hoped that it wasn't too serious. He deactivated Sparky's gyros, locked the gun in standby mode, and went to the distant end of the *Krepost* to investigate. He floated into the control area and audited the gauges on the station's power status panel. The discrepancies he saw gave him cause for alarm. For some inexplicable reason, the fuel cell was not charging the station's batteries.

He returned to the docking hub and floated headfirst down the access tunnel into the cramped dry compartment of the freighter. Since he and Travkin had transferred the dry compartment's contents—food, spare parts and other items—from their aluminum storage racks to the *Krepost*, the free space around the fuel cell was mostly empty, except for trash—used food containers, primarily—they had packed into the module to be jettisoned.

Holding a small flashlight with his teeth, he painstakingly inspected the wiring harness that led into the docking hub and then toward the power distribution module within the *Krepost*. Then, returning to the freighter, he methodically checked the exterior of the fuel cell and its connections. Since everything seemed to be in good order, he tugged out several trash bags packed around the fuel cell casing, and then traced the cryogenic hoses, swathed in insulating blankets, that snaked from the wet compartment. The cryogenic conduits supplied the all-important components of liquid oxygen and liquid hydrogen, warmed into their gaseous state by the fuel cell's residual heat, which the device then combined to generate electricity and water. Perplexed, he shook his head; everything appeared to be in kilter. So why wasn't the fuel cell running?

He returned to the docking module to ponder the puzzle. Although the fuel cell's failure was aggravating, it was by no means calamitous. Although he hadn't gained absolute confidence in the fuel cell, he did have to admit that it had chugged along for over two weeks with very few problems. Since he had not detected any physical damage to the cell, and the cryogenic pressures looked nominal, he felt secure that the engineers at Control could provide him with a set of procedures to coax it back on line.

Until the fuel cell was returned to service, the *Krepost* still had three sources of electrical power. The first was the station's main batteries, which were normally replenished from the fuel cell. He had just checked their gauges in the control area; all of the batteries were completely charged and in excellent repair. By themselves, even in the worst of circumstances, they could fully power the station for roughly four days, which he could stretch to almost a week or even longer by carefully allocating the power to the most critical systems.

The *Krepost* was also equipped with a pair of solar panels, protruding from the hull like large wings. These were also in good order, and when oriented correctly at the sun, generated power to trickle-charge the main batteries. Unfortunately, the solar array were extremely dependent on another system—the station's stabilization gyros—which kept the *Krepost* appropriately oriented, not just to keep the panels pointed at the sun, but also to maintain several other directionally dependent systems—communications antennas, Sparky, Sparky's detection and acquisition radars, etc.—in correct alignment. Without the stabilization gyros, the *Krepost* would gradually begin to wobble and then eventually just drift. So, while the solar panels were important, approximately half to three-quarters of the auxiliary power they generated was eventually lost to the gyros.

Finally, there was yet another entirely redundant set of back-up batteries within the independent module that housed the Egg. Those batteries, which were intended specifically to ensure that the Egg could be deployed even in the event of a catastrophic power failure on the *Krepost*, were not even connected to the station's main power grid. They were only placed into operation when the Egg was activated. Since the batteries and their connecting bus were physically separate from the main station, Vasilyev could not gain access to them, even in an emergency, to supplement the station's power.

8:50 a.m. GMT
GET: 30 Days 8 Hours 5 Minutes, REV # 485

Hovering in the control area, Vasilyev watched the mission clock and waited patiently for the next contact window. As the remaining seconds ticked off, he donned his headset and listened for the call. Clear as a bell, it came through: "*Krepost, Krepost*, this is Control. Are you monitoring?"

Speaking calmly, he curtly reported: "Control, this is *Krepost*. Be aware that the fuel cell has failed."

A few seconds elapsed before Control replied: "*Krepost*. We acknowledge that you are experiencing difficulties with your fuel cell. Is it charging the main batteries?"

Vasilyev groaned. *Is it charging the main batteries? Why, no, it's not; because I'm not simply experiencing difficulties with the fuel cell, the damned thing has FAILED. Weren't they listening down there?* He took a breath, keyed the microphone, and slowly said: "*Nyet*, Control. The fuel cell is not charging the main batteries. The fuel cell has *failed.*"

"*Krepost*, are you alleging that the fuel cell has ceased to function?"

Am I alleging that the fuel cell has ceased to function? Why, yes, I am. Is there some other definition of failed? "Affirmative," he reported. "The fuel cell has *failed.*"

Two minutes passed, and all he heard in his earphones was the faint crackle of static. Impatient, he keyed the microphone and said: "Control, this is *Krepost*. I request guidance concerning this situation." In composing his transmission, Vasilyev chose his words very carefully. Under *no* circumstances would he allow himself to utter the word 'emergency.' Arbitrarily declaring an emergency implied that he was losing control, and if his bosses assessed him as even the least bit prone to panic, he probably would never be permitted to fly again. No more flight bonuses. No more perks. *Better to die a hero than create a fuss.*

As he waited, Vasilyev gritted his teeth. He had hoped that the engineers would have already anticipated all potential problems with the fuel cell and would already have a contingency plan ready for him to diagnose the shortcoming and bring the device back on line. Unfortunately, if their prolonged silence was any clue, that didn't appear to be the case.

Finally, after two minutes of pregnant silence, he heard a voice on the radio: "*Krepost*, this is Control. Be advised that we will provide you with a restart protocol on the next contact window. Do you understand?"

Vasilyev took a deep breath, let it out, and replied: "This is *Krepost*. Affirmative. I am standing by for the restart protocol." Moments later, he heard the faint warbling tone that indicated that he was losing contact, following a persistent buzz of static.

Now, things were becoming more serious. While the situation was not yet an emergency, it was coming dangerously close to that territory. Vasilyev assessed his situation and decided that it was an opportune moment to set priorities for power consumption and to scale back non-essential systems to extend the batteries as long as possible, just in case the fuel cell predicament took a while to resolve.

He still considered defending the *Krepost* to be his foremost priority. While most of the defensive systems—Sparky and its attendant target acquisition radar—were fairly power-hungry when activated, they could remain in standby for the time being. Only the proximity alarm radar needed constant power, so that circuit would remain open. Sparky—the automatic cannon—and several other key systems, including the highly directional communications antennas, were effectively useless if the *Krepost* was not kept in stable alignment, so he had little choice but to let the station's stabilization gyros continue running.

After he identified those items that absolutely had to remain operational, he drifted slowly through the station, switching off lights, air circulating fans, and any other equipment that wasn't absolutely essential. As he cut off the lights in the docking hub, he considered the potential consequences of a worst-case scenario, one in which the fuel cell could not be restarted and the main batteries were eventually depleted. He was extremely conscious that the three docking mechanisms were dependent on electrical power; without adequate power, even if a freighter or relief mission was launched, their Soyuz spacecraft would be unable to dock.

Of course, if worst came to worst, a Soyuz rescue vessel would not have to dock in order to effect a rescue. Since the docking hub could be depressurized, with provisions available for an airlock, theoretically he could use it to transfer to a rescue vehicle. But although it was *theoretically* possible, it was also largely irrelevant as an escape option. He and Travkin had been fitted with the very rudimentary SK-1 spacesuit, and while the SK-1 was considered adequate to contend with a catastrophic loss of cabin pressure, there was no absolute assurance that they could provide sufficient environmental protection outside the spacecraft. His SK-1—now safely tucked away in a storage locker in the docking hub—was provided with a chest pack that would provide oxygen for roughly twenty minutes, but it was specifically intended to ensure that they could evacuate the *Krepost*. Solely intended for internal use within the station, it *might* work if he was compelled to take a stroll outside, but that was certainly not an excursion that he looked forward to.

After switching off all non-essential systems, he returned to the control area. He consulted the list of surveillance targets he was supposed to observe, and realized that it would be a few hours before the first of them came into view. His stomach growled and he was tired. Lacking anything useful to do, he closed his eyes and tried to snatch a quick nap until the *Krepost* spun back around for the next communications window.

10:20 a.m. GMT
GET: 30 Days 9 Hours 35 Minutes, REV # 486

He cursed his bosses for placing such reliance on untested fuel cell technology. Even the Americans had experienced problems with the temperamental cells and their associated cryogenic tanks. In fact, only last year, they had come very close to losing the three-man crew of Apollo 13 when a liquid oxygen tank abruptly ruptured.

As he nibbled on one of the vitamin-enriched wafers that he had once deplored, he used a hand-held microfiche viewer to examine schematics of the fuel cell. Letting go of the battery-powered viewer so that it floated before him, he wished that Travkin were still here. A consummate problem-solver, Petr would have rectified this issue hours ago, and they would already have been well into another marathon of durak or chess.

He heard a flight controller's voice in his earphones: "*Krepost, Krepost,* this is Control. Are you ready to copy the restart protocol for the fuel cell?"

Am I ready? Why, yes I am. Vasilyev grabbed his pencil and pad. "This is *Krepost,*" he replied. "Go ahead with the protocol."

It wasn't a simple procedure; he spent at least ten minutes frantically copying down the complex protocol that Control read up before he put it into practice. After verifying all of the circuit breaker settings and switch positions, he came to the last step of the instructions. "Now?" asked Vasilyev. His right index finger was poised over the master switch for the fuel cell.

"*Da,*" replied Control. "You may turn it back on."

Vasilyev threw the toggle switch and heard a series of disconcerting metallic clicks, followed by a faint rattle. He checked the gauges on the power status board. *Nothing.* The needles had not moved, not even a flicker.

"Now that the cell has re-started," said Control, "re-set breakers one through nine..."

"*Nyet,*" he snapped. "The fuel cell did *not* re-start."

The radio emitted mild static for several seconds. Vasilyev could picture the earthbound engineers scratching their heads, flailing with their slide rules, and jabbering nonsense at one another. Finally, after almost thirty seconds, there was an anxious voice from Control: "The fuel cell did not re-start?"

"That's what I said," replied Vasilyev, trying mightily to conceal his seething anger. Their scheme had cost him precious time.

"Nothing?" asked Control.

"*Nothing.*"

As if they had been waiting for him to say something, Control immediately responded: "*Krepost,* we will provide additional instructions on next orbit. We are collaborating with the designers of the fuel cell, and they will coach you on shutting down and restarting it."

Vasilyev cursed. *The designers of the fuel cell? So, were they were calling someone in Connecticut, where they had stolen the plans from the NASA contractor who built it?* It certainly wasn't the guidance that he expected or wanted. Obviously, continuation of the Egg mission—and *not* necessarily his survival—was their current priority. He was already over four hours into this potential crisis and had already expended his own limited array of diagnostic tools, so now he was at a crucial decision point.

The next step was not one that he could take lightly. With his resources dwindling and a deadline looming, he was compelled to finally admit that the fuel cell outage was not just a passing annoyance, but a genuine emergency. Despite this, once he formally declared a crisis, his chances of a return trip to space were essentially nil. It was a foregone conclusion that his every decision would be questioned, and that he would be hereafter perceived as a timid soul who would snivel and retreat before reaching the absolute brink. The RSVN was not the least bit lenient concerning cowardice or even perceived cowardice.

As he mentally inventoried the various tools stored in a locker in the docking hub, he heard a voice over the radio and recognized it as Travkin's: "Pavel, this is Petr. Listen, the engineers are positive that they can formulate a solution for the fuel cell. They want you to wait another revolution, and they will have it for you, I promise."

Vasilyev cursed. Maybe the mission planners believed that Travkin's soothing voice could somehow lull him into calm complacency, but he didn't appreciate the sly ploy. Although he knew that he could be candid with Travkin, there were others listening in on the comms loop, so he resisted the impulse to speak with frankness.

He couldn't believe that they expected him to waste another ninety minutes just so they could have another engineering séance. *Can't they comprehend that the damned fuel cell is broken? And if those asinine engineers know so damned much about the proper care and feeding of the fickle fuel cell, then why couldn't they have designed and built one of their own, instead of filching plans from the Americans?*

"*Pavel?* Did you hear me, brother?" asked Travkin from Control. "The engineers will have a solution for you on the next pass."

"*Da*," replied Vasilyev, gritting his teeth in anger but feigning a tone of optimism. "I concur. I understand that I am to wait for instructions and I *will* comply."

He took off his headset and let it float, tethered by its cable. Seething with anger, he pounded his left fist against a bulkhead until his knuckles bled. Next, he drifted through the station and shut off *everything* that was drawing power, except for the most vital systems. There wasn't an established protocol or procedural checklist for the shut-down, but he just didn't give a damn at this point.

Rubbing his swelling left hand, he coaxed himself into believing that it was essential to remain calm. He heard a buzzing sound and pivoted to look at the Channel Three recorder. Its blue light flickered, indicating that it was automatically receiving an intelligence update. The wire recorder whirred for a few seconds and then stopped. Vasilyev thought it very odd, since the recorder normally took at least three to five minutes to completely capture a single compressed message. Assuming that the recorder might be malfunctioning, he went through the steps to process and decrypt the burst transmission so he could listen to it.

He donned his headset and plugged its jack into the Channel Three recorder. Expecting to listen to a report concerning yet another calamitous event threatening peace on Earth, he was surprised to hear General Abdirov's distinctive voice: "Comrade Major Vasilyev, I regret to inform you that the international security situation has deteriorated to the extent that the General Staff of the High Command has authorized me to issue you an Independent Action Code. That code is Seven-Six-Eight-One-Zero-Seven-Two-Three. Again, your Independent Action Code is Seven-Six-Eight-One-Zero-Seven-Two-Three. Because of exigent circumstances, assuming that a major conflict is imminent, you are now authorized to deploy the warhead at your discretion. This code will remain in effect until the security situation has stabilized and the High Command determines that exigent circumstances no longer exist. Comrade Major, I trust you to exercise your best judgment."

Vasilyev quickly scribbled the eight-digit code on a scrap of paper and stashed it in the chest pocket of his coveralls. Stunned, he rewound the message and replayed it several times. The world was obviously descending into chaos; otherwise the High Command would have never entrusted him with the incredibly powerful Independent Action Code. It just seemed somewhat bizarre that they would transmit it when he was on the brink of a potential emergency with the fuel cell. Of course, they might have abruptly sent it in the expectation that communications might be severed at any moment.

He contemplated the sobering implications. If thermonuclear war erupted below, even if he did not deploy the Egg, it was a foregone conclusion that he would be stuck up here forever. Without doubt, Kapustin Yar and Baikonur would be priority targets for American warheads, so there might not even be the necessary infrastructure to send up a Soyuz to fetch him home. Even if the launch sites miraculously escaped destruction, organizing a mission to rescue a single cosmonaut would hardly be high on the list of priorities.

Although he did not doubt that the code was legitimate, there were significant discrepancies in the manner it was conveyed to him. Why would Abdirov personally transmit the message? After all, as powerful as he was, the weapons deployment codes were still under the domain of the *Perimetr* leadership. Moreover, why was the code transmitted on Channel Three? Logically, if the High Command had authorized the Independent Action Code, it should have been transmitted on the dedicated *Perimetr* channel. Had something gone wrong with *Perimetr*? Had the Americans found some way to interdict the much-feared "Dead Hand" network? Could there have been a clash between the High Command and the *Perimetr* bureaucracy? If that was the case, could the High Command have purged the *Perimetr* leadership? In the relatively short history of the Soviet Union, stranger things had certainly happened.

Vasilyev remembered his conversation with Gogol, on the night the cosmonaut came to his cottage in search of vodka. Considering that Gogol had insisted that they would deploy the Egg during their mission, it seemed more than a coincidence that Vasilyev was on the verge of doing just that. If Gogol had been granted such warning, could this be part of a massive and long-planned scheme arranged by the High Command? Could they have somehow orchestrated the chain of international incidents to force a nuclear confrontation with the West?

17:50 p.m. GMT
GET: 30 Days 17 Hours 5 Minutes, REV # 491

Fitfully dozing in the darkened control area, Vasilyev was rudely awakened by a squawking alarm. Tumbling and flailing in the dark, he struggled to find the satchel that contained his emergency respirator, but eventually found it floating near one of the portholes. He placed its canvas strap around his neck, pulled his flashlight out of his pocket, and headed to the forward area of the *Krepost*.

He noticed a radiant orange light at the far end of the station, like the flickering glow of a fireplace on a dacha's timbered wall, and assumed that something—probably the defective fuel cell—was burning. Fighting the urge to panic, he grabbed a fire extinguisher and scrambled toward the docking hub. Briefly pausing at the entrance to the access tunnel that led to the freighter, he donned the awkward respirator mask.

He quickly floated down the tunnel to assess the situation. There was an undulating orange blob of fire, but it was outside the fuel cell's casing, so the cell itself was apparently not directly involved. He hoped that the fire would snuff itself out like Gogol's match on his first mission, but for some reason, the expanding blaze did not seem the least bit starved for oxygen.

Quickly abating the blaze with his fire extinguisher, he yanked off the respirator mask and moved in for a closer examination. Fanning away the smoke with his hands, he glimpsed a dense clump of charred debris in the narrow gap between fuel cell casing and the pressure hull of the dry compartment, and recognized that the combustion had been sustained by the discarded food packages that he and Travkin had earlier tamped around the casing.

Floating globules of firefighting foam splattered his coveralls. Thinking that he had resolved the crisis, he breathed a sigh of relief, but his reprieve was not long-lived. He heard an ominous noise, a shrill sound somewhere between a whistle and a hiss, and suspected that an oxygen line had somehow been damaged, perhaps crimped just enough to facilitate a pinhole leak. There was a sudden flash of light, accompanied by a loud whooshing noise, as the fire abruptly reignited.

Reacting instinctively, Vasilyev braced himself against the pressure hull and sprayed the fire extinguisher at the growing inferno. It had virtually no effect, and the bottle was quickly depleted. He rushed back into the docking hub to locate another extinguisher.

As he started back down the access tunnel, he saw that the fire was *still* growing. He smelled an unmistakable acrid odor and realized that the fire was now not only stoked with oxygen-saturated trash, but that metal was also being consumed. Few people—except perhaps welders and crash-rescue firefighters—even comprehend that metal could be ignited, but not only *could* metal burn, it typically burned with an incredible intensity. Aluminum, as an example, burned at a temperature exceeding three thousand degrees Celsius. Worse, the lattice of storage brackets within the dry compartment offered more than an ample supply of the flammable metal.

Certainly, a garbage fire was bad enough, but now, given an almost inexhaustible stock of other volatile materials, the fire could rapidly engulf the dry compartment. If not subdued, the expanding fire could soon impinge into the docking hub and then into the main compartment, threatening the entire *Krepost*.

Vasilyev forced himself not to yield to panic. When faced with an uncontrollable fire, the station's emergency protocols mandated that he physically isolate the involved module and conduct an emergency depressurization of that compartment to snuff out the blaze. Unfortunately, isolating the freighter was not nearly as simple as wrestling the hatch shut and dogging it down. Three conduits—an electrical supply cable, roughly the thickness of a man's thumb, a water supply hose and a flexible oxygen duct—ran from the top of the fuel cell, through the Soyuz's forward hatch, through the access tunnel, and into the docking hub. The oxygen duct provided a steady stream of the essential gas, preheated by the fuel cell's residual heat, to supplement the station's life support system. Vasilyev valiantly tried to undo their connections at the top of the fuel cell, but was quickly forced back by the terrible heat of the inferno. If Travkin were still aboard, one of them could have fought the fire while the other loosened the connections, but Vasilyev did not have the luxury of an extra set of hands.

The noise grew progressively louder and now was like the shriek of a factory whistle. With his head pounding from inhaling toxic smoke, gasping for breath, Vasilyev scurried back up the access tunnel. If he could not disconnect the cables at the fuel cell, his only option was to break the links somewhere else. Save for the ones on the top of the fuel cell, there were no other quick disconnect fittings, so he would have to somehow chop or saw through the cables and hoses. He opened the tool locker in the docking hub, but found nothing adequate for the job. There was a hefty spanner wrench, which might be handy if he could somehow bash his way out of this conundrum, but no saw, knife or axe.

Then he remembered Gogol's satchel of survival gear, and recalled that he had stashed it in the other docking tunnel. He quickly found the green bundle, unlatched it, and hastily groped through its tightly packed contents to find a wire saw. The wire saw, a short length of titanium cable embedded with serrated cutting barbs, was intended for cutting tree branches and amputating limbs, amongst other things. It would be the perfect tool to quickly slash through the cables and conduits. Soon, the confined space of the docking hub was filled with floating items— bandages, a tourniquet, compact rations, a compass, a pocket knife, a signaling mirror, a small flare gun—that had spilled out of the kit, but Vasilyev could not find the elusive wire saw. He saw Gogol's hatchet floating end over end and decided that it would have to do.

Clutching the hatchet to his chest, he scrambled back up into the docking hub. He made short work of the oxygen duct, but the electrical cables—manufactured of heavy gauge copper wiring—were not as yielding. Swinging the hatchet like a man possessed, he hacked at the cables at the point that they passed through the hatchway and into the docking hub. He desperately yanked at the severed cables, shoving their loose ends back into the docking tunnel. By now, the freighter's dry compartment was fully involved, and the undulating blob of fire was climbing up the access tunnel. He wielded the hatchet again, chopping the water conduit apart with one well-placed swing. A jet of stagnant water spurted from the slender conduit, but quickly subsided as the residual pressure was exhausted.

Severing the conduits had cost precious time. Since the growing fire was now rising up into the access tunnel, he could not close the freighter's front hatch. If he was able to close the hatch, even if he was unable to contain the fire, he could still jettison the freighter. Since jettisoning the freighter was no longer an option, the *Krepost* was now limited to two docking ports rather than three, which would severely constrain future operations, assuming that they could continue operations at all.

Since he could no longer isolate the freighter, his next course of action was to preserve the docking hub. Thinking that the access tunnel's hatchway was clear, he slammed the hatch closed but realized that some small obstruction precluded the disk from fully sealing. In moments, blobs of flame and tiny rivulets of smoke seeped around the unsealed portal.

Using the screwdriver tip of his pocketknife, he desperately tried to dislodge the miniscule object—a fragment of insulation from the electrical cables—to clear the hatch. The circular hatch was mounted so that it swung into the docking hub; now, the swelling pressure within the freighter and access tunnel prevented it from completely closing. To gain some leverage, Vasilyev planted his feet on the opposite wall of the docking hub before lunging at the hatch's handles to wrestle it closed. Yowling as the hot metal burned his hands, he realized the blunder of not wearing his gloves.

As the pressure continued to mount inside the freighter, a plume of dense smoke spewed into the docking hub. Vasilyev heard a roaring noise like a blowtorch burning out of control, so he assumed that the pinhole break in the oxygen line had probably worsened.

Chest heaving, he swung the hatch partially open, then slammed it shut and latched it closed. The air was filled with choking smoke, so he was forced to don his cumbersome rubber respirator. Convulsing with pain and struggling to catch his breath, he heard a muted thump as the freighter's pressure vessel spontaneously ruptured. The entire station shuddered with the muffled blast.

He scuttled out of the docking hub and into the main station. He worked his way down to the control area, and then looked out a porthole to witness a billowing cloud of debris spewing into the vacuum. Although he could see the aft end of the freighter, the part jutting furthest away from the docking hub, he couldn't view the dry compartment at the front end of the freighter, so there was no way to fully assess the damages and diagnose the impact. In any event, it was clearly obvious that he was in an extremely tenuous situation.

Just when he believed that things could not possibly get any worse, they did. He suddenly heard a steady stream of popping noises, and realized that the station's maneuvering thrusters were firing almost constantly. Eyes burning from the smoke, he quickly moved to another porthole to assess the situation outside; gazing outside, he realized that he should be seeing the earth, but was instead looking at the sky, and then saw the earth come into view. As best as he could tell, gases flowing from the breach in the cargo ferry's ruptured pressure vessel acted essentially like a low-powered thruster, imparting a very slow but constant spin to the *Krepost*. The stabilization gyros detected this torque and tried to dampen it by automatically activating the small maneuvering thrusters.

The thrusters' almost incessant firing was not a significant issue, because although there was a shortage of virtually everything else, there was no dearth of maneuvering fuel, since he and Travkin had pumped every single drop from the freighter's wet compartment to the Krepost's storage tanks.

Although there was an ample reserve of maneuvering fuel, the same couldn't be said for his remaining reserve of electrical power. Left to their own ends, the gyros and maneuvering thrusters would constantly labor to compensate for the station's slow spin. It was similar to driving a car with badly aligned wheels down a highway, so it was only a matter of time—and not much time at that—until the station's batteries were entirely exhausted. Consequently, even though it meant entirely sacrificing the defensive and communications systems, he reluctantly yanked out the circuit breakers for the stabilization gyros. The crippled *Krepost* was now entirely adrift, like so much flotsam and jetsam that already whirled around the earth.

Right now, his immediate concern was to dissipate the smoke in the cabin. Even with his respirator on, the acrid fumes irritated his eyes to the extent that he could barely see, and his throat and lungs burned like he was inhaling directly from a bottle of acid. The only logical solution was to reactivate the air exchangers, but they could quickly deplete his precious power reserves. But if he didn't clear out the asphyxiating smoke, he wouldn't survive long, so he was forced to turn on the power-hungry scrubber fans to filter the air. He reluctantly threw the switch and the dormant fans rumbled to life, immediately drawing the air into motion. Weary, he used a thick strap to anchor his body to a bulkhead, so that his face was positioned next to an exhaust vent that blew out filtered air. He slowly tugged off his respirator mask, took a deep breath of the purified air flowing from the vent, and then fell fast asleep from sheer exhaustion.

10

STRANDED

Krepost Station, On Orbit
10:47 a.m. GMT, Sunday, October 15, 1972
GET (Ground Elapsed Time): 31 Days 10 Hours 2 Minutes, REV # 502

Vasilyev dreamed that he was nestled tightly against Irina, clinging to her welcome warmth, and then awoke to the soft jangling of the alarm clock he had salvaged from the docking hub. He used it to warn of imminent events where he had to take action, like communications windows or passes over surveillance targets. Shutting off the alarm, he used his flashlight to consult his handwritten schedule and saw that a surveillance target would soon come into view.

Shivering, he jammed his hands under his armpits to stay warm. To stave off the chill, he wore two sets of his own coveralls, as well as makeshift hat and mittens he had fashioned from Gogol's extra socks. He had also improvised a sleeping bag from the two previously unused cocoon-like sleep restraints. Preferring to float free as they slept, he and Travkin had foregone using the restraints. He had slipped one inside the other and then stuffed the space between them with extra clothes and insulation battens he scrounged from storage compartments in the galley.

To conserve his ever-dwindling supply of electricity, he had shut down virtually any and all equipment that wasn't absolutely vital. With the environmental systems now dormant, the ailing station's atmosphere was dank, cold and stale. A film of moisture, composed primarily of his exhaled breath, gradually formed on the metal walls and bulkheads, eventually evolving into a thin layer of frost.

The cold sapped his strength. His temples throbbed with a headache. Since the fire and explosion, he had acquired a persistent cough and a chronic pain in his chest. That wasn't a surprise, since the station's stale atmosphere reeked with acrid fumes of scorched metals. Since the fire in the freighter involved burning aluminum, Vasilyev suspected that the station's atmosphere still contained minute metallic particles, because the conflagration's dense smoke had vented into the docking hub for several seconds before he was able to wrestle the hatch shut to isolate the compartment.

Although the air was slightly more tolerable, lingering fumes still stung his eyes and made him slightly lightheaded. He considered donning his respirator but decided that he preferred dizziness and blurred vision over wearing the claustrophobic rubber mask.

Groggy, he slipped out of his sleeping bag and floated over to a porthole. Since the fire and explosion, the *Krepost* rotated like a spitted chicken on a rotisserie, slowly turning on its long axis. As a result, it was almost futile to monitor the surveillance targets, since it was essentially a matter of luck that the designated landmark would actually appear in the porthole at the appointed time. Even so, he stayed at it; his sense of duty kept him on watch.

Although he would be content to remain in his snug cocoon to doze, the *Krepost* would pass over an American SAC—Strategic Air Command—bomber base in slightly less than five minutes. Other surveillance targets would march underneath the station every few minutes, so he had to remain awake and alert for a while.

The most debilitating aspect of the station's rotation was its impact on the power supply. In order to produce electrical power, the two arrays of solar panels had to remain oriented at the sun. But even though they didn't function anywhere close to capacity, the arrays regularly generated a modest amount of electricity to charge the batteries. Unfortunately, according to Vasilyev's calculations, the amperage he expended each day exceeded what the panels yielded, so he was still operating at a loss. It also didn't help that a substantial amount of the harvested electricity was automatically diverted to the Egg's inaccessible power system, like an insidious tax that must be paid regardless of the circumstances.

He longed for some hot tea, but since water and electricity were now precious commodities, that luxury was no longer available. With the fuel cell out of service, his water supply was limited to what remained in the station's reservoir, and he wasn't overly confident that it would last even as long as the residual power in the batteries. Besides depriving him of his treasured refreshment, the water shortage also restricted his food intake, since the *Krepost's* pantry was mostly stocked with dehydrated and freeze-dried fare. Without ample water to rehydrate them, the desiccated foodstuffs were useless, so he subsisted primarily on the vitamin-enriched and nearly tasteless "energy" wafers. But as hungry as he was, he had set aside most of his assorted goodies as a special larder. In his mind, there would soon come difficult times when he would need to boost his spirits, and a timely treat might serve to do so. And since it was so likely that he would consume his last meal up here in this freezing can, he was determined that his final repast would be a feast.

He checked his wristwatch and then used his pocketknife to slit open a tea pouch. Hoping that they might jolt him awake, he crunched on the brown crystals as a stimulant. Checking his wristwatch, he saw that he had several minutes remaining until his next major task, a contact window over several radio sites strategically spaced through the Soviet Union.

The station's constant rotation, combined with the unique properties of the *Krepost's* antennas, severely hindered his ability to communicate. Because they received signals from ground stations where transmitting power was not an issue, the radio receivers employed omnidirectional antennas. Consequently, regardless of the *Krepost's* orientation, these antennas constantly gathered signals sent up from the ground, so he heard a steady stream of message traffic as he passed over the Motherland.

On the other hand, his capacity to transmit was hamstrung. To conserve power and to mitigate the chances of interception, the antennas for the *Krepost's* transmitters were all unidirectional. To function properly, the unidirectional antennas relied on the station's stability; they only worked when the *Krepost* rotated so that the station's "underside" was oriented towards the earth. As a result, during any given contact window, he might be able to squeak out a few seconds of messages to Control. He had long ago memorized the locations of the communications stations back on Earth, so he was able to use the porthole as something of a bombsight, only transmitting when those stations came into view.

Because the radios and associated equipment gobbled power, he only energized them during contact windows. Despite the agonizing fiasco following the fuel cell's initial failure, he had to hand it to the men at Control. In very short order, they had devised a shorthand procedure to adapt to the lop-sided communications flow. During every contact window, Control slowly read up a list of key questions. Each question was numbered, and for brevity's sake, required only a brief transmission—Yes, No, a simple statement, or a number—in reply. *One – How many amps remain in your batteries? Two – How many minutes per day do you operate your air scrubbers? Three – How many liters of water remain?* So when he was able to speak, he responded to each question with the appropriate but abbreviated answer.

But even though he could convey information to Control, there was little they could do to alleviate his growing concerns. Other than his surveillance chores and contact windows, there was little to occupy his waking hours, so he regularly calculated how much time he theoretically had left, based on the remaining stores of the simple stuff—oxygen, water and electricity—necessary to keep him alive.

He felt like a schoolboy, struggling to balance a complex equation, only to discover that the variables refused to remain constant. He had long since surmised that electrical power was the key factor. But since there was no way to effectively predict how much power he might have to expend on any given day, he was compelled to constantly revise his estimates. As an example, the air scrubbers consumed a relatively enormous amount of power when he ran them, but he had little choice but to do so, since their oxygen regenerators filtered out the carbon dioxide that was the natural byproduct of his breathing. So, he regularly monitored the build-up levels of the hazardous gas, and only activated the scrubbers when the accumulation crept toward dangerous limits. Oddly, as he traded precious electricity to cleanse the air, he was also rewarded with a slight amount of water, since the scrubbers also filtered out and collected condensed moisture for reuse.

Numbers, numbers, numbers. The remainder of his life was parceled out in fractions, and with every hour that passed, his critical resources were gradually and irreversibly diminished. He dreaded the unavoidable moment when the last of the gauges ran completely down, conscious that his end would be more agonizing than he could possibly imagine.

Reflecting on his plight, Vasilyev sighed as he switched on the radios. His patience was frayed. He felt like a doomed sailor aboard a sinking ship, trapped in a compartment already deep underwater, anxiously listening to the hissing and pops of breaking rivets as he waited for the straining bulkheads to collapse under pressure.

On the other side of the darkened control area, the faintly glowing lights of the Egg taunted him. Supplied with an independent collection of batteries, out of reach behind tamper-proof shields, the weapon's systems were fully operational and would likely be active for several days after the *Krepost* had finally given up the ghost. Vasilyev mused that he would soon be dead, but the Egg would still be incubating, ready to hatch oblivion.

Krepost Project Headquarters
7:50 a.m., Monday, October 16, 1972

Yohzin listened as the flight director briefed Abdirov concerning the *Krepost's* current status. Mildly stated, the situation did not appear overly promising. With every hour that passed, it looked ever less likely that a rescue mission—if Abdirov even approved such an effort—would reach the stranded cosmonaut before he expired. As much as Yohzin admired Vasilyev and wanted to see him returned safely to Earth, a failed rescue—or perhaps no rescue attempt at all—would likely result with the *Krepost* abandoned and the overall effort shelved. Perhaps the program might be resumed in the future, but certainly not under Abdirov's control. At this juncture, that was certainly an outcome that Yohzin favored.

The flight director's reedy voice interrupted Yohzin's thoughts. "So even if a Soyuz does make it to the *Krepost* in time, it still might not be able to dock."

"How so?" asked Abdirov.

"We have been fortunate enough to focus some very powerful tracking radars on the station," explained the flight director. "The radars detected a constant and consistent change in aspect ratio, which verifies Vasilyev's observations that the station is rotating around its long axis."

"This rotation is not being counteracted by the stabilization gyros and thrusters?" asked Abdirov.

"*Nyet*, Comrade General. There is far too much torque to overcome, so the gyros and thrusters are ineffective. Vasilyev has shut them down. Unfortunately, unless the rotation is somehow abated, we cannot rendezvous and dock another vehicle."

"Refresh my memory: Can Vasilyev not manually fire the thrusters to compensate for the rotation?" asked Abdirov.

"The station's thrusters were never meant to be fired manually," replied the flight director. "The *Krepost* was expressly designed to be automatically controlled, without any interaction from the crew. The stabilization gyros control the thrusters during routine operations. When and if the warhead is deployed, the Egg's targeting computer takes over and properly aligns the *Krepost* for weapons delivery."

"So, is there not anything that can be done?" asked Abdirov.

"We don't think so, sir, at least for the time being," answered the fastidious flight director. "But on a positive note, our radar studies show that the rotation is gradually slowing. Granted, it's a preliminary finding, but it lends credence to Vasilyev's theory that venting gases are effectively acting as a thruster. Eventually, the escaping gases will be completely exhausted, if they have not already been depleted, and then it's just a matter of time before the inertia is degraded to the extent that the gyros and thrusters can effectively dampen the rest of the rotation."

"That sounds promising," said Abdirov.

Removing his black-framed eyeglasses, the flight director shook his head and said, "Unfortunately, according to our projections, even in the best-case scenario, Vasilyev's electrical power will be gone long before the rotation can be brought under control. Granted, he can likely stretch out his power if he uses it judiciously."

"I have to imagine that he is using his power judiciously right now," said Yohzin. "After all, Vasilyev is a very practical man."

"True," said Abdirov, studying a graph that depicted power expenditures. "So, at this point, when would be the earliest we can execute a rescue mission?"

"Given the best case scenario, where absolutely everything lines up in our favor, we still wouldn't be able to launch a rescue mission for at least another week," said the flight director. "On our side, we are developing options to execute the rendezvous so the least amount of electrical power is expended on the *Krepost*. We are confident that we can track and control the entire rendezvous from the ground, to minimize reliance on the *IGLA* rendezvous system on the station.

"Ideally, we should be able to get the Soyuz in close proximity, so that all that remains would be to actually dock. Unfortunately, though, the rescue Soyuz could not dock unless the docking equipment is powered. So, if there's no power remaining at that crucial moment, the whole effort would be pointless."

"Agreed," said Abdirov. "But doesn't the docking mechanism have an option to facilitate remote activation?"

"*Da*, Comrade General," replied the flight director. "Good point. If that option is active, and the *Krepost* has adequate residual power in the batteries, then the docking mechanism can be switched on from the ground or from the approaching Soyuz."

Yohzin remembered the remote activation circuit was included in the early designs, when it was assumed that the *Krepost* would be briefly vacated between crews. After the High Command insisted that the *Krepost* be occupied at all times, the vestigial circuit remained in the overall design, primarily because it was simpler to leave than to remove. As he looked over Abdirov's shoulder to examine the power expenditures graph, Yohzin drew his slide rule from its leather scabbard to verify some of the flight controller's numbers.

"So, theoretically, even if Vasilyev was dead or otherwise incapacitated, the Soyuz could dock and another crew could board, correct?" asked Abdirov.

"*Da*, Comrade General," replied the flight director. "But—"

Abdirov quickly interrupted him. "And even though the power situation is currently less than optimal, the batteries are still being charged, correct?"

"*Da*. They are being charged to a marginal extent, Comrade General," answered the flight director. "But at the rate Vasilyev is expending electricity, the batteries will not accumulate…"

Scowling, with his single eye twitching, Abdirov held up his disfigured hand to silence the flight director. "Colonel, I think that you're misplacing your priorities, or at least you're failing to appreciate *my* priorities. From what you have described to me, the *Krepost's* batteries will continue to charge, at least to a limited extent, and as the rotation slows, they will accumulate progressively more power as the solar panels are exposed to the sun longer. Correct?"

"*Da*, Comrade General."

Abdirov held out the power expenditure graph towards the flight director. "And according to your graph, the systems that consume the most power are the radios and the environmental systems, correct? If I order Vasilyev to shut those systems down, then it's almost a certainty that the batteries would accumulate sufficient power to facilitate docking, correct?"

"*Da*, that's correct, Comrade General," answered the flustered flight director. "But if he couldn't communicate…"

"Again, Colonel, you're missing the point. Right now, short of telling him what systems to shut down and instructing him to remain patient until we climb up to rescue him, there's little that can be accomplished in maintaining regular communications with him. So, shutting off the radios now would conserve more power for the docking mechanism later, agreed?"

"*Da*, Comrade General," answered the flight director. "But you must be aware that the power levels would still be marginal. There *might* be enough to actuate the docking mechanism, but there is no guarantee. Moreover, it's imperative that Vasilyev runs the oxygen regenerator, at least periodically, or carbon dioxide will eventually kill him."

"Enough," stated Abdirov emphatically, tapping one of his few remaining fingers on the table. "Here's what I want. Draft a message to send up to Vasilyev on the next contact window. Tell him that we have thoroughly studied his power situation. We will limit future communications to one very brief contact window per day.

"He is not to activate his environmental systems, including the air scrubbers and oxygen regenerator, unless we specifically instruct him to do so. Instruct him to check the settings in the docking hub and double-check to ensure that the remote activation circuit is active. Make it abundantly clear that he should reduce his activities to the absolute minimum, so that he is essentially hibernating until we get up there to bring him home. I'll be here for the next contact window, and I want to talk to him personally, to convey a sense of urgency."

"As you wish, Comrade General," replied the flight director, scribbling down notes.

"Two more items, Colonel," said Abdirov. "Assemble your best people to cobble together some kind of manual mechanism to power the oxygen regenerator. Perhaps they can improvise some sort of bellows to pump air through the lithium hydroxide filters, to scrub out the carbon dioxide. That will keep Vasilyev decisively engaged and maybe lend him a bit more oxygen to inhale.

"Also, immediately begin preparing a profile to send up *one* man in a Soyuz, rather than two. He and Vasilyev might be up there for at least a little while until the next freighter is ready, so plan on stuffing all available free volume with consumables."

"*Da*, Comrade General," replied the flight director. "Shrewd plan, sir."

"*Go!*" ordered Abdirov. "Time is wasting. The next contact window will be soon upon us."

As the flight director obediently scurried toward the front of the room, Yohzin quietly asked, "A word, Rustam?"

"Certainly."

"May I speak freely?" asked Yohzin.

"Of course, little brother."

"I checked those numbers," stated Yohzin, holding up his slide rule. "You're correct that the batteries will accumulate sufficient power to actuate the docking mechanism, given a week's time, but Vasilyev will run out of breathable air if he can't run the oxygen regenerator."

"Perhaps, but did you not hear me instruct the flight director to come up with a manual solution to scrub the carbon dioxide?" asked Abdirov.

"That's a lot of volume to scrub with a manual system. Honestly, Rustam, I don't think it's a plausible option. I just don't understand why you want to lend Vasilyev false hope."

"False hope is better than *no* hope," replied Abdirov.

Yohzin was astounded at his friend's callousness. For whatever reason he might have, he was effectively sentencing Vasilyev to die.

"Look, Gregor, this hypothetical rescue mission still has to be approved by the General Staff of the High Command," said Abdirov. "If I was to tell them that Vasilyev would likely be a stiff corpse by the time the Soyuz made it upstairs, don't you think they would balk at flinging a Soyuz and R-7 into orbit on a fool's errand?"

"Then if that's the case, Rustam, why send it at all?" asked Yohzin. "If Vasilyev is so likely to be dead, what's the damned use?"

"And what exactly makes you believe that I *care* whether Vasilyev is alive or dead when the Soyuz arrives?" asked Abdirov. "For once, Gregor, be pragmatic. So long as that Soyuz is able to dock, and we can stick a man aboard, I could care less whether Vasilyev survives. Granted, I cannot admit that to the High Command, so I am compelled to paint this endeavor as a rescue mission."

Krepost Station, On Orbit
7:46 a.m. GMT, Monday, October 16, 1972
GET: 32 Days 7 Hours 1 Minutes, REV # 516

As he waited for the impending contact window, Vasilyev patiently transcribed a Channel Three intelligence update that had been automatically recorded during the previous communications pass. His fingers were so numb that he could barely hold his pencil.

Studying the transcription, he whistled softly through his badly chapped lips. The news was still plenty dire. Throughout the world, all strategic forces of all nations were on high alert. ICBMs were fueled and ready for immediate launch. Bearing nuclear payloads, strategic bombers were either in the air, in holding patterns as they awaited orders, or on strip alert with ready crews aboard. Lurking undersea, missile submarines maintained combat stations, many hovering close to the coastlines of their enemies. It was just a matter of time, perhaps only hours, before open hostilities ensued.

Shivering, he gnawed on a half-frozen energy wafer as he glanced down through the porthole. He half-expected to see the world aflame below him, but saw no clear evidence of impending war. *How could this be?* Vasilyev did not doubt that the intelligence updates were accurate, but this paradox was exasperating. With such limited access to information, it was like seeing and hearing the world through a tiny keyhole. Perhaps if he had some means to monitor regular radio channels, particularly those of international news stations, then maybe he would verify the veracity of the intelligence reports. If nothing else, if all-out war loomed so close, then surely governments would broadcast public information alerts to grant their citizens ample time to make necessary preparations.

Since his own world was now effectively confined to the control area, he adjusted the layout of his modest habitat to achieve the most efficient use of the space and his resources. He rarely left the meager comfort of his improvised sleeping bag, but now wore it as a sort of padded garment to fend off the frigid temperatures. Most often, he stationed himself close by the porthole, tying a length of cord around his waist as an anchor, so he readily could look at the earth below.

He sloshed as he moved. Concerned that his remaining water would freeze in the reservoir, he had salvaged several used tea pouches from the trash and then painstakingly filled them with the remaining water. He stuffed the pouches into his respirator bag, which he now wore like a bandolier around his neck, and rationed himself one pouch a day.

He switched on the bank of radio receivers, allowing a few extra minutes to lend their circuits adequate time to warm, and then was immediately beset with a prolonged series of agonizing coughs that originated deep in his chest. He regained his composure, then grimaced as he clamped bitterly cold earphones over his tingling ears.

As the *Krepost's* revolutions took it over Soviet territory, the station passed over a network of radio outposts. Connected by microwave relays and telephone lines, the transmitters at the outposts were carefully synchronized to broadcast the multiple frequencies originating from Control, to ensure an uninterrupted flow of communications. Approximately two minutes after he powered up the radios, the receiver warbled slightly, and then he heard a single message, previously recorded, repeated several times. Since the *Krepost* initially came into reception range in the middle of the message, Vasilyev had to wait almost half a minute for the message to cycle through to the beginning.

The lengthy message began with the tracking number—122—and was comprised mostly of technical instructions on shutting down specific systems to conserve electrical power. Vasilyev was slightly alarmed when he heard Control order him to switch on his radio receivers for only one contact window per day, and not to switch on his transmitters—including the circuits for data telemetry—unless they specifically instructed him to do so when they initially established contract on any given window, or unless he was experiencing an emergency situation. Keeping the transmitters powered off did make plenty of sense, since they consumed a considerable amount of power. Moreover, any data transmitted on the telemetry frequencies was largely irrelevant now.

The communications limitations bothered him, but far more disconcerting was that Control directed him not to activate the station's air scrubbers. Fortunately, Control also stated that they were devising an alternative method to exploit the lithium hydroxide in the oxygen regenerator, so that the device could cleanse the atmosphere of carbon dioxide without drawing on electrical power.

The recorded message concluded with General Abdirov's distinctive voice: "Comrade Major Vasilyev, although we have suffered a setback, your mission has not changed. You are a Soviet officer, and you will act accordingly. Speaking on behalf of the General Staff of the High Command, I expect you to remain vigilant. Given our current difficulties with communications, I expect you to exercise your best judgment in the absence of orders.

"You must be cognizant that given the current world climate, we will not abandon the *Krepost* under any circumstances. At this very moment, we are preparing a mission to reinforce you. To reiterate, until such time that you are reinforced, you must be prepared to exercise your best judgment in the absence of orders."

It was difficult to assess Abdirov's tone, particularly since his regular speech bordered on a lisp. He wasn't sure whether Abdirov sought to soothe or perhaps reassure him, but if nothing else, the most exalted General made it abundantly clear that Vasilyev's lot was not to cry and whine about his circumstances, but to obediently stay at his post until relieved. *As if I have a choice*, thought Vasilyev, scratching his numb nose.

He remembered the slip of paper in his pocket, the note that bore the eight numbers of the Independent Action Code, and understood that in the absence of orders, as Abdirov had emphasized, he must be ready to deploy the Egg if he witnessed any evidence that the thermonuclear war had commenced.

Confident that he had captured the salient details of the transmission, Vasilyev prepared to reply. He mused that it was unfortunate that his communications would now be restricted, since he had devoted many hours to developing a predictive model to determine which ground radio sites would be "visible" during any given contact window. He wrapped a towel around his face so that his exhaled breath would not inadvertently create frost on the porthole. He consulted his graph, checked his wristwatch, and patiently waited a few minutes until he saw the familiar terrain of Irkutsk, a Siberian city just west of Lake Baikal, gradually rotate into view. He picked out the communications site, paused until it was almost centered in the glass, and then keyed the microphone button for his terse message: "Control, I acknowledge your message number One-Two-Two. Again, I acknowledge One-Two-Two. Will comply."

Seconds later, he heard Control's equally terse reply: "*Krepost*, we acknowledge your receipt of message One-Two-Two. Control Out."

And that is that, he thought, switching off the radios. Perhaps Control would be more forthcoming on their next planned contact window, which was slightly less than twenty-four hours away. In the meantime, it was a considerable amount of information to absorb. As he reviewed their instructions, he struggled to understand their ultimate intent.

He tried his best to be upbeat, but still could not shake a lingering sense of doom. No matter how much he juggled the numbers, he couldn't foresee a favorable outcome. Although their instructions appeared to contradict his calculations, he had to assume that there was some brilliant method to their madness, and that they had divined some mysterious protocol to squeeze out the maximum benefit from what little power remained. Moreover, they obviously knew how quickly a relief mission could be mounted, and he did not, so that was a critical variable that he could not account for. But surely their intent was to keep him alive long enough to greet his reinforcements, so their plan certainly *had* to be the best potential solution. After all, he was one man with one brain, with his thoughts muddled by cold and fatigue, and they were Control, a conglomeration of some of the brightest minds in the storied RSVN. Surely, they knew what was best, so as miserable as he was, fully cognizant that his conditions would progressively deteriorate, he would obey their orders and remain vigilant, ever ready to execute his most sacred duties.

Krepost **Project Headquarters**
12:03 p.m., Monday, October 16, 1972

"I have a mission for you, Gogol," said Abdirov, gesturing for the squat cosmonaut to take a seat. "I trust that your foot has healed sufficiently for you to fly into space."

"It has, Comrade General, but even if it had not, a single swing with an axe would resolve the issue. Am I to assume that I will be going up with Major Travkin?"

"*Nyet*," answered Abdirov. "This may be slightly awkward for you, but you will be taking this journey by yourself. I'm sending only you up to the *Krepost*."

"Excellent, Comrade General," said Gogol. Grinning from ear to ear, he was obviously elated at the prospects of returning to orbit. "I will not disappoint you."

"Of that I'm sure, but you should be aware that it might be a week or even longer before we have a vehicle ready and a suitable launch window. There's much work to done in the meantime. Control is revising flight procedures so that you can ascend alone, and as soon as they are ready, you'll go into the simulators to practice.

"This is all tentative, of course, because even though the hardware and procedures are being readied, we are still awaiting final approval from the General Staff of the High Command. Moreover, when and if you fly, we are not even sure that the *Krepost* will be adequately stable for you to dock."

"And my mission, sir?" asked Gogol.

"In theory, you and Vasilyev will occupy the *Krepost* and wait for another freighter and relief crew."

"In theory, Comrade General?" asked Gogol.

"*Da*," said Abdirov, handing him a slip of paper. "Between you and me, here's your real task."

"Sir?" asked Gogol, examining the note. It simply contained eight digits: 76810723.

"That's an Independent Action Code. Memorize it. As soon as you are aboard, I want you to deploy the Egg on the juiciest target that becomes available, preferably a large American city."

"As you wish, Comrade General," replied Gogol. He glanced at the paper again, folded it, and nonchalantly handed it back to Abdirov. "It will be done."

"And you don't have any qualms about doing that?"

"None whatsoever, Comrade General," replied Gogol. "But I do have a question, sir."

"Go ahead."

"What about Major Vasilyev, Comrade General?"

"What about him?" answered Abdirov. "The chances are that he will be long dead by the time you get up there."

"But what if he is not, Comrade General?" asked Gogol. "If he is still alive, can I expect him to be cooperative?"

"*Nyet*," answered Abdirov. "You probably shouldn't expect him to be cooperative in the least, so my previous instructions remain in effect. Be prepared to subdue him if necessary, and it's better that you act sooner than later, while you still might have the element of surprise."

2:10 p.m., Monday, October 16, 1972

"Gregor, listen to me," said Abdirov quietly. "I'm stuck in quite a dilemma here."

"What is wrong, Rustam?" asked Yohzin. "How can I help?"

"I've talked to Gogol. As soon as he boards the *Krepost*, he will deploy the Egg," revealed Abdirov. "He will use the code that you pinched from the *Perimetr* people."

Yohzin closed his eyes and swallowed. *So, this is it* he thought.

"Did you not hear me, Gregor?"

Yohzin opened his eyes and replied, "I did, Rustam. I know that such a decision must weigh heavily on you."

"Listen, little brother, dropping the Egg is not my dilemma. Being unable to fulfill my promise to you is what troubles me so."

"How so?"

"Do you remember that I told you that I would allow you and your family to visit Luba's parents when this time was near?"

"I do."

"Then I regret to tell you that I might be unable to make good on my commitment," explained Abdirov. "The GRU is still looking for the leak, and until they find the culprit, you're still under suspicion. I can request authorization to facilitate travel for Luba and your sons, but I am concerned that if I request the same arrangements for you, then none of you will be able to leave Kapustin Yar."

Yohzin considered the implications. If there was even the slightest chance that Luba and his sons could be delivered into safety, even if he might perish, then his decision was simple. "If that's the case, then I will stay here, Rustam," he declared. "And Luba can take the boys to see her parents, if you would be so kind as to request their travel documents. Luba is a very capable woman, and I am confident that she can travel by herself."

"And you, Gregor?" asked Abdirov.

"You can count on me to remain at my post, Comrade General," answered Yohzin. "I will remain at my duties until the very end. If my fate is to die for the Motherland, then it will be my honor to be at your side when my end comes."

Residential Complex # 4, Znamensk (Kapustin Yar-1)
8:12 p.m., Thursday, October 19, 1972

Yohzin composed his weekly message to the Americans. It was simple and to the point. He stated that the *Krepost* was badly damaged but still occupied by one cosmonaut, and a rescue mission would be launched no earlier than four days from today. He also requested that the extraction plan for his family be executed as planned.

He went through the familiar process of encrypting the message and then transcribing it to the thin paper strip that he would insert into the capsule to be "transmitted" by Magnus tomorrow night. He shook the plastic vial containing the capsules, and realized that there was only one remaining; it was no matter, because he had two more vials filled with capsules of similar composition, except that they were colored dark gray rather than dull brown. He prepared the capsule, placed it on his desktop, and then carefully gathered and cached his secret writing tools. In just a few minutes, he would drop the capsule in Magnus's kibble, and just a few minutes after that, the message would be on its day-long journey through the dog's alimentary canal.

Two days after the new message was passed, he would return to his study just after midnight, to set up a miniature shortwave radio to listen for a coded confirmation message that the extraction plan would be executed as he requested. He knew from a previous message that his American handler—"Smith"—had already approved the plan and made the necessary arrangements, so now it was just a matter of setting those wheels in motion.

Confident that he had done what he could do to safeguard his family, Yohzin focused his thoughts on another matter of grave consequence. He was tormented by the knowledge that if he had not stolen *Perimetr's* secret code and passed it to Abdirov, he wouldn't be in this terrible predicament. He had to stop Abdirov's plot, regardless of the terrible costs he might incur. It was not merely a matter of sacrificing himself to save his family, but he was compelled to do so to save the world.

Now, he had a plan. Once he was sure that Luba and his sons had safely departed Kapustin Yar, he would diligently attempt to persuade Abdirov not to follow through on his scheme. If he failed, and if it was coming down to the wire, he would simply go to the GRU and divulge the truth, ideally before Gogol was launched. Of course, he had no doubt that both he and Abdirov would be arrested and subjected to horrendous ordeals, but if that sufficed to spare the world from thermonuclear destruction, then so be it. So long as Luba and the boys were safe, he could endure any agony that the GRU could see fit to mete out.

Krepost Station, On Orbit
22:30 p.m. GMT, Friday, October 20, 1972
GET: 36 Days 21 Hours 45 Minutes, REV # 590

For at least the fifth time in less than an hour, Vasilyev was jolted awake from a terribly realistic nightmare. He dreamed that he was drowning; as he awoke, his heart pounded furiously in his chest, and he gasped for breath.

He finally concluded that the recurring nightmares were actually an instinctive mechanism of his unconscious brain, and that they served to warn him of a very real danger. The distressing sensations he experienced were symptoms of carbon dioxide poisoning.

Without the air circulation normally provided by the scrubbers, the station's atmosphere was effectively stagnant. Even if the cabin was theoretically saturated with oxygen, as he breathed, he drew in oxygen and exhaled carbon dioxide. Without some mechanism to stir the air into motion, even gently so, a pocket of carbon dioxide would eventually accumulate in the vicinity of his mouth. The situation reminded him of his first flight into orbit, when Gogol tried to convince him that it was safe to smoke. Gogol had lit a match, the matchhead burned for a few seconds, and then was snuffed out as the surrounding oxygen was depleted. Likewise, Vasilyev's life would be extinguished almost as abruptly if there was no oxygen around his mouth and nose, regardless of how much there might be just a half-meter away.

Even though Control had ordered him not to switch on his scrubbers, he had to do something about this situation, because it was only a matter of time before he failed to wake up.

How would Travkin handle this? he thought. He studied the dimly lit control area, and considered the potential resources available to him.

Gripping his penlight between his teeth, he wielded a screwdriver to unfasten and remove a protective panel from the instrument console, and then extracted a small electric fan that was intended to cool a cathode ray display. He wired the miniscule fan to a spare flashlight battery, and in short order had improvised a ventilation fan to churn the air around his face as he slept, to dispel the build-up of carbon dioxide. *And that's what Travkin would do!* he mused, grinning. Now, once he became accustomed to the faint buzzing sound, he would be able to sleep without interruption.

But if his sleep was not interrupted by carbon dioxide poisoning, there were plenty of other things to keep his mind from rest. He could not help but feel despair about his probable fate; he still kept up with his estimates, and he just could not see how he could be still be alive by the time a rescue mission arrived. He had come to the conclusion that their priority was to charge the batteries so they would have adequate power stored to actuate the docking mechanism later, but unless Control arrived at a timely solution to run the oxygen regenerators without power, it wouldn't matter if the docking mechanism worked or not, because he would no longer be alive to rescue.

That wasn't all that concerned him. With every day that passed, he grew ever more skeptical about the intelligence updates. Although he regularly stared at the earth until he was in danger of hallucinating, he still could not detect any discernible evidence that the dispatches were valid. If the updates were false, why would Control subject him to such a charade? He considered an array of their possible motives. On one end of the spectrum, they may have been concerned about his capacity to remain alert after Travkin departed, and employed bogus intelligence reports to motivate him toward a state of enhanced vigilance. Another possibility, certainly reflecting a much darker agenda, was that erroneous reports were part of a devious scheme to trick him into deploying the Egg, now that he was entrusted with an Independent Action Code. Finally, it dawned on him that he really did not know whether the Independent Action Code was truly valid, so perhaps it was all part of an elaborate reliability exercise to determine if he would actually follow through on deploying the Egg if ordered to do so.

If only he possessed some means to double-check the intelligence updates, then he could at least know if they were legitimate or not, and if they weren't, he might have a clearer insight into Control's intent.

11

MISSION TWELVE

Launch Complex 41
Cape Kennedy Air Force Station, Florida
6:30 p.m., Saturday, October 21, 1972

Clad in their space suits, reclining in a pair of matching beige Barcaloungers, Carson and Ourecky waited impatiently in the cramped suit-up trailer. It was the same Airstream camper that they had occupied for their prolonged stint on Johnston Island last summer, anxiously waiting for word that the Soviets had launched their massive Proton rocket. In the hours just before Hurricane Celeste came ashore to demolish the PDF launch complex, the trailer—along with a few tons of other specialized equipment—had been hastily evacuated to Hawaii aboard a C-130 turboprop transport.

For at least the tenth time this hour, Ourecky checked over his flight plan notes. "Man, I'm *starving*," he grumbled. His stomach growled audibly. "You would have thought they'd plan for some chow. I haven't eaten since breakfast, and that was almost twelve hours ago."

"Thirty minutes," announced a suit technician, holding up a walkie-talkie radio. "They're getting ready to remove the shroud."

Ourecky sat up and twisted around so he could look through the small window behind his right shoulder. The sun was setting. In the fading light, he glimpsed Launch Complex 41—LC 41—in the distance, slightly less than half a mile away from where the suit-up trailer was parked, near the base of the skyscraper-like Vertical Integration Facility. The launch pad's main structure, a Titan umbilical tower, was positioned atop an enormous concrete "flame bucket." The umbilical tower was surrounded by four tall metal structures that resembled radio broadcast towers; they were literally gigantic lightning rods, to prevent stray bolts from destroying the fragile electronics of a rocket on the pad.

The fiberglass shroud had been fitted on the Titan II to disguise it as an unmanned payload for a limited notice operational test. Theoretically, the purpose of the test was to validate ICBM emergency firing procedures under extremely realistic conditions; as far as the outside world—which included local civilians and NASA workers—was concerned, a stand-by ICBM alert crew would be issued launch orders on very short notice, so the Titan II could blast off at any time, day or night. The shroud would be removed and lowered immediately after the sun was completely below the horizon. Most the complex's lights would remain off, cloaking the remaining launch preparations in darkness.

"Hey," declared the technician, "if you're that hungry, Scott, I think there's a case of C-rations stashed in the van. The pad techs like to keep some extra chow around in case they're working late and can't leave. Want me to call them?"

"I'm not *that* hungry."

"We'll wait for Virgil," said Carson, reviewing notes on a clipboard. "He said he would make it back on time, so we'll just have to trust him."

The men heard gravel crunching as a car pulled up outside. "I think the chuck wagon has arrived," noted Ourecky.

Seconds later, the trailer's aluminum door swung open, and Virgil Wolcott entered, ducking his head to avoid banging into the low doorframe. He bore two brown paper bags. "Catsup, extra onions?" he asked, digging into one of the bags to produce a large hamburger wrapped in wax paper.

"That's me," answered Ourecky. He quickly unwrapped the sandwich and commenced to eat.

"I guess the other's mine," noted Carson, taking the other hamburger that Wolcott proffered. "You're a lifesaver, Virgil. Thanks a million."

"Glad to oblige," drawled Wolcott. "You boys are mighty lucky. Those came straight from the Sunrise Diner, best danged burgers in town."

"How much, Virg?" asked Carson, talking around a mouthful of burger.

"Seven bucks for the both of you, pard."

"I'm a little short at the moment," declared Carson, patting his chest pocket. "I left my wallet back in Ohio. I'll pay you when we get back, provided we get back. Besides, it's legal for me to put meals on my travel voucher, right?"

"Danged good point. If you're stickin' it on your voucher, why don't we call it ten bucks even, then. Hell, I should rate at least a sawbuck for pre-flight caterin'." Wolcott sat on a wooden stool between the two recliners where Carson and Ourecky reposed, and asked, "Are you gents comfortable with what you've got?"

"I don't like the thought of flying the intercept without the radar," replied Carson, squirting catsup into a small paper sack of French fries. "I know we have plenty of support from the ground, and everything is supposed to be perfectly synchronized, but I'm just not comfortable relying on someone else's fix. We've done this deed countless times in training, and seven times for real, and we're used to the radar."

"Duly noted, pard," said Virgil, snitching one of the catsup-drenched French fries. "But even though that durned *Krepost* is supposed to be unoccupied, we just can't afford to take any chances. And you, Ourecky? Any major reservations on your part?"

"Other than riding a rocket to work?" asked Ourecky, crumpling the hamburger wrapper and hurling it toward a cardboard box that served as a trash repository. "None to speak of, Virgil. I have to admit, I'm with Drew concerning the radar, but other than that, I'm anxious to fly this thing, kill the target, and then get back home so I can move on with my life. You're sure that General Tew intends to make good on his promises?"

"Absolutely, son, absolutely," replied Wolcott, nodding vigorously.

"Then, like you say, Virgil: if you need someone to go upstairs to sweep out the attic, then we're your huckleberries."

Eating quickly, enjoying their last minutes on Earth, the two men were silent for a few minutes.

"The van is outside," announced the technician, listening to his walkie-talkie. "Time to go, guys."

"Guess you fellers have to skedaddle," said Wolcott, patting Ourecky on the shoulder. "Best of luck to you. We're countin' on you two to get this deed done."

"Thanks," answered Ourecky. "We've had plenty of practice runs, so it's great to finally go after the real deal."

"Thanks, Virg," said Carson, just before finishing the last bite of his hamburger. "We'll do our best."

Ourecky sat upright, cinched the elastic thigh strap on his knee-board, and then jammed his slide rule's leather holster into his right calf pocket. He stood up, bent over to retrieve his helmet, and then announced, "I'm ready to roll."

"Give me a minute," replied Carson, squeezing Colgate toothpaste onto his index finger. He briskly rubbed his teeth with the finger, and then swished out his mouth with water from a paper cup. "Okay, buddy, let's hit the road."

8:41 p.m.

Ourecky slid down his clear visor, locked it into place, and then did a quick scan of his instrument panel. All readings looked nominal for launch. He checked the computer console again, ensuring that its knob was set to 'ASC' for Ascent mode and made sure that his kneeboard was securely in place on his left thigh. Confident that he was ready, he granted himself a minute to close his eyes to think about Bea and Andy before saying a quick prayer to request a safe flight.

For various reasons, this was a ground-directed mission, in which he and Carson would have very little input into the intercept process. Their orbital track would be paced by a virtual armada of tracking stations based on ships and airplanes, supplemented by several ground-based tracking locations.

Powerful radars at the tracking stations would carefully monitor the orbital position of their Gemini-I as well as their target: the Soviet *Krepost* space station. The tracking information would be quickly routed to the Mission Control facility at Wright-Patterson, where a bank of high-speed computers would analyze the data to produce very precise instructions for the intercept profile. Those instructions would be relayed back to the tracking facilities, where they would be uploaded by data link to the Gemini-I's on-board computer.

Compared to previous launches, there was a lot less chatter in his earphones than normal. Like the intercept process, the communications procedures for this mission were also considerably different than their eight previous missions. Although the ground-based CAPCOM— Capsule Communicator—would keep them informed with a running commentary of flight-related issues, Carson and Ourecky would maintain radio silence unless there were circumstances that dictated an emergency abort. They would keep quiet even during the launch, since in theory, anyone—which typically included Soviet intelligence trawlers sitting offshore in international waters—eavesdropping on the radio frequencies would be compelled to believe that the Titan II's payload was unmanned, as advertised.

"Vehicle is transferring to internal power," stated the CAPCOM. "Stand by for engine gimballing."

"Gimbals," muttered Carson on the Gemini-I's intercom loop.

"Copy gimbals," responded Ourecky.

"T-minus one minute and counting," declared the CAPCOM. "Five seconds to Stage Two fuel valves....Thirty seconds...T-minus twenty seconds...."

Ourecky felt the almost overwhelming sensations that had become so routine in the past three years: the vibrations of the powerful turbo-pumps raced through his spine.

If but only to each other, both men chanted out the final countdown: "T-minus 10, 9, 8, 7, 6, 5, 4...first stage ignition."

Gritting his teeth, Ourecky felt a shudder as the first stage engines roared to life. Two seconds later, right on schedule, the explosive hold-down bolts cracked, releasing the rocket to the sky. They were on the way, hoisted by 215,000 pounds of thrust.

"Here we go!" declared Carson. "Lift-off and the clock's running. Last time, buddy?"

"*Definitely* the last time."

Gemini-I, On Orbit
3:50 a.m. GMT, Sunday October 22, 1972
GET: 3 Hours 5 minutes / REV # 2

Ourecky was bored, if it was truly possible for a man to be bored while travelling at 17,500 miles per hour through the cold vacuum of space. It wasn't that spaceflight had become commonplace to him, but rather that he was accustomed to being so busy that he scarcely had time to even steal a glance out the window. Of course, the mission planners had asserted—repeatedly—that although the initial approach was virtually automated, save for manually executing the maneuvers specified by the computer, he and Carson were absolutely critical for the final phase of the intercept.

As they passed through their communications window, Ourecky watched as the upload light blinked on their on-board computer. He scrolled through the computer display's screens, copying down the key pertinent information on an index card. He tugged his slide rule out of its holster and then painstakingly verified the mission controller's calculations.

Curious, Carson unwrapped a package of Fig Newtons as he watched the right-seater. He handed two of the fruit-stuffed wafers to Ourecky, ate two, and then tucked the remainder into his side storage pocket. "Uh, Scott, you do recall that this is predominately a ground-directed mission, right?" he asked. "We're only supposed to cook the recipe they beam up. No deviations or extra spices."

Ourecky floated the cookies in front of him, nodded, finished his computations, stowed his slide rule, and then compared his numbers to the ground's. "Looks good," he commented, double-checking an entry from the computer's digital read-out.

"So, is there any particular reason that you feel compelled to rework those problems?" asked Carson.

"Well, Drew, I suppose that old habits are hard to break."

"I guess. So, what's next on the agenda?"

"Another minor phase shift in thirty-four minutes."

"Thirty-four minutes? That seems like a day away. I wish that I had brought a book up."

"Well, you do have time to grab a nap if you want," replied Ourecky, leaning forward to grab a slowly tumbling Fig Newton between his teeth. "I'll cover while you're down."

"Thanks. The controls are yours, Right Seat."

"Right Seat has the controls."

Mission Control Facility, Aerospace Support Project
1:25 a.m., Sunday, October 22, 1972

As the mission was in progress, Mark Tew kept vigil from his habitual vantage point during flights, the glass-enclosed Executive Observation Area located at the rear of the Mission Control facility. He was happy, if not still apprehensive. So far, the sortie was proceeding entirely as planned, but they were still a long way from climbing up to the *Krepost* and destroying it.

Tew glanced at the mission clock on the front wall, past the first row of controller consoles. Just starting their third orbit, Carson and

Ourecky had been up for slightly less than five hours. A few minutes ago, Virgil Wolcott had called from Flight Operations here on the base, to announce that he had returned from Cape Kennedy. Admiral Tarbox was at the Pentagon, ostensibly to keep General Kittredge and other high-ranking officers appraised of the mission's progress.

Watching the controllers at work, Tew felt his pulse at his neck. His heart had fluttered several times in the past few days, accompanied by stabbing pains in his chest. His current medications didn't seem to be having much therapeutic effect. If all went well, this mission should be concluded in forty-eight hours or less, so he promised himself that he would report straightaway to his cardiologist the moment they had confirmation that the boys were safely back on the ground. In the meantime, there was still much to be done and little time to squander, so his health be damned for the moment.

He heard Wolcott's distinctive Oklahoma twang and looked down to see his counterpart, still dressed in his flight suit, entering the facility through one of the lower side doors. Ted Seibert, also attired in sage green Nomex, was with him. The two had obviously come straight from the flight line without changing.

Wolcott customarily made his rounds on the floor before coming up to the observation area, but today he strolled directly up the sloping aisle between the rows of consoles. Wearing concerned expressions, he and Seibert entered the observation room together.

"I think you need to grab a chair, Mark," declared Wolcott, as the door swung closed behind him. "Ted has some hot news that just came over the wire, and it ain't soundin' good."

Tew took a seat behind one of the unoccupied desks, took a deep breath, and said, "Go on."

"We apparently have an ugly situation developing, General," confided Seibert, wiping sweat from his forehead with his sleeve. His face bore a faint outline of his oxygen mask, and his hair—usually perfectly coiffed—was matted down from his flight helmet. The intelligence officer was normally fastidious about his appearance; Tew had never seen him even slightly unkempt.

"Something that could affect the mission?" asked Tew.

"Most definitely," answered Seibert, nodding solemnly. "We just received this week's message from the source at Kapustin Yar. He indicated that the *Krepost* might still be manned."

Tew gasped. "*What?* I thought we had solid verification that a Soyuz had undocked from it, reentered, and the crew was successfully recovered."

Seibert cleared his throat and said, "He claims that they only sent *one* man home."

"Oh."

"Like I said, Mark, this ain't good," interjected Wolcott. "It don't bode well at all."

Seibert bent down, unzipped his left calf pocket, and pulled out a briarwood pipe and a leather pouch. He stuffed the pipe's bowl with tobacco and started to light it with a wood match, but stopped when Wolcott faintly shook his head.

Nodding in affirmation, Seibert put away his matches and continued. "There's more. According to the source, the *Krepost* may be damaged in some manner. The Soviets apparently are planning to launch a mission to rescue the man on board. The source at Kapustin Yar also apparently suspects that he is in danger of compromise, because he has requested an emergency extraction for his family and himself."

"That ain't good," declared Wolcott.

Nervously toying with his pipe, Seibert stated, "We also have information from *another* source that indicates that the Soviets know that we are aware of the *Krepost* and that it is armed with nuclear weapons."

"So the Soviets know that we're conscious of the *Krepost*?" asked Tew angrily. Convulsing slightly, he clutched his chest and began to wheeze. As casually as he could manage, he reached into his desk drawer, found a flat tin containing nitroglycerine tablets, and slipped one in his mouth.

"Probably," answered Seibert.

Tew felt the nitro tablet dissolve on his tongue, and then said, "This is terrible news. How about our boys upstairs? At this point, we have to assume that the Soviets must at least suspect that we might target the *Krepost*. Do you have any indication that they know that we have a mission underway? Are Carson and Ourecky in imminent danger?"

Seibert shook his head and replied, "General, there is no indication that they're aware that we are currently executing a mission, nor do we think that they even know about this Project. On a negative note, they are obviously being more vigilant than usual, so we suspect that they might be anticipating an attack on the *Krepost* at some point, but they still seem fixated on the notion that we're employing an unmanned intercept platform."

"So, if the *Krepost* is vacant, our men should be relatively safe, right?" asked Tew.

"Honestly, we have no way of knowing," answered Seibert. "The man still up there could very well be waiting in ambush. On the other hand, assuming that the information from our Kapustin Yar source is accurate and the *Krepost* is damaged, there's no way of knowing if their defensive systems are still functional. Obviously, we should probably err on the side of caution and assume that they are."

"Agreed," replied Tew. His head was spinning as a result of the nitroglycerine tablet, and he struggled to maintain his composure.

"So, pard, I s'pose you want to call those boys home now, right?" said Wolcott, loosening the laces of his flight boots. "I'll go and get Gunter so we can start puttin' the appropriate plans in place."

Tew shook his head. "Let's not be too hasty. As far as I'm concerned, the situation really hasn't changed that much. This *Krepost* is still a critical target that needs to be knocked down. Yes, I'm extremely inclined to order those two home, but their past performance has shown that they are going to do as they wish, regardless of what directions we give them."

"So you ain't goin' to order them down?" asked Wolcott, frowning as he raised his eyebrows.

"No. We will give them all the pertinent information that we have, ensure that they are apprised of the risks, and grant them an opportunity to exercise their initiative. I would be more than happy if they decided to come home forthwith, but I strongly suspect that they won't.

After all, we put Carson and Ourecky in that cockpit for a reason, didn't we?"

"That we did, brother," answered Wolcott, gazing through the glass at the controllers at their consoles. "Indeed, that we did."

Gemini-I, On Orbit
6:10 GMT, Sunday October 22, 1972
GET: 5 Hours 25 minutes / REV # 3

"Crypto's locked in," announced Ourecky, verifying the green light that indicated the voice scrambler had accepted the cryptographic variable that he had just keyed in.

"I copy that the crypto is locked. Standing by for transmission," said Carson. "Hey, Scott, just to be absolutely safe, why don't you switch off the transmitter on the voice side, just so we don't accidently break the rules?"

"Will do," replied Ourecky, throwing a series of switches. "Voice transmit is disabled."

A few seconds passed before they heard the voice of the Mission Controller, who was physically located in a tracking station in California. Because he was transmitting through a scrambler, his voice bore a cartoonish distortion that made him sound almost like Daffy Duck. "Scepter Twelve, this is Track West. Stand by to copy critical traffic."

Already prepared with index cards and pencils, Carson and Ourecky looked at each other. "Critical traffic?" asked Ourecky. "Wonder what this is about?"

He had their answer soon enough. "Scepter Twelve, this is Track West. Be advised that I have a lot of information and will prioritize from most important to least important, in case we lose contact. Be advised that the next station will pick up where I leave off...break...Specific instructions from Golf Mike Tango: You may continue to execute the intercept at your own discretion...break...I say again: Specific instructions from Golf Mike Tango: You may continue to execute the intercept at your own discretion."

Track West continued: "If you elect *not* to continue intercept, you are cleared to reenter for PRZ One-Nine on your fifth rev. Current weather is six thousand scattered with twelve miles of visibility. Winds out of Nine Zero at six, gusting to nine. Current altimeter is Two Nine Six Seven. TACAN is Channel Three. I say again, if you elect *not* to continue intercept, you are cleared to reenter for PRZ One-Nine on your fifth rev. All other contingency recovery zone data remains in effect.

"This is current intelligence concerning your target. First, assume that it has sustained some form of damage, extent unknown, but is likely still manned and crew will probably aggressively defend target if they detect your approach. Stand by...stand by...stand by..."

As they waited for more, Carson whistled. "Isn't this something? This definitely isn't the walk in the park we anticipated, Scott."

Ourecky nodded.

"Scepter Twelve, this is Track West. Be aware that we have some last breaking developments to pass on. Stand by to copy."

Seconds later, Track West announced: "Tango Two-One has reported a significant anomaly with the target vehicle." Ourecky knew that Tango Two-One was a ship-based tracking station that employed a powerful radar to monitor the *Krepost* and accurately determine its position in space. Like the other tracking radars that would guide their rendezvous, Tango Two-One had been switched off until this phase of the mission, since the *Krepost* was in such a predictable orbit, but was now illuminating the station in short duration "snapshot" pulses of radar energy.

"As of an hour ago, Tango Two-One indicated that there are now multiple radar targets in the immediately vicinity of the target, as well as multiple targets trailing it in the same orbit. Their observations have been confirmed by another terrestrial station as well as an airborne station."

Track West continued: "Earlier guidance remains in effect: You may continue to execute the intercept at your own discretion. Regardless of your decision, you should proceed with extreme caution. Acknowledge this transmission by breaking squelch three times, and then switch circuit for data upload on my mark."

"Scott, switch to voice," said Carson.

Ourecky rotated a knob and stated, "Voice on."

Carson keyed his mike three times.

"Copy acknowledgement," replied Track West. "Switch circuit to data upload on my mark…three…two…one…mark."

Ourecky checked the computer display and verified that the DCS data upload light was lit.

"Whew," said Carson. "That was a lot to absorb. What's your take, Scott?"

"Honestly? All that sounded pretty interesting, and it was really nice of General Tew to extend us the courtesy of deciding whether to continue or not, but it really doesn't change things."

"How so?" asked Carson, unwrapping a stick of Juicy Fruit.

"Even though we were told that the *Krepost* would be unmanned, we've always assumed that they might have some remote or automatic system for firing the gun. To me, the only difference is that they apparently know or suspect that we're coming."

"Point taken."

"The new development is what concerns me. If there are additional radar returns, that either means that they've experienced some sort of catastrophic accident, like Ground implied, or…"

"They're dumping some sort of chaff to spoof our radar," interjected Carson. "And that's actually a good sign, if they believe we need to light up our radar to make our approach. Chances are that they also have a gizmo like ours to detect radar frequencies. So if we creep in with a cold nose, they might not even notice us until it's too late."

"DCS upload is complete," observed Ourecky, watching the computer console. "The computer has accepted the DCS data. So, Drew, it sounds like you're intent on proceeding. Is that the case?"

"You know I can't order you to do this, Scott," answered Carson. "You have a wife and family to go home to, and I don't. Personally, I would just as soon go ahead and execute, but it's your call. What say you, Ourecky?"

Ourecky thought for a moment. He thought about Bea and Andy at home, and wanted nothing more than to return to them as quickly as he could, but he also thought of the millions of people who would be living under this monster's menacing shadow if they failed to act.

"They sent us up here for a reason," he said. "If it's all the same to you, Drew, I would prefer to follow through."

12

THE LONGEST SUNDAY

Krepost **Station, On Orbit**
7:10 a.m. GMT, Sunday, October 22, 1972
GET: 38 Days 6 Hours 25 Minutes, REV # 612

S waddled in his makeshift sleeping bag, Vasilyev wedged his body against the frigid aluminum facing of the Egg's control panel so he could use its faint glow to illuminate the calculations penciled in his notebook. Two days ago, Control had notified him that Gogol would reinforce him, and that his launch was scheduled for tomorrow. While he did not fancy Gogol's company, he had been enamored with the notion that he would at least have the means to return to earth once the Soyuz was docked. Unfortunately, the timing was definitely not in his favor, since it was highly unlikely that he would survive long enough to welcome Gogol with salt and bread.

Oxygen. Oxygen was *the* issue. There was actually plenty of oxygen in his midst, but unfortunately, at the molecular level, those plentiful and precious oxygen atoms were securely bonded to carbon atoms, and the resultant gas—carbon dioxide—was the thing that would soon kill him. The lithium hydroxide in the oxygen regenerator could break the atoms' clinging grasp on one another, but Control had forbade him from switching on the power-hungry air scrubbers. They insisted that they were still diligently working on a solution to manually operate the oxygen regenerators, but surely they recognized—as Vasilyev did—the essential devices were buried deep inside the service module and were all but inaccessible.

Updating his daily projections, he clumsily manipulated his slide rule with stiff fingers, squinting at its tiny numbers in the faint light. Not satisfied with his results, he processed the variables a second time and then a third time, but his answers were the same on the subsequent attempts. His best-case estimate was that he would be alive for twelve more hours. On the worst-case end of the equations, he could quite possibly expire within the next six hours.

He let go of the slide rule, allowing it to float before him, and shivered from the cold as he studied the numbers and grappled with his despair. Certainly, he could activate the air scrubbers against Control's orders. But even in the best-case scenario, before the batteries lost power altogether, the oxygen regenerators could only process enough carbon dioxide to yield roughly twenty-four hours' worth of life-giving oxygen. So, although he might postpone his demise for another day, there was no way to evade the inevitable: he would still be dead before Gogol arrived. As painful as it was to admit, Control was correct in their decree; since Gogol could not come up any faster, little could be gained by swapping power for oxygen. Unless he elected to take his own life, and perhaps that was an unspoken part of their plan, he was doomed to eventually die from asphyxiation.

Almost delirious from the pervasive cold, he attempted yet again to regulate his breathing to conserve oxygen. He had read that Hindu mystics could meditate with such intensity that they could literally slow their metabolism, so that their hearts rarely beat and they might need to draw just one breath a minute. In the past few days, he had tried to replicate their feats, but his attempts had met with little success. Exhaling a cloud of steam, he gasped, finally coming to the realization that his impromptu breathing exercises accomplished little.

He felt his tiny world collapsing in upon him. Despondent, he knew that his fate was irrevocably sealed; he reconciled himself with the idea that it was just a matter of time before he drew his last breath. The crippled *Krepost* was a deathtrap.

He wriggled toward the main instrument panel, and checked the power gauges, in the vain hope that the solar panels might now be generating more power. They weren't. The gauges reflected that there was almost an adequate charge built up in batteries to operate the docking mechanism, but that was the extent of it. As he studied the fluttering needles, Vasilyev contemplated the supposed rescue mission. He surmised that Abdirov's ultimate intent had nothing to do with rescuing him, but was focused more on getting Gogol aboard the *Krepost* to ensure continuity of the mission. It made sense. Like Gogol, Abdirov did not seem to be encumbered with a conscience or sense of pity.

His aching stomach growled like a bear waking from hibernation, and he suddenly gained enough presence of mind to realize that while most of his precious reserves were dwindling, he still possessed a considerable amount of food. He had been depriving himself to stretch his larder, but now there was clearly no reason to die of starvation.

And that wasn't all. When Travkin departed, Vasilyev had vowed not to drink until his mission was over and he was safely home, but it was now abundantly clear that his mission *was* over, since he would soon be dead. Why should he spend his few remaining hours wallowing in misery? He decided to abandon civility and go out on a well-deserved bender. After all, he still had a copious volume of liquor on board, and it made little sense to let it go to waste. If he accomplished nothing else, then he could at least embalm himself.

He soared up to the galley, opened a cabinet and yanked out a rubber hot water bottle filled with cognac. Unsealing it, he pressed it to his lips. He didn't drink; he guzzled. He drank like a stranded man happening upon an oasis pool in the searing heat of the Saharan Desert.

Then he ate. Drooling over the culinary possibilities, he foraged through the goodies he had stashed for this moment. Ravenous, he crammed his mouth with dense, half-frozen fruitcake. He voraciously gobbled cookies and chocolate bars. He opened a tin of fancy beluga caviar and savored the salty fish eggs, relishing the buttery aftertaste they imparted as he crunched them between his teeth. He slathered raspberry jam on dark bread that was now stale and hard, and enjoyed it as if it had just emerged fresh and hot out of the oven.

After he sated himself with food, at least for the moment, he opened another flexible container of liquor and swigged some more. Very soon, he was all but numb to his woes and lost all track of time.

Gemini-I, On Orbit
8:27 a.m. GMT, Sunday October 22, 1972
GET: 7 Hours 42 minutes / REV # 5

Not knowing what lay in store, their final approach was painstakingly slow and cautious. They couldn't advance during the forty-five minute night phases of their orbits, for fear of accidentally colliding with the *Krepost* in the darkness. As orbital dawn arrived, Carson and Ourecky commenced the last stage of their hunt. Glancing out the window, Carson saw that they were passing just to the east of Brazil, and then he rolled the spacecraft a few degrees to watch for the *Krepost*. Just as planned, it was almost three miles away, directly ahead of them.

"My Mark One eyeball has us at about five thousand meters," declared Carson, nudging the maneuver controller as he studied the faint object in the distance. "Of course, it's way out of calibration. Does it look about right to you?"

"Yeah," answered Ourecky, holding up his hand to shield his eyes. "I think that's accurate."

"We're well into the danger area if that gun is active. Radar detector?"

"No tone on any bands. Nothing at all."

"How about your radiation detectors?"

"Still nothing but background," replied Ourecky, looking down to examine the sensor. He looked back up and scrutinized the *Krepost* through powerful binoculars.

"Any indication that it's been transmitting?"

"None," answered Ourecky, verifying the "ferret" scanner that swept radio frequencies. "Not since we're been in range, anyway. Either no one is home, or they're not very talkative."

"Okay." Carson verified the pressure on the fuel and oxidizer tanks that supplied the four one-hundred-pound main thrusters. "Pressure is still good on the mains. Edging forward."

The pilot kept his hand poised on the maneuver controller, ready to immediately ignite the powerful main thrusters at a split-second's notice. As they closed the gap, it was readily apparent that the *Krepost* was canted out of alignment and slowly spinning around its long axis.

The two men periodically saw sparkles as sunlight glinted off pieces of debris that trailed the station. Obviously, their target was badly damaged, but it was like stalking a wounded but still very dangerous beast. The radar warning alarm had been silent through the entire approach, so they were much less fearful of some automatic system triggering the cannon. Besides, the automatic cannon would obviously be reliant on the station's alignment, and every indication was that the *Krepost* was completely adrift.

As they drew still closer, they were astonished by the eerie sight. The much-feared *Krepost* looked less like a formidable space station, and more like the shattered remnants of a child's broken toy. There was a gaping maw in the mangled forward module of the Soyuz cargo ferry. Scorched tangles of wire and loose chunks of debris floated out through the jagged and charred edges of the breach. A large semicircular piece of the Soyuz's pressure vessel drifted alongside the *Krepost*, tumbling in slow motion. Carson kept his normal distance and tacked on a very liberal safety margin; it was just too treacherous to maneuver any closer.

"Ugh," noted Carson, gingerly navigating around the derelict space station. "That definitely looks like the cargo ferry, still docked. There's no crew transfer Soyuz, just like we were told. If I were to hazard a guess,

I think that if someone was left aboard, they're probably dead by now. That thing looks as dead as a doornail."

"Agreed. What's your plan?" asked Ourecky, snapping photographs.

"My plan? I think we need to complete our documentation as quickly as possible, deploy the Disruptor, and then skedaddle the hell out of here as fast as we possibly can," replied Carson, pausing to squirt some water in his mouth. He followed the water with a stick of gum. "I'm not at all comfortable with the idea of maneuvering with all this damned wreckage in the vicinity. Granted, it's relatively stationary in relation to us, so it shouldn't cause any major problems, but I don't feel like taking any chances."

Carson continued. "Besides, even though the main block is wrecked, the warhead module looks intact. There's always the possibility that it could be controlled from the ground, and I don't want to be in the neighborhood when those retros fire."

"Amen to that," noted Ourecky, changing film on his 35-mm Hasselblad camera. He stowed the exposed roll in a metal film can, checked to see that it was correctly labeled, annotated it in his photo log, and then tucked the film can into his right-side storage pouch. "But that thing is still rotating, Drew. How are we going to snare it with the Disruptor?"

Shaking his head, Carson replied, "I haven't figured that out yet. It sure would have been sweet if the eggheads had delivered that torpedo they promised. That would have been the hot ticket for this job. Anyway, you keep snapping pics and watching the detectors and ferret gear, and I'll keep us clear. Once you have all the shots you need, we'll figure out some way to smoke this cigar."

Krepost Station, On Orbit
9:41 a.m. GMT, Sunday, October 22, 1972
GET: 38 Days 8 Hours 56 Minutes, REV # 613

After gorging himself from the pantry, Vasilyev dozed for a while, floating in the galley. Still very intoxicated, he awoke with a start and wondered where he was in relation to the world below. Clutching his queasy gut, he poked his arm out of his cocoon to consult his watch. He was several hours away from his next contact window. Not hungry anymore, but still plenty inebriated, he meandered down to the control area to look at the earth.

Peering through a porthole, he immediately saw that he was in a dark phase of an orbit, but he recognized the northwest coast of the United States. He oriented himself; there were the familiar lights of Seattle, and over there Tacoma, Boise, Missoula, and Billings. Then he glimpsed something that made him gasp; in the vicinity of Great Falls, where the Americans had buried scores of Minuteman ICBMs in subterranean silos, he saw brilliant flashes of light in the darkness. Instinctively, he rushed to the Egg's console to switch on the targeting computer.

Wheezing, he opened the code safe to secure the target book, and then stopped short. He realized that he was operating entirely in the blind. Lacking the regular weather report normally sent up from Control, he had no way of knowing whether the bright flashes were missile launches or merely lightning strikes from a thunderstorm passing over the barren hills and hinterlands of central Montana. Granted, the dire situation reported in the intelligence updates suggested the former, but he still was not sure.

Staring at the dimly lit console, rubbing his sore eyes, Vasilyev realized just how effectively he had been conditioned to do the terrible deed without even pausing for thought. He was no longer a man, but merely a component of the machine.

As he thought about how quickly he was willing to trigger annihilation, he recalled a story from the Manhattan Project, the Americans' crash program to develop the atomic bomb. Before the Americans dropped their nuclear weapons on Japan, they detonated a test device in the high deserts of New Mexico. As he witnessed the test, Robert Oppenheimer—the lead designer of the bomb—was simultaneously awed and horrified at the destructive potential he had wrought. Afterwards, he was inspired to write that he had become Shiva—Death—the Hindu Destroyer of Worlds. Reflecting on the story, Vasilyev laughed. That intellectual

Oppenheimer was nothing but a self-deceiving fool, because although he had created the means for such wanton destruction, he was not burdened with the knowledge that he could wield it at his own whim.

The General Staff of the High Command had sent him up here to babysit this terrible Egg, to obediently do their bidding. *But if thermonuclear war is now so inevitable, then why not just get on with it? What is the sense in waiting?* At this juncture, Vasilyev was probably the sole man in the universe endowed with his own personal nuclear weapon, and it made little sense for him to exercise restraint. After all, even if he *wasn't* going to be dead in the coming hours, what was there to live for anyway? His wife and daughters had been taken from him, and although he would rejoin them soon, maybe this terribly destructive act could be their legacy. By unleashing one bomb, he might teach the world the futility and fallacy of mutually assured destruction.

Without hesitation, he reached into the code safe, fumbled for the target book, and selected a city to be destroyed. Squinting in the pale light, he entered the target point—the coordinates for New York City—and summarily sentenced millions of civilians to a fiery demise. But it was just as well: the world would be thrown into turmoil, but they would finally have to face the reality of thermonuclear war once and for all. They could no longer hide from it, since Vasilyev would not let them.

With tears welling in his tormented eyes, Vasilyev entered his eight-digit Independent Action Code into the interlock's keypad. What was once purely theoretical was now harsh reality; as the computer acknowledged his code as valid, he became the most powerful human in the universe. Only one simple step remained now, and millions would die.

As his finger hovered over the Arm button that would lock in the coordinates and initiate the arming sequence, he was startled by a screeching noise, which he recognized as the long-dormant stabilization gyros spinning back to life, followed by cacophony of thrusters popping. To his ears, the welcome sounds were something like treasured notes from a favorite symphony. In his drunken state, he had entirely forgotten that once a target's coordinates and the Independent Action Code were entered, the targeting computer would assume control of the station to optimally align the platform for weapons deployment. In this instance, the targeting computer obviously saw fit to counteract the *Krepost's* rotation. In mere minutes, the *Krepost* was stable again. *The rotation was gone!* It was truly a serendipitous event.

The pink glow of orbital dawn streamed through the portholes. Struggling to contain his exuberance, he rushed to the main console and checked the power gauges. With the *Krepost* now stable, the solar panel arrays now functioned at full capacity, so the batteries were charging normally again. He did some quick calculations; it would still be touch and go, but so long as the sun continued to shine and the *Krepost* held stability, he should be alive when Gogol arrived in the Soyuz.

And then grim reality set in. Only moments ago he was overjoyed, but now he was consumed with fear as he considered the consequences of his actions. Granted, it was an extraordinary set of circumstances, but it was still highly probable that he could hauled before a tribunal and charged for attempting to deploy the Egg without express authorization. Certainly, he might be granted some modicum of leeway since he had been issued an Independent Action Code and explicitly stated instructions to use his best judgment, but the fact was that he might have formed his decision after witnessing a string of ICBM launches or observing a bad thunderstorm. The truth was that he could have readily unleashed hell based on a simple weather event. And to cap it all off, he strongly suspected that Abdirov would be just as angry with him for *not* following through with the Egg deployment. Anyway, there was just no way of knowing if he might be punished for this gross infraction.

He realized that since all the transmitters were switched off, Control had no means of knowing what had transpired. Feeling warm and flush, despite the bitter cold, he decided to tidy up after himself. First, fidgeting with the keypad, he cleared the data from the targeting computer. The he shifted his attention to the interlock. He tried inserting his key and entering various sequences of arbitrary numbers, in the hope it would reset the interlock, but his desperate efforts were to no avail. The interlock remained disabled.

Gemini-I, On Orbit
10:02 a.m. GMT, Sunday October 22, 1972
GET: 9 Hours 17 minutes / REV # 6

To remain safely clear of the slowly rotating *Krepost*, Carson had parked the Gemini-I "ahead" of the station, a few hundred yards forward of the warhead end. In this position, they were well clear of the constellation of debris that trailed the derelict station. They had been watching it for over an hour, struggling to come up with a viable solution to deploy the Disruptor.

The two men were discussing a diagram that Ourecky had drawn, depicting a potential approach, when Carson happened to glance out the window as the sun rose above the horizon. "Damn it, he's maneuvering!" he declared, instinctively throwing switches to prepare the main thrusters to fire. "That thing certainly isn't dead."

"Checking ferret gear," stated Ourecky.

"Nominal pressure on the fuel and ox," stated Carson. "We can bolt if need be."

"Good," answered Ourecky.

"Any radio activity over there?" asked Carson. He tugged a set of binoculars out of his left storage pocket and studied the *Krepost*.

"I'm seeing nothing on the ferret," replied Ourecky, adjusting the dials to gradually sweep through radio frequencies. "He's not transmitting now, and we didn't hear any alarms earlier."

"We have to go in *now* with the Disruptor," announced Carson, stowing the binoculars. "This might be as good as it gets. We might not have another opportunity."

"Aren't you being a little hasty?" asked Ourecky. "You don't want to watch him for at least a few minutes?"

"Look, we don't know how long he's going to be stable and stationary," said Carson. "Besides, I can think of only one logical reason…"

"He's getting ready to deploy that warhead," blurted Ourecky.

"Yep. Ready?"

"Let's go for broke," replied Ourecky. "It's why we're up here."

"Definitely not an occupation for the fainthearted," replied Carson, grasping the hand controller. "I'm going to nudge forward. After we've closed the gap and nulled out our residuals, you'll pop up the boom and get it ready, just like we've practiced, except this time we're obligated to move a lot faster. Okay?"

"Okay on my side."

"Maneuvering forward," said Carson calmly. "Fasten your seatbelt and secure any loose items, because this ride could get a mite bumpy."

In just a few minutes, Carson had moved the Gemini-I into position. "Hey, I know we're not supposed to work in the dark, but today we're going to push through."

"I'm good," answered Ourecky. "Light or dark, rain or shine, let's do it."

Switching on the floodlight, Carson said, "Deploy the Disruptor."

"Deploying the Disruptor," replied Ourecky. He threw a switch to initiate the chain of events that would swing out the Multi-Function Disruptor's boom from the adapter section in the rear of the spacecraft. Powered by compressed gas cartridges, the large boom would extend out and lock into position. A large hoop, constructed of metal tape, was attached to the end of the boom. Once the boom was locked in place, Ourecky would activate a motorized drive to expand the hoop until it was slightly larger than the diameter of the target. Once the hoop was properly sized, Carson would deftly maneuver the Gemini-I to center the hoop around the target, much like placing a ring on a finger. After it was in position, Ourecky would reverse the motorized drive, gradually cinching the hoop until the Disruptor's target engagement head was snug against the target. Once this was accomplished, Ourecky would fire a pyrotechnic charge to sever the boom, and then Carson would carefully back away from the target.

Aptly named, the Multi-Function Disruptor incorporated three potential options to interdict a suspect satellite. First, the Disruptor's target engagement head contained an explosive charge capable of demolishing a target outright. Second, it was fitted with a solid rocket motor that could cause its target to prematurely fall out of orbit. The third means of attack was probably the most devious: a miniscule "needle" thruster, powered by a tank of compressed nitrogen, which would force the target into a slow spin that interrupted key functions.

At this point, the Disruptor had proven itself time and again. Carson and Ourecky had successfully used it to attack and destroy seven Soviet satellites. But today, there was a problem: the vaunted satellite killer just wouldn't deploy.

The hoop was cradled in a series of small supports in the Gemini-I's adapter section. Ourecky looked up through the viewport in his hatch, but did not see the boom protruding from the adapter. He rotated a knob to check the various functions on the Disruptor's control panel. The "STOW" light on the Disruptor's control panel flashed green, indicating that the hoop hadn't budged, so the boom had not deployed.

Ourecky tapped the switch, then toggled it again.

"Not good," he observed. "No joy on the Disruptor. The boom must be stuck back there."

"Ouch," replied Carson. "Keep trying."

Ourecky persisted for a few minutes, but to no avail.

"Buddy, we need to stop for a quick chat," said Carson. "Although we're going to continue working the problem, we have to consider the worst-case scenario, in case we have to act immediately."

"Okay. What's on your mind?"

"Right now, we have to assume that this thing is being controlled from the ground, since we've seen no evidence that anyone is alive inside there," said Carson. "If they're preparing to deploy that warhead, we have to be prepared to immediately react. Here's my take: if the warhead physically separates from the station, we should light the main thrusters and chase after it."

"And then what?" asked Ourecky, rechecking the Disruptor's function lights.

"Ram it."

Ourecky swallowed and replied, "I was afraid you would say that." The two of them had discussed this very situation in the past, but had never considered a serious option. Of course, they never even contemplated that the Disruptor might fail, since it had functioned flawlessly on seven previous missions.

"If we were forced to ram it, I'm sure that we could pick up sufficient velocity to disable it," noted Carson. "And although we probably wouldn't survive, there's still a chance that we might. Our nose would bear the brunt of the impact, and there's nothing up there we really can't live without, since we're flying this job without radar. Provided that the pressure vessel isn't compromised and we're intact enough to make reentry, we could probably still limp home."

"But it's still pretty damned unlikely that we would live, right?" asked Ourecky.

"Affirmative. So, I need to know now: Are you in or are you out?"

"I'm in," answered Ourecky. "If it comes down to that, we'll do what we must."

"Good. Now, like I said, let's keep working the problem. We're going into darkness in a few minutes, so I'm going to back away slightly while you continue working with the Disruptor. If we can't fix it before we go back into daylight, then we'll punt toward doing an EVA. We'll fly a close-in all-around inspection to see what's immediately vulnerable. After that, we'll put you outside as soon as possible to break stuff."

"I suppose I should clip their antennas first, and disable the docking mechanisms?" asked Ourecky.

"Good call. After you've done sufficient damage, your next priority will be to head back to the adapter to see if you can determine why the Disruptor is jammed up. If you can fix it, we'll deploy it after you climb back in."

"One question," said Ourecky. "What if the warhead breaks away while I'm outside?"

"Always with the negative waves, Ourecky. Always with the negative waves."

Krepost **Project Headquarters, 3:15 p.m.**
Sunday, October 22, 1972

Looking for Bogrov, Yohzin strolled into the mission control facility. He sat at the table at the rear of the room and watched.

On the Control side, the mission controllers were rehearsing launch procedures for tomorrow's Soyuz mission. Yohzin was aware that Gogol was in a simulator somewhere else in the building, so that he was able to practice his ascent in concert with the mission controllers. On that side of the big room, the atmosphere was calm and unhurried, like a skilled construction crew applying the finishing touches at the end of a long project.

On the *Perimetr* side of the room, however, things were immensely different. The usually complacent *Perimetr* engineers had stashed away their chessboards and novels, and were huddled around a television monitor. They seemed frantic; something was obviously very wrong.

Wearing his customary grey suit, poorly fitting and rumpled, Bogrov walked up to Yohzin's table. The *Perimetr* engineers had obviously chosen him as their emissary.

Yohzin noticed that Bogrov was sweating profusely, and the *Perimetr* engineer seemed to be in somewhat of a daze.

"What, Aleksey?" asked Yohzin. "Why did you ask me here?"

"Comrade General, can you tell me if your Control people have received any telemetry from the *Krepost*?" asked Bogrov.

"You called me here to do your bidding?" asked Yohzin. "You couldn't just ask them yourself?" It seemed so idiotic that the *Perimetr* people could never see fit to leave their well-appointed enclave, even though only a waist-high wooden partition separated their side of the room from Control's. The simple divider might as well have been the formidable Berlin Wall, festooned with broken glass and razor wire. He just could not comprehend how two teams of professional officers could function in such close quarters, day after day, and yet allow so much animosity to separate them.

"Bear with me, *puzhalsta*, Comrade General."

Yohzin grudgingly pushed himself out of his seat, walked away from the table, and waved over the flight director.

"I know that this may sound absurd," he said. "But have you received any telemetry from the *Krepost*?"

The flight director shook his head, slipped on his reading glasses, referred to a clipboard, and then replied, "Comrade General, the last contact window was roughly twenty minutes ago. It wasn't a full pass over Soviet territory, but just involved an abbreviated pass over the communications site on Sakhalin Island."

Yohzin knew that the site was their furthest station to the east, a remote facility occupied by just ten men. Sakhalin Island was in the Sea of Okhotsk, roughly sixty kilometers north of the Japanese island of Hokkaido. "Did you have any contact with the *Krepost*?" he asked.

"*Nyet*," replied the flight director. "We did not direct Vasilyev to turn on his transmitters, so we received no transmissions from him."

"No data telemetry?" asked Yohzin.

"*Nyet*," replied the flight director. "Begging your pardon, Comrade General, but in accordance with General Abdirov's orders, we instructed Vasilyev to keep his transmitters off, to conserve power, unless we specifically directed him to turn them on. That includes the telemetry transmitters. Has General Abdirov amended his orders?"

"*Nyet*. He has not."

As Yohzin started to turn, the flight director asked, "Do you know what's going on over there, sir? They seem to be acting very peculiar today."

Yohzin shook his head, and answered, "Hopefully, I'll find out soon enough. If I can, I'll let you know." He walked back to the table and sat down next to Bogrov.

"To answer your question, there have been *no* transmissions from the *Krepost*," stated Yohzin. "Including telemetry. Vasilyev is obviously abiding by our orders and is not energizing his transmitters, or he's dead, in which case the orders are rather irrelevant. Aleksey, you look like you've seen a ghost. What the hell is going on?"

"Comrade General, can we speak in confidence?" asked Bogrov.

"Of course. Now, what is it?"

"*We* received telemetry when Vasilyev passed over Sakhalin," confided Bogrov quietly. "The telemetry indicates that the interlock is disengaged. We don't know how it happened, but we suspect that there might have been some sort of malfunction, possibly from a power surge."

"The interlock is disengaged?" asked Yohzin incredulously. "Did I hear you correctly?"

"*Da*," answered Bogrov meekly. Tilting his face toward the floor, he looked like a mongrel puppy anticipating a brutal beating for soiling an heirloom rug.

"*Wait.* This makes absolutely no sense. If Vasilyev hasn't turned on his transmitters, then how could you be receiving telemetry?" asked Yohzin.

"Our communications equipment and frequencies are entirely independent," explained Bogrov. "And they are on a separate power source. Vasilyev cannot activate them, nor can he deactivate them."

"Of course. That completely slipped my mind."

"There's more."

"More? How can there possibly be *more*?"

"You remember that the targeting computer assumes control and aligns the platform once a target has been entered and an authorization code has been verified?"

"*Da?*" asked Yohzin.

Bogrov swallowed deeply and then admitted, "As best as we can tell, Comrade General, those things occurred as well. We received telemetry from the Egg for the entire duration of the contact window, about a minute's worth. Theoretically, that could only be possible if..."

"The platform was aligned, and the stabilization gyros halted the station's rotation," interjected Yohzin. "What the hell is going on? I sat with you for *weeks*, and we tested that damned interlock every way from Sunday, and as skeptical as I was, I'll be the first to admit that your design was bulletproof."

"We were sure that it was," said Bogrov, wearing a pained expression. "But obviously it wasn't."

"So the interlock is disengaged," said Yohzin. "What are the implications?"

"Since it's unlocked, there's absolutely nothing to stop your guy from dropping the Egg on his own," answered Bogrov. "He wouldn't need an Independent Action Code or any other code."

Shaking his head in sheer amazement, Yohzin said, "Is there any chance that Vasilyev is aware of this discrepancy?"

"We doubt it. From what we have heard eavesdropping over the partition, we suspect that he is probably just clinging to life at this point, if he's not already dead, so I doubt that he would apply too much energy to the Egg."

"Agreed," said Yohzin. "Aleksey, you do know that we're going to have to broach this matter with General Abdirov, right?"

Bogrov nodded. "*Da*. I am really not looking forward to that."

Krepost **Station, On Orbit**
13:47 p.m. GMT, Sunday, October 22, 1972
GET: 38 Days 13 Hours 2 Minutes, REV # 616

As he waited for the next contact window, Vasilyev did some quick calculations. Although he would still be compelled to keep the station in a largely dormant state, he now had ample power to activate the scrubbers. Now, he had to decide what to tell Control, because he obviously could not conceal the situation indefinitely. Badly dehydrated, his head pounding from his bender, he forced himself to concentrate on the task at hand.

The moment of truth had arrived. Just a minute prior to the scheduled contact window, he switched on the receivers and donned his headset. In a few moments, a voice cut through the static; he recognized it as Major General Yohzin. "*Krepost*, turn on your voice transmitter and acknowledge," ordered Yohzin.

As directed, Vasilyev switched on the transmitters. Swallowing deeply, he knew that he must be in truly dire straits if Yohzin was assuming the role of communicator. Speaking slowly and deliberately so as not to slur his words, he keyed the mike and said, "This is *Krepost*."

"Good," answered Yohzin. "We weren't absolutely sure that you were alive."

"I am, sir, but just barely. I need to ask about the scrubbers, I…"

Interrupting him, Yohzin asked, "*Krepost*, has your rotation stopped?"

"It has," he answered. "The platform has stabilized, and the batteries are now charging normally."

"And why did you not inform us of this development earlier?" asked Yohzin.

"I was following Control's orders. I was instructed not to activate the transmitters."

"Correct answer, *Krepost*. Admirable," stated Yohzin. "Now that you have sufficient power, I want you to operate your scrubbers for one hour. We will advise on further actions. We have nothing further to discuss, unless you want to report any other developments up there."

"Control, this is *Krepost*. I have nothing else significant to report."

"Then except for the scrubbers, all existing guidance and instructions remain in effect. Switch off your radios."

Vasilyev switched off the radios, as ordered. The contact window was a puzzling turn of events, and certainly not the exchange he expected.

Krepost **Station, On Orbit**
14:26 p.m. GMT, Sunday, October 22, 1972
GET (Ground Elapsed Time): 38 Days 13 Hours 41 Minutes
REV # 617

Suddenly, he was distracted by a brilliant flash of light sparkling through a porthole. He chuckled. Perhaps an angel had miraculously arrived to rescue him. Curious to see the source of the light, he lightly pushed off the control panel and drifted toward the window.

Either as a result of the alcohol or sheer exhaustion, his tired eyes were slow to focus, but he gasped at what he saw. A brilliant dazzling blob of intense light floated outside. *It could not possibly be!* Mesmerized, he thought that he surely must be hallucinating. His senses must be deceiving him. He shook his head and closed his eyes, resolved to the thought that it would be gone when he looked again. He rubbed his eyes and slowly opened them. It was still there! Only a moment ago, he scoffed at the idea that the captivating light might be an angel sent to save him, but that seemed to be the only logical explanation. It was *real*, a genuine angel.

As the lighting conditions gradually improved and his vision was less obscured, what he thought was an angel more clearly came into view. An American Gemini spacecraft was seemingly hovering motionless less than a hundred meters away. The shimmering light he saw moments earlier was a reflection from the grey-painted adapter section.

An American spacecraft? If an angel's presence was miraculous, this occurrence wasn't too far behind. He tried to comprehend how it came to be here, and his only logical explanation was that it was a rescue mission, launched on extremely short notice. While it made little sense that the Americans would send up a spacecraft to save him, he was very aware that a joint Soviet-American mission was being planned to test rescue procedures.

With their immense wealth and enormous resources, perhaps the Americans had already developed a rescue system; maybe the Gemini was dedicated to the task and had been maintained on constant pad alert. But a Gemini? As a two-seater, it didn't seem logical as a rescue vehicle. On the other hand, the Gemini spacecraft was proven technology. Besides, the vessel out there might have been flown by one man, with the other seat left unoccupied for Vasilyev's return. As he studied it, he realized that the spacecraft wasn't exactly the version that NASA had flown a few years ago; for one thing, this Gemini's nose was considerably larger than NASA's version.

He attempted to look through the windows, but the reflections off their glass, combined with his blurred vision, made it virtually impossible to determine how many men were aboard. Without question, the plan surely was for him to don his SK-1 suit, exit the *Krepost* via the airlock in the docking hub, and then transfer over to occupy the vacant seat. It would be an extremely risky gambit, since although his SK-1 suit was equipped with an emergency oxygen supply, its duration was roughly fifteen minutes at best.

Since he was marginally fluent in English and knew Morse code, he should be able to communicate with the Gemini pilot, even if they could not converse over the radio. First and foremost, he had to make sure that the Gemini was here to rescue him. He checked his penlight and composed his message. It had to be extremely clear, concise, and leave no room for error.

As he waited for an opportune moment, the Gemini gradually maneuvered until its truncated nose was pointing directly at him. He could clearly discern the faces of *two* astronauts looking through the windows. Dismayed that it was likely not sent to deliver him back to Earth, he sent his terse message anyway. Slowly blinking the flashlight in Morse letters, he spelled out RESCUE followed by the Morse punctuation signal—Dot-Dot-Dash-Dash-Dot-Dot—that denoted a question mark.

Minutes passed as he anxiously waited for a reply. Finally, he watched a light blink from the left side window, and quickly scrawled down the letters. Like his message, the Americans' reply was short and simple: NO.

Gemini-I, On Orbit
14:35 p.m. GMT

Carson and Ourecky had been shocked to see the cosmonaut's face in the window, but probably just as confusing was his apparent belief that they had been sent up on a rescue mission. Ourecky's ferret equipment showed two recent transmissions, but they suspected that the signals were associated with some automated telemetry system. It was almost impossible to believe that someone could be alive over there.

As they waited for a reply, they discussed alternatives to deal with the unfolding situation. "Drew, you're the boss up here, but are you sure we're doing the right thing?" asked Ourecky. "Should we have been so quick to let him know that we're not on a rescue mission?"

"I don't think so," said Carson. "Look, let's review the things we know. We have to assume that this guy's no dummy, or he wouldn't be up here. He knows Morse code. He obviously knows at least some English. His ship is broken and he has no way home. Agreed?"

"Agreed."

"He has to be smart enough to know that we're not in a position to give him a lift," said Carson. "So he probably has long since figured out that we didn't come up here to sell him encyclopedias. Most importantly, he still has a nuclear warhead at his disposal, and we're essentially dead in the water, at least for the moment."

"So what do we do?" asked Ourecky.

"We *bluff*," answered Carson, drafting a message on his kneeboard. "How's this?"

"Strongly worded, but certainly gets the point across."

Krepost Station, On Orbit
14:38 p.m. GMT

Vasilyev read the Americans' message: DO NOT DEPLOY NUCLEAR WEAPON OR WE WILL DESTROY YOU, followed by the Morse three-letter "procedure word"—QSL—directing him to acknowledge. He copied down the letters; the message certainly appeared to be a threat, but Vasilyev wasn't confident enough in his English to be absolutely sure.

He remembered the phrase books and translation guides that Gogol had stashed aboard the freighter. He raced to the docking hub. Gripping his penlight between his teeth, he rummaged through Gogol's satchel. He remembered that Gogol had stowed his maps and books in a mesh sack that he had jammed in the bottom of the satchel. Digging down, he pulled out the sawed-down Kalashnikov assault rifle, the bag of medical supplies, the multi-band radio receiver, the sextant, Geiger counter and sundry other items, and finally found the flimsy sack containing the phrase books and other references. Clutching the sack to his chest, he rushed back to the control area to make a more definitive translation.

Carefully translating each word to ensure absolute clarity, he read the message. The Americans weren't mincing words; it was clearly a threat. The question was whether their threat was empty or they had some means to follow up on it. If they weren't dispatched to rescue him, then they certainly hadn't come up here for a picnic, so it was a logical assumption that they did have some sort of weapon to attack the *Krepost*. In Vasilyev's mind, that explained why their spacecraft—with its enlarged nose—did not look like a "normal" Gemini. Moreover, since their nations were at a state of war, or at least on the verge of war, then it made perfect sense that the Americans would attempt to destroy the *Krepost*. Given the timing, assuming that they were aware of the *Krepost's* mission, it was a threat they could not possibly ignore.

How should he respond to this potential peril? He thought that the answer rested in the last radio transmission he had received from Control, about an hour ago, in which General Yohzin had spoken to him directly. Although he wasn't always sure what to make of General Abdirov, and was sometimes concerned that the horribly scarred officer had lost his equilibrium, Vasilyev trusted Yohzin emphatically. Equally respected by the *Perimetr* engineers and *Krepost* program officers, Yohzin was the calming influence who seemed to keep everything in balance at Kapustin Yar.

In Control's last transmission, Yohzin plainly specified that all guidance and instructions remain in effect. Consequently, Vasilyev had to infer the dire reports and instructions transmitted on Channel Three were valid, so even though he had not personally detected any tangible signs of conflict, he had to assume the intelligence was valid and that he had to remain ready to deploy the Egg.

Before replying to the Americans, he went to the Egg's controls. Shivering, he opened the code safe, took out the targeting book, and leafed through its pages to find the coordinates for New York City. He entered the city's coordinates—40 degrees 46 minutes North by 73 degrees 58 minutes West—into the targeting computer. Now, if the Americans made a threatening gesture, all he needed to do was press the green ARM button, and the Egg would be on its way. Exhaling his breath in a mist, he sighed. He might be dead hours before the Egg navigated its way to its final destination, but his death would be avenged.

Grimacing, referring to the Russian-English translation guide, he composed his reply to the Americans' threat. He held the penlight against the porthole, clicked out the message, and then sent it again for good measure.

Gemini-I, On Orbit
14:52 p.m. GMT

"And here it is," said Carson, copying down the message: OUR NATIONS ARE AT WAR. I WILL FOLLOW MY ORDERS. YOU FOLLOW YOURS. QSL

"Our nations are at war," read Carson. "I will follow my orders. You follow yours."

"At war?" asked Ourecky. "Did I hear you right? Are we missing something here?"

"Well, we had better reply," said Carson. "Any thoughts?"

Ourecky scribbled down a message and showed it to Carson. "How about this?" he asked.

"Hmm. Scaling back the rhetoric is a good start for negotiations," he said. "Let's go with it."

Krepost Station, On Orbit
14:55 p.m. GMT

Vasilyev watched the blinking light and wrote down the message: OUR NATIONS ARE NOT AT WAR AND WE DO NOT WISH TO START ONE. DO NOT DEPLOY YOUR NUCLEAR WEAPON. WE DO NOT INTEND TO CAUSE YOU HARM. QSL

Vasilyev took a few moments to translate. He was perplexed by their message. *What? The Americans mean me no harm? And they claim that there is no war?* What a strange situation. Surely, they were attempting to deceive him.

As a Soviet officer, he had long since been conditioned to blindly obey orders and to accept reports without question, but the intelligence reports still troubled him. There should be at least some correlation between the Channel Three reports and what he witnessed through the portholes, but there was none. He wished that he had some way to verify the reports. And then he realized that he *did* possess the means to do so. He held his flashlight against the window and blinked out a one-word message: WAIT.

He made his way up to the docking hub, where he found Gogol's multi-band radio receiver. He returned to the control area, switched it on, but it didn't work. Clumsily, he pulled out the batteries from the radio and warmed the batteries under his armpits. He shook uncontrollably; the frigid batteries felt like ice cubes. He waited a few minutes; hoping that they had thawed at least somewhat, he inserted them back into the radio and tried again. He heard static, and then the faint sounds of a piano concerto.

Slowly rotating the frequency knob, he periodically paused to listen to different radio stations, most broadcasting in English. The news he heard absolutely contradicted the information he had received on Channel Three. As he had earlier suspected, the intelligence reports had to be bogus. The Americans *were* telling the truth; except for some isolated pockets of conflict, the world was clearly at peace. His own countrymen—Yohzin included—were lying to him, apparently using him as an unwitting stooge to instigate a war with the West. He was distraught. With this knowledge, how could he ever go back home?

As he listened to the BBC international news broadcast, Gogol's notebook floated by his face. He picked it out of the air, opened it, and perused Gogol's intricately detailed reentry instructions. Flipping through the pages, an idea came to his mind.

Gemini-I, On Orbit
15:32 p.m. GMT

Over thirty tense minutes had elapsed since the last message. As they waited, they decided to prepare for Ourecky's EVA. Ourecky had pulled the ELSS chest pack out of its bracket and had wedged it between his thighs to inspect it. Munching on a Fig Newton, Carson S-rolled the right seater's umbilical as he kept an eye on the *Krepost's* porthole.

"There's a light," blurted Carson. He quickly scribbled down the message: I WILL NOT DEPLOY WEAPON. I NEED ASSISTANCE. WILL YOU HELP? QSL

"Well, this is certainly interesting," noted Ourecky.

Without hesitation, Carson held his flashlight to the teardrop-shaped window and transmitted: YES. WE WILL HELP.

"He's sending another message," said Carson moments later, jotting down text. "It looks like his radio frequencies."

"No good," declared Ourecky, looking at the numbers written on Carson's kneeboard and shaking his head. "These are outside our range. The ferret gear can show us when he transmits on these frequencies, but we can't hear him or talk to him."

"Well," said Carson, unwrapping a stick of gum. "There's just one logical solution, then. Let's pass this ball downstairs and let them put it in motion."

Gemini-I, On Orbit
17:40 p.m. GMT

It had only been two hours since they had passed the information to Mission Control at Wright-Patt, but Carson was definitely aware that the ground had established contact with the *Krepost's* cosmonaut and that plans were already beginning to gel. At this point, he and Ourecky had effectively been pushed to the sidelines, and there were plenty of indications that Tew might call them home sooner than later, even before the situation with the *Krepost* was entirely resolved. After all, there was little that they could contribute to the situation now.

"I want to go outside," announced Ourecky.

"You've ridden in this bucket *how* many times?" asked Carson. "I would have thought you were well past your claustrophobia by now."

"Well, let me be more specific," said Ourecky. "I think I can fix the Disruptor. That's why I want to go outside on EVA."

Carson shook his head as he contemplated Ourecky's comment. *It just wasn't worth the risk.* Over the course of eight missions, *nine* including this one, they had witnessed more than their share of technical glitches, and had been very fortunate to have survived a few of them. Merely opening the hatch entailed a huge amount of risk; if it failed to close or properly seal after Ourecky returned from his jaunt outside, they could not safely reenter. The simple fact was that the longer they stayed up here, the more likely it was that something would go wrong. It might start as something minor, like a failed relay or a stuck solenoid, and then quickly escalate into something catastrophic. So, if granted the opportunity, Carson would gladly light his retros and go home at the earliest possible opportunity. In his mind, the time for orbital heroics had long passed, and he was anxious to hear Tew order them to reenter.

"Nope," said Carson. "It's obvious that the Disruptor's not necessary anymore. The ground has this game now, and all there's left for us to do is be passive observers. Even then, we're still pretty damned superfluous, because whatever is going to happen is going to happen whether we're here or not."

"Maybe, but everyone is assuming that our comrade will follow through on his promises," said Ourecky. "What if he doesn't? Right now, if he decides to deploy that weapon, there's nothing we can do to stop him, short of ramming it. Fixing the Disruptor would at least give us an option if worse comes to worse."

"Agreed," replied Carson, looking out the window at the ominous *Krepost*.

"Moreover, we don't know if this thing can be controlled from the ground," said Ourecky. "If that's the case, it really doesn't matter how much help we're able to lend our new friend over there, if his bosses decide to drop that bomb without his knowledge.

"Excellent point," observed Carson. "But let's be realistic, Scott. Tew is not going to approve an EVA. It's just too dangerous."

"I agree, but I also know that as hazardous as it might be for us, there are still millions of lives at risk downstairs. I'm guessing that something simple has the Disruptor jammed up, and I'm confident that I can fix it. Once it's fixed, we can deploy it on that warhead next door, and Ivan would never know, but doing that would give us a hole card if this deal suddenly goes south. Agreed?"

"Agreed," answered Carson.

"Well then," said Ourecky, reaching back to unstow the ELSS chest pack. "I guess we just have to resort to our normal way of doing business. On the next comms window, we ask for permission to do the EVA, and if they shoot us down, then we go ahead and do it anyway. Sound good to you?"

"Yep."

"Then if you don't mind, you want to prep my umbilical? The faster we make this happen, the better."

Gemini-I, On Orbit
19:03 p.m. GMT

Less than ten minutes since he opened the hatch to venture outside, Ourecky was hovering in the adapter at the far rear of the Gemini-I spacecraft. He took great care to remain absolutely clear of the Disruptor's boom, since they had already triggered its deployment; a combination of compressed gases and high tension springs strained to swing it out to its deployed position, but all that potential energy was apparently held in check by some sort of malfunction in the latch mechanism. If the boom spontaneously deployed while he was back here, his helmet's visor could be shattered, his umbilical could be snarled or any number of other misfortunes could befall him, killing him immediately or at least preventing his safe return to the snug sanctuary that was the Gemini-I's cockpit.

Resisting the urge to rush, Ourecky moved slowly to keep his respirations and heartrate under control. It had been weeks since he had walked in space to transfer from the Gemini-I to enter the MOL to rescue Russo, and even longer since he trained for EVA while immersed in the massive Tank in New Orleans, but he was abundantly conscious of the need for patience and restraint.

His suspicions were confirmed as he examined the latching mechanism that was intended to hold the boom in check before deployment. The latch had not completely opened; a tiny sliver of metal, perhaps a sixteenth of an inch wide, prevented the Disruptor boom from deploying.

Ourecky gradually made his way to a makeshift "toolbox" mounted in the adapter, selected a implement that resembled a metal pry bar, and floated back to the latching mechanism. He checked again to ensure that he and his umbilical were safely clear of the boom's path, carefully inserted the tool's blunt tip into a gap in the latching mechanism, and gave it a slight twist.

Now set free, the boom immediately sprang to life, swinging out of the adapter and properly locking into place. The rest of the Disruptor's components followed suit.

"Drew?" he asked over the intercom. "See it?"

"I do," replied Carson. "Now, get back in here before I have to come get you. Good work, Ourecky."

Krepost Project Headquarters
7:40 p.m., Sunday, October 22, 1972

Several key personnel met in Abdirov's office to brief the general on the newly revised flight. Yohzin saw that it was patently clear that no one had previously expected Vasilyev to survive to this point, and now they were compelled to make rapid adjustments to accommodate his continued participation in the mission

Clearly terrified of Abdirov, Bogrov was the only representative from *Perimetr;* obviously he was being offered up as the sacrificial lamb.

Abdirov called the group to order, and then asked, "Is Vasilyev aware of the issues with the interlock?"

"We don't think so," answered Yohzin. "He still has his hands full dealing with other issues, since the crisis is far from being resolved. He might be aware that the targeting computer malfunctioned and caused the spontaneous alignment of the platform, but we don't believe that he understands that the interlock is currently disengaged."

"But we have talked to him since this interlock issue surfaced, correct?" asked Abdirov.

Yohzin answered, "Comrade General, be aware that I personally spoke with him on the last contact window. He didn't mention the interlock, and I avoided calling his attention to it. I have instructed the others in Control to follow suit."

"Very good, Gregor," said Abdirov. "So, once he has docked and gained access to the station, Gogol's first priority is to reactivate the interlock. Correct?"

"That's affirmative, Comrade General," answered Bogrov, clearly avoiding eye-to-eye contact with the disfigured officer. He held up a screwdriver, pointed at its unique tip, and explained, "In order to gain access to the power board, Lieutenant Colonel Gogol will need this special tool to undue the fasteners on the Egg's panel. Beyond that, it's a relatively simple procedure. We will ensure that he is ready before he departs tomorrow."

"Good. And you are comfortable with this?" asked Abdirov, pivoting toward Gogol.

Gogol answered with a grin, displaying his collection of stainless steel caps, and a nod.

"But, Comrade General, can we trust this man to reactivate the interlock?" asked Bogrov timidly. "After all, we could be placing the fate of the entire world in his hands."

"I trust him, and it's *my* decision," blurted Abdirov. "But if you want to question *his* reliability and *my* judgment, then perhaps we can stuff you in the Soyuz in his stead, and shoot *you* up there to mend your damned gadget. Would that suit you?"

Cowering like a lamb before a wolf, Bogrov trembled as he meekly answered, "*Nyet*, Comrade General. I fully trust Lieutenant Colonel Gogol to execute this task."

"That's good, because if you recall, you idiots claimed that your damned interlock could pose no threat to my crew!" he bellowed. "That was clearly erroneous, since my people are now compelled to clean up after your mess."

Abdirov obviously relished the opportunity to berate Bogrov. The previously untouchable *Perimetr* people had been abruptly knocked down from their high pedestal, and Abdirov had resumed his dominance. It had only been hours since the interlock had failed, but there were already rumors that the unreliable *Perimetr* equipment might be scrapped from future *Kreposts*.

"Since our circumstances have changed so drastically in the past twenty-four hours, what is the rest of our plan?" asked Abdirov.

The flight director answered, "Until the next freighter arrives, Gogol and Vasilyev will inhabit the Soyuz, and will enter the *Krepost* only when absolutely necessary. To continue the mission indefinitely, we will need to clear the docking port that is currently obstructed by the damaged freighter. My engineers are fabricating some special equipment for this, which will be stowed on the next freighter.

"Once the new freighter arrives, Gogol and Vasilyev will seal the docking hub from their Soyuz and the *Krepost*. Comrade Gogol will be suited up and will remain in the docking hub. He will evacuate the docking hub's atmosphere and then conduct an internal spacewalk to enter the damaged freighter and close its nose hatch. He will mount a small remote-controlled rocket motor, which our engineers are currently fabricating, to the nose hatch.

"Once the docking hub hatch is sealed, Gogol will repressurize the docking hub, and then he and Vasilyev will over-pressurize the access tunnel and then jettison the freighter. The over-pressure should shove the freighter well clear of the station, and then Gogol will remotely fire the rocket motor to brake its orbit. Afterwards, they will transfer cargo from the new freighter, activate its fuel cell, and we should be able to resume normal operations."

Yohzin studied Gogol as the flight director recited the revised flight plan. Picking at his fingernails, Gogol wore a bored expression. Yohzin suspected that the stocky cosmonaut had heard the new plan so many times that it was entirely second nature, or he might not be paying attention because he knew the plan would never actually be implemented. Yohzin hoped for the former, but resigned himself to the knowledge that Gogol was still programmed to drop the Egg.

"So, we are prepared to execute this mission?" asked Abdirov. "Do you have any significant discrepancies to report?"

"We are experiencing some minor problems with the R-7 launch vehicle," answered the flight director. "The pad crew is seeing some erratic readings on the first stage oxidizer lines."

"Will that preclude you from launching on time?" asked Abdirov.

"Probably not, Comrade General, but although the optimal launch window is tomorrow, we still have an acceptable window the following day. Since Vasilyev is in good health and the *Krepost* is currently stable, we would like your permission to delay the launch one day."

Abdirov's face instantly turned red. It looked as if he were on the verge of detonation. "That is unacceptable!" he shouted angrily. "We will *not* delay! You will be ready *tomorrow*, or you will be digging for turnips in Kamchatka by the end of the week. *Do you understand?*"

Yohzin was momentarily perplexed at Abdirov's unusual behavior, and then remembered that within the Soviet space program, it was expressly forbidden to launch on October 24. The date was considered incredibly unlucky because an R-9 rocket had caught fire, killing nine pad workers, on October 24 in 1963. More importantly, October 24 was the anniversary of the infamous Nedelin catastrophe, in which the second stage of an experimental R-16 ignited prematurely, instantly immolating scores of the Soviet Union's most proficient rocket scientists. Abdirov incurred his terrible injuries on that tragic day, so the date's significance was certainly more than merely symbolic to him.

Grabbing the flight director by the back of his lapel, Yohzin yanked him down into his seat. He leaned over his shoulder and whispered, "Shut up, idiot. Look at your calendar."

"Anything else?" asked Abdirov.

No one spoke.

"Then get back at it," he said.

As the others departed, Abdirov gestured for Gogol to remain.

"This new *Perimetr* development is intriguing," said Abdirov quietly, placing his disfigured hand on Gogol's shoulder. "But your ultimate task has not changed, not in the least. Understood?"

"Completely, Comrade General."

"Are you ready to leave tomorrow? All of your affairs are in order?" asked Abdirov.

"Entirely, Comrade General."

"Then we will see you off tomorrow evening. Get some rest tonight."

Yohzin waited for Gogol to leave and then asked, "A word, Rustam?"

"What is it, Gregor?"

"Rustam, surely you know that I have more hands-on experience on the R-7 than anyone else assigned to this project, right?"

"I'm sure that's true."

"Since I am at a lull with my other chores, maybe I could lend the launch crew some assistance with resolving the issues with their rocket."

"Excellent idea, little brother!" declared Abdirov. "I know that I can always count on you. But, listen: Luba and the boys leave tomorrow, don't they? On the noon train, correct?"

"That's right, Rustam. Thank you for your help in expediting their travel documents. Your signature carried a lot of weight with the KGB. They approved the papers without delay."

"Excellent," said Abdirov. "Listen to me, little brother: don't stay here all night working. Go home to see your family. I am confident that you can fix this rocket tomorrow, and then we can get on with what needs to be done."

Residential Complex # 4, Znamensk (Kapustin Yar-1)
11:45 p.m.

Try as he might, Yohzin couldn't sleep. Just before midnight, he slipped out of bed and quietly padded to his study. He locked the door behind him and lit a cigarette. He retrieved his miniature shortwave receiver, strung up its wire antenna, and switched it on to await the message.

As he waited, he thought about the events that would unfold tomorrow. In less than twelve hours, certified travel documents in hand, Luba and his sons would board the train to begin an arduous four-day journey to the west. With several stops and transfers, the railroad would take them as far as Odessa, from which they would proceed further west by bus or car.

The final leg would be by a donkey-drawn cart, proceeding along a rough dirt track, to a rustic single-room cabin in the wilderness. Luba's parents resided in the remote setting because her father served as a watchman of sorts for a large natural gas pipeline. He had one simple chore, to daily read a set of gauges at a pressure monitoring station; if the readings were out of range, he would use an ancient Navy wireless radio to contact the pipeline's control station at Chisinau. And that was it. Beyond that, the old couple had plenty of time to tend to their vegetable garden, hunt wild game, and enjoy their solitude. A wagon occasionally came by to deliver supplies and the mail, but that was the extent of their contact with the outside world and other humans.

If the truth be known, the assignment was simply a make-work posting, awarded to him because of his valiant military service. Luba's father had been a highly decorated tank officer who had risen from platoon commander to division commander over the course of the Great War. He hadn't been a brilliant tactician, by any means, but his incredible personal bravery inspired his men to victory in several momentous engagements. He always led from the front, taking the lead tank into battle. He was devoted to his men and they were devoted to him; they would follow him anywhere, straight through the portals of hell if need be. Since his tank division was largely comprised of former militia units from his region, he knew many of the men personally. He had grown up alongside them, and counted scores of them as friends.

In the course of several battles, his division was brutally decimated during the summer counteroffensive in 1944. Hurled against overwhelming German firepower, all of his tanks were destroyed and over ninety percent of his men were killed. Terribly wounded, he should have died with them, but his ever-faithful subordinates would not let him perish, despite his pleadings to leave him be. He was evacuated to a field hospital and spent years after the war recuperating. He was entirely broken by his wartime experiences. Ghosts followed him everywhere he went, and he fell down weeping whenever he encountered widows of his subordinates. His behavior deteriorated to the extent that he could no longer be around other people, and his wife—Luba's mother—begged the government officials for some sort of relief. And that's how they came to settle in the wilderness, since he had served the Motherland with such bravery.

Yohzin's sons had never met their grandfather, and they were anxiously looking forward to the trip as if they were embarking on a grand adventure. They were, of course, but certainly not what they expected.

Tomorrow, as his family travelled, Yohzin intended to finally resolve the crisis he had caused. His plans had drastically changed in the past few hours. Until he learned of the problems with the R-7 booster, Yohzin had intended to wait until his family was safely away from Kapustin Yar and several hours to the west, and then he would go to the GRU before Gogol launched. He would confess to stealing the secret code from *Perimetr*, and describe how Abdirov was plotting to use the code to drop the Egg. In short order, in Yohzin's mind, the GRU would immediately inform the General Staff of the High Command, and the launch proceedings would be halted immediately.

Yohzin knew that certainly he wouldn't emerge unscathed in the aftermath, since he was still guilty of treason, no matter how the onion was sliced. He would probably join Abdirov on the gallows when the accounts were settled and this ordeal finally ended.

Today's problems with the R-7 had yielded a serendipitous opportunity, one in which he might be able to derail Abdirov's scheme and yet still survive. After visiting the pad earlier and conversing with the launch crew, Yohzin knew exactly where the problem resided, and also how to quickly correct it. Tomorrow, under the pretense of helping the launch crew, he would bide his time until he announced that he had found and corrected the problem. While he was at it, he would sabotage the R-7. In his long experience with the venerable booster, he knew well its idiosyncrasies and vulnerable points. He was confident that he could place a tiny snippet of wire in a critical valve in such a manner that it would be entirely undetectable, but it would cause the rocket to explode moments after it left the pad. And the best part was that if his new plan came to fruition, the only casualties would be Gogol, whose demise could certainly not be counted as a great loss to humanity, and Vasilyev, since it was highly unlikely that another mission could be mounted quickly enough to save him.

If for some reason he was unable to sabotage the R-7, he could still resort to his original gambit—confessing to the GRU—as a fallback. He sighed, knowing that while he could not predict his own fate in the coming days, he was confident that the world would not be pitched into a thermonuclear war.

Yohzin checked his watch, stubbed out his cigarette in the ashtray, switched the radio on, adjusted the dials, and listened as an announcer read a repeating stream of illogical phrases and seemingly nonsensical strings of numbers. In a few minutes, his heart jumped as he heard the droning voice announce the set of words and digits intended specifically for him. His plan was *approved*. The die was set. All that mattered now was to send Luba and the boys on their way.

He took down the aerial, carefully stashed the radio, and returned to the bedroom.

He pulled the blanket up to his chin, turned toward Luba, and held her close. "Wake up," he whispered, hoping not to wake the boys. "I need to tell you something important."

Yawning, she stirred. "What, dear? What is it?"

"No matter what happens, you *must* leave this place tomorrow," he confided. "It's not safe to remain here. Promise me that you will be on that train at noon."

"Of course, darling, I promise. What in the world has you so bothered? You've been tossing and turning all night, so I know that you haven't slept a wink."

"Listen," he said. "I have done something terrible, and I must undo it tomorrow, or many people will suffer."

"What could possibly be so terrible?"

"I cannot tell you, but please just trust me, Luba. No matter what happens, leave here tomorrow. It will be too dangerous to stay. Don't be surprised if you see people acting strangely. If they do, stay your course and make sure that you and the boys get all the way to your parents' place before you stop. That is the only safe place for you. Will you do that?"

"I will, Gregor."

"Promise me," he demanded.

"I promise."

There's something else," he said. "In the coming days, at some point, a man will come to see you. He will ask you what it is the fastest way to travel to Tiraspol. That's how you will know that he is the right man. You should answer that he can travel by train, but an airplane would be faster. That's how he will be sure that he should talk to you. No matter where you are, no matter how ludicrous it might seem, that's how you should answer."

"That sounds so silly," she said. "You are causing me to worry, dear. Please stop it."

"Remember that he will ask you the fastest way to travel to Tiraspol," he reiterated.

"Tiraspol," she said.

"And how do you reply?" he demanded.

"I will tell him that he can travel by train, but an airplane would be quicker."

"Good. You can trust him. Do whatever he says, no matter how crazy it might seem. He will be there to protect you and the boys. You *must* do what he says, without question."

"But how about you?" she asked.

Closing his eyes, he swallowed deeply and replied, "I'll follow along as soon as I can."

"You know I love you, Luba, right?" he asked. "You and our sons are my entire world. Please do these things for me."

"I will. I love you too."

Mission Control Facility
Aerospace Support Project, Wright-Patterson Air Force Base, Ohio
6:53 p.m., Sunday, October 22, 1972

The large room had been overtaken by a flurry of activity since Carson and Ourecky had established contact with the Soviet cosmonaut roughly eight hours ago. Even as they intently monitored the ongoing Blue Gemini mission, Heydrich and his mission control team orchestrated a crash project to render assistance to the Soviet pilot, who had since identified himself as Major Pavel Vasilyev.

The cosmonaut was communicating directly with the Project's far-reaching web of communications sites, planes and ships. To ensure greater clarity, the Project was being augmented with a battery of Russian linguists, borrowed from several Air Force Security Agency sites around the world. Other technical specialists were being drafted as well. As they arrived and were assigned to compartmented work areas, the reinforcements were told exactly what they needed to know to perform their jobs, and nothing else.

As they watched the hectic proceedings, Tew and Wolcott waited in their glassed-in office at the rear of the facility. A visitor—Major Ed Haney—had just arrived from NASA's Manned Spacecraft Center in Houston. Haney was an Air Force officer notionally assigned to NASA as a military liaison. The nephew of a prominent New England senator,

Haney possessed an impeccable social pedigree, and it was a virtual certainty that he would be eventually selected as an astronaut. Glancing through his personnel file summary, Wolcott was confident that Haney intended to use his pending astronaut experience as a springboard to a political career. Of course, mused Wolcott, Haney might have to be very patient as he bided his time, since NASA didn't appear inclined to hire on any new hands, particularly since their manned spaceflight program was on the verge of stalling.

Wolcott grinned at the frantic Gunter Heydrich and then spit tobacco juice into an empty Mountain Dew bottle. He generally despised guys like Haney, but they were presently compelled to deal with the sycophant because of a unique aspect of his background. By happenstance, Haney was a quasi-authority on the Soviet Soyuz spacecraft. To occupy his time, since neither NASA nor the Air Force made any significantly pressing demands on his work hours in Houston, he was actively attending the joint NASA/Soviet meetings associated with the ASTP—Apollo-Soyuz Test Program—effort, which was intended to develop an international space rescue capability. According to Wolcott's contacts at NASA, Haney was also taking Russian language lessons, in hopes of eventually flying on the ASTP test mission, in which a specially configured Apollo spacecraft would dock with a Soyuz in Earth orbit. His was a very ambitious goal, thought Wolcott, since he hadn't even been fitted for a space suit yet.

Still clad in sweaty Nomex, Haney had come straight from the flight line after landing his NASA T-38. "Gentlemen, I'm Ed Haney," he announced, throwing up a half-hearted salute before plopping an aviator's chart box on the table and casually taking a seat. "I didn't get a briefing before I left Houston. How might I be of help?" Haney was a couple of inches shy of six feet, with an athletic build and handsome features. He wore his dark brown hair in a crew cut.

"Simply stated, Haney, we're in contact with a Soviet cosmonaut aboard a Soyuz spacecraft," said Tew. It was obvious that Tew was equally disgusted by the would-be astronaut. "He has requested our assistance, and time is of the essence."

Haney chuckled, and then said, "Obviously there's been some mistake, General. The Soviets don't have anyone in orbit right now. The last Soyuz mission they sent up was in June of last year, and those three fellows died during reentry. Besides, I'm confident that you're aware that if the Soviet space agency needed American assistance, they would be communicating with NASA."

"No, Major, there has been *no* mistake," replied Tew. "Moreover, this cosmonaut apparently wants to terminate his flight here in the continental United States, so we need your technical expertise to make that happen."

"Oh," said Haney, laughing again. "This is *so* rich! I get it now, General. This is some sort of exercise, right? No one told me that I was coming here for an exercise."

Since Tew appeared ready to spontaneously explode, Wolcott answered. "It ain't an exercise, bub," he declared impatiently. "Now, I know you ain't acquainted with the way we do business here, but we need you to suspend your disbelief and focus on what General Tew is tellin' you. The faster you latch on to that concept, the faster we'll resolve this situation so you can hop back on your white pony to zoom back to Houston to hang out with your high falootin' NASA buddies."

"Yes, sir," replied Haney.

Moreover," added Wolcott, "the next time you see fit to laugh at something the General says, unless he's obviously tellin' a joke or relating an amusing anecdote, which ain't ever goin' to happen, I'll be obliged to rap my knuckles across your forehead. Savvy, pard?"

"Understood, sir. Sorry that I got off on the wrong foot." Haney's eyes grew big as he glanced at some of the pictures on the wall behind Tew and Wolcott.

Tew handed him a Teletype transcription of a translated exchange between Vasilyev and a Russian linguist aboard an EC-135E "Snoopy" ARIA tracking aircraft flying over the North Pacific. "This is extremely awkward," said Tew. "It appears that our friend upstairs seems intent on dictating his landing site to us."

"He seems to fancy New Mexico," noted Wolcott, spreading out a large-scale topographic map of the southwestern state. "That's where those coordinates would have him come to Earth."

Haney seemed transfixed on the reentry instructions in the Teletype print-out. "Your cosmonaut is reading this down?" he asked. "Sir, this makes no sense whatsoever."

"How so?" asked Tew.

"General, prior to each Soyuz mission, a set of mission reference books are packed aboard the spacecraft," explained Haney. "One of the books contains detailed instructions for reentry and recovery, for their primary landing sites as well as their alternative sites in case they experience an emergency.

"The format of *these* instructions corresponds with how they are formatted in the reference books, but I can assure you the Soviets haven't yet designated any primary or emergency landing sites in the United States. They might do so in the future, depending upon how the ASTP program shakes out, but at present I think they would prefer that their guys die outright than endure the humiliation of seeing one of their spacecraft land on American soil. I think that they would rather put a Soyuz down in the ocean than surrender one to be picked apart and studied."

Haney studied the map of New Mexico and added, "Moreover, sir, if the Soviets did plan for an emergency landing site in the United States, you would have to assume that it was intended for extreme circumstances, and that they would notify us so that we could deploy the appropriate assets to locate their crew, right?"

"That makes sense," replied Tew.

"Agreed," added Wolcott.

"If that's the case, sir," said Haney, pointing at a spot on the map, "why in the world would they want their guys to come down into a remote wilderness area in the middle of nowhere? Wouldn't it make more sense for them to drop into an area where there was a greater likelihood that they would be quickly found?"

13

GAS LEAK

GRU Internal Security Office, Kapustin Yar Cosmodrome
2:50 a.m., Monday, October 23, 1972

Federov was terribly concerned, so much so that he ached in the pit of his stomach. In the month he had been assigned at Kapustin Yar, he had completely reorganized the Internal Security Office's surveillance operations, but his carefully drawn net had yielded no fish. He had endured countless hours of painstakingly analyzing surveillance films and photographs, but he hadn't yet uncovered anything even remotely suspicious.

He clenched his fists and groaned. He was *absolutely* convinced that some traitorous malcontent was passing secrets to enemies of the State, and he would remain steadfast at his station until he found the culprit, but the question remained how long he would be allowed to pursue the task.

His superiors at the Aquarium were demanding results, and he had nothing to show for his efforts. It was only a matter of time before he was called back to Moscow—yet again—and his return would not be triumphant. There would be no glorious posting to the Bureau of Special Cooperation; in fact, he would be fortunate if he wasn't dispatched to some miserable gulag in Siberia. It was true: if he wasn't successful here, then he might spend the remainder of his military career guarding political prisoners. Or worse, he might find himself huddling with the hungry wretches within the frozen wire, looking out instead of in.

This whole episode was an incredibly frustrating experience. To a large extent, his hands were fettered, since he was only permitted to monitor the scientists and engineers in the residential area located within the cosmodrome. He had repeatedly demanded authority to conduct surveillance in the work areas and test sites, but his requests were met by steadfast resistance. If his bosses really wanted results, why didn't they give him full rein to do those things that needed to be done?

It was late, and his eyes ached from scrutinizing film, but he vowed to relentlessly study the materials until he uncovered a clue. Of the ten men suspected by the GRU, only the four currently stationed at Kapustin Yar were still being investigated. To use his time effectively, Federov would intensely focus on one subject at a time. Since he might be at this for days, he decided to concentrate his initial efforts to eliminate the *least* likely suspect, Major General Yohzin of the RSVN, before he moved on to the other three potential culprits.

Early in their investigation, the Internal Security Office had considered Yohzin as their *most* likely suspect, primarily because of his previous affiliations with German rocket scientists, but the cloud of suspicion had largely passed away from him. Yohzin was the picture of anyone *but* a despicable traitor. The general labored like a tireless automaton; he normally worked seven days a week and rarely took a day off. He and his wife didn't socialize with other couples. His two boys diligently applied themselves to their academic studies; they were so dedicated to their books that they didn't even participate in sports or games.

Once he came home from work, he only emerged—usually with his spouse—to walk his dog. Throughout the course of their walks, Yohzin made no furtive gestures, and his hands were always clearly visible. And that was the principal reason that Federov wanted to quickly eliminate him as a suspect and move on to the others. Federov had an innate proclivity for catching people in the act of passing secret messages, and he knew that it was virtually impossible to pass a message except by either handing it to someone else or physically placing it in an intermediate location—a dead drop—where it could be later retrieved. He had studied the films of Yohzin for hours, and the general did neither. Likewise, he had watched films of Yohzin's wife, Luba, as she went to the market and other locations during the day, and it was clear that she was not passing messages on his behalf.

As a high-ranking officer, Yohzin was able to travel more frequently than most, so there was always the possibility that he somehow passed messages when he and his family occasionally went to Moscow. Consequently, the GRU had subjected them to intensive surveillance while there—literally deploying squads of counterespionage agents to shadow the general and his family—but none of them had done anything even the least bit suspicious.

Yawning, Federov threaded an 8-millimeter film into a hand-cranked editing machine, which would allow him to examine the film frame by frame, if need be. Yohzin was so uncompromising in his routine that it was painfully boring to watch the films. It was more monotonous than watching potatoes grow.

He taped a large sheet of graph paper to a table. Using a red pencil, he painstakingly mapped the route of the Yohzins' evening stroll. Once he had defined the couple's predictable circuit, he went into the next step of his analysis. He would watch the films of other subjects—people entirely unrelated to Yohzin or the other three suspects—to look for the intersection, the point at which Yohzin's path coincided with someone else. If Yohzin was passing messages—which Federov strongly doubted at this point—the intersection would be the place where the message drop would be found.

Through the night, he cranked through film after film. He could only imagine the exorbitant cost of this project, which had been in full swing long before he arrived on the scene. Thousands of rubles must have been expended solely to purchase and process the seemingly endless library of movie films. If nothing else, if the project was audited, he should not have to account for those expenditures.

In the wee hours of the morning, he was studying one of the last individuals who routinely traversed the commons area where Yohzin daily walked his dog. The subject was an elderly woman who was brought here to clean the domiciles of several high-ranking officers. A war widow from Minsk, she lived alone in a servant's room in the attic of one of the apartment buildings. Illiterate and somewhat feeble-minded, a stranger in a place where strong intellect was so greatly prized, she kept mostly to herself. Reviewing the report that accompanied the films, Federov noted that the maid did not clean the Yohzins' apartment, nor did she have any contact with the other three suspects.

Not one to leave a stone unturned, Federov sat down at his desk and steeled himself for the task. Once he cleared the maid, he could turn his attention to someone other than Yohzin. Through his window, the rosy glow gave notice that the new day was arriving. He groaned, rubbed his weary eyes, and slowly turned the crank to study the film.

Like Yohzin, the housekeeper was a creature of highly engrained habit. Also like the general, she routinely worked seven days a week. Once leaving her meager quarters—bearing a mop, broom and pail—she went from one apartment to the next, rarely emerging into the daylight until her day was done.

Like most servants, especially the most accomplished at their humble duties, the old woman was essentially invisible, but for one minor exception. Every day, roughly at noon, she walked into the commons area, sat on a concrete bench, and ate her simple lunch. Federov studied her actions carefully; in this instance, the bench was such an obvious place for to conceal a message drop, but while the housekeeper was seated at the bench, her hands were always in plain sight. Besides, in all of the films he had studied, Yohzin and his wife had never occupied the same bench.

After the housekeeper finished her lunch, she habitually tore apart a scrap of bread and cast out a handful of crumbs to feed the doves.

As he spooled through the films, watching the crone's identical actions from day to day, Federov felt almost like ripping out his eyes out of sheer boredom. After innumerable iterations, he knew exactly how she untied the faded kerchief in which she wrapped her meal, and how she anxiously glanced from side to side before turning down her face to offer thanks for her blessings.

Straining to keep his eyes open, he was on the verge of moving on to one of the other suspects when he saw it, a gesture so subtle that it surely would have escaped the attention of anyone who lacked his highly refined powers of observation. On one of the films, after the maid threw out her crumbs to the pigeons, she stood up, slowly walked out a short distance into the grass, and then crouched down to pick up something. Federov gasped in astonishment; in his impatience, he had almost missed it.

He rummaged through the batch of reels that documented the housekeeper's activities, examining each film in turn, quickly advancing to the segment where she fed the pigeons before going back to work, and determined that she only took her abbreviated post-lunch stroll on one specific day of the week—Saturday—and each time, she knelt in the very same manner at the very same spot in the grass.

Grinning, he scrutinized his hand-drawn map of Yohzin's daily walk; he had identified the intersection, the key piece to the puzzle, so all that remained was to patiently find and assemble the rest of the pieces. At this juncture, he had due cause to haul Yohzin *and* the maid in for interrogation, but what fun would that be? There was so much more to be gained if he bided his time and played his cards carefully. The Bureau of Special Cooperation would be *his*; he knew it, as surely as the sun was now rising over the steppes.

Federov heard the door open in the outer office and knew that his desk sergeant was reporting for duty. The sergeant was the same man that he had brutally punched when he first arrived at Kapustin Yar. Despite his slovenly appearance, the hefty sergeant had proven to be quite an effective organizer and record-keeper, so Federov had elected to keep him here at the office, working at his same tasks, rather than putting him on permanent detail as a latrine orderly or stable minder.

Invigorated by his momentous discovery, Federov shoved open the door and strolled out of his office with a triumphant air. Perching a pot of water on a paraffin burner, the sergeant almost fell over as he looked up.

"Pull *everyone* in this morning," ordered Federov. "We'll meet in an hour."

Snapping to attention and throwing his best salute, the sergeant replied, "As you wish, Comrade Colonel, but most of the men will already be headed to their surveillance assignments. Are you sure that you mean *everyone?*"

"*Everyone.*" Uncharacteristically, Federov winked and grinned. "Today, we're taking a holiday from our regular spook work. Instead, we're going fishing, not with a net, but with a *hook*. And sergeant..."

"Sir?"

"If you would be so kind, *puzhalsta,* fetch me a glass of tea once you get it on the boil."

Wide-eyed, seemingly ready to topple over, the sergeant was speechless. Finally, he found his words and falteringly answered, "As you wish, Comrade Colonel. It would be my greatest pleasure."

Residential Complex # 4
Znamensk (Kapustin Yar-1), Astrakhan Oblast, USSR
10:10 a.m., Monday, October 23, 1972

Luba Yohzin stuck three pans of brown bread in the oven and closed the door. The loaves were for her and the boys to snack on during their journey. She checked their travel papers yet again, folded them neatly, and placed them carefully in her purse. Their two suitcases were in the living room, waiting by the front door.

A few minutes later, startled by the piercing wail of a warning siren, Luba Yohzin ran to the kitchen window, tugged back the curtains and peered outside. Initially, listening for an explosion and looking for a tell-tale plume of black smoke, she was horrified that there might have been an accident at the cosmodrome. *Could this be the terrible thing that Gregor had warned about?*

Her heart thumped as she thought that her beloved Gregor might be injured...or worse. Thinking of Gregor's terribly disfigured friend, Rustam, she remembered that there were some fates even worse than death. She prayed that if her husband ever was in an accident—today or any other day—that he would not have to suffer in the same manner as Rustam. As much as she loved Gregor, she knew that he would prefer a quick death over a life of prolonged misery.

With no discernible evidence of a mishap at the cosmodrome, she watched as people--mostly women and young children—streamed out of other apartment buildings to gather in the commons area below. Cupping her ear, she heard the sounds of heavy footsteps on the stairs and fists rapping at doors. She suspected that it might be just another routine emergency drill, except that the block captain usually provided at least some forewarning of drills. Just to be sure, assuming that she would soon join the crowd milling in the commons, she grabbed her coat and purse.

She heard a tapping at the front door. She promptly opened it to see a broad-shouldered, redheaded man in gray cotton coveralls. A blue metal tool box rested by his feet. "Gas leak, ma'am," he announced politely. "Somewhere in the complex. We're evacuating all of the buildings, just as a precaution, and then our fitters will ferret out the leak. You'll need to go outside to the commons area and wait for the all clear signal before you can come back in. It shouldn't take any more than an hour or so."

"Certainly," she calmly replied, tying a dark kerchief over her hair.

"Before you go, make sure that you've cut off the gas line to your stove. I'll show you where the valve is located, if you don't know."

Luba frowned. "I do know where the valve is," she said. "But I have bread in the oven. I'm afraid it will be ruined."

The tall man sniffed the tantalizing odor. "Sorry," he said. "Switch your oven off and make sure that the pilot is extinguished. And ma'am?"

"*Da?*"

"Is there anyone else here now?"

"*Nyet,*" she replied. "My husband is at work and my sons are at school."

"Good. This shouldn't take very long. I'm sorry about your bread, though. Oh...is there any chance that you have a pet here? Perhaps a cat or a dog? We've had problems before, when dogs have mistaken our inspectors for intruders."

She shook her head. "We do own a dog, but it's with my husband. He takes it with him to work every day."

Nodding, he smiled. "This shouldn't take too long, and you'll be able to get back to your chores. I'm really sorry about your bread. It certainly smells delicious."

As directed, she switched off the oven. *I guess we'll just be a little hungry today*, she thought. Stepping out of the kitchen, she glimpsed the suitcases waiting by the front door. Smiling, she casually picked them up started through the door.

"Wait," blurted the man, gesturing towards her valise. "What are those?"

"They are evacuation suitcases," she replied. Actually, that was a half-truth; the larger suitcase *did* contain their emergency supplies and her clothes, but the other just held her sons' clothes and a few of their books.

"Evacuation suitcases?" he asked.

"*Da,*" she replied. "They're for emergencies." Luba raised her eyebrows; she was very surprised that he didn't know about the rule, particularly if he worked with the gas department. "It's prescribed in the emergency protocols. Every family is required to have them, in case we are evacuated to a temporary shelter."

"Oh," he replied. "Well, we shouldn't be very long, so you can just leave them here. I'll see to it that they are not disturbed."

Standing on the stair landing, she pointed down, toward the growing crowd in the commons area. "But everyone has theirs. I'm afraid that my block captain will be checking, and my husband will receive a bad report if I don't have it with me. It's procedure, you know."

The man nodded reluctantly and replied, "*Da*. You're right. Bring them with you. I certainly wouldn't want for your husband to land in hot water."

"*Spasiba*," she replied. As she made her way down the stairs with the clumsy suitcases, she was very conscious that something was amiss. She was familiar with most of the maintenance workers who serviced the apartment complex and hadn't recognized the man who came to her door, and it was extremely suspicious that he wasn't aware of the rule about the evacuation suitcases. Moreover, she was almost certain that there were only two gas fitters who worked at the complex, and right now a small fleet of sedans and military trucks were parked on the street in front of the building, disgorging a veritable phalanx of "gas workers." People were definitely acting strangely, just as Gregor had predicted.

She recalled Gregor's stern admonition that she must leave Kapustin Yar and make her way to her parents' cabin in the wilderness, no matter what happened. To that end, she concocted a plan. She would blend into the swelling crowd of housewives and small children evacuated from their apartments, wait until things settled somewhat, and then walk to the train station, which was about two kilometers distant. A sedan from the school would deliver the boys at 11:30, and they should board the train shortly afterwards.

10:21 a.m.

After he spotted Yohzin's wife walking across the street, Federov stepped into the doorway, opened his tool box, extracted a portable "walkie-talkie" radio—about the size and weight of a large brick—and switched it on. As he waited for the radio to warm up, he cursed himself for not being more aware of the evacuation plans for the apartment complex. Extending the radio's antenna and speaking into the microphone, Federov summoned his technical search team.

As he waited for his men to arrive, he slowly strolled through the apartment to gather an appreciation of the layout. It was certainly not a palatial residence by any means. There were two bedrooms, a small living room, the kitchen, a small dining room, and a room that Yohzin apparently used as his study. He found it odd that Yohzin was satisfied with quarters that were more appropriately suited to a junior colonel or perhaps even a senior major; as a major general, he was entitled to a much more spacious unit, as well as a housekeeper and cook. It made little sense that he didn't requisition quarters commensurate with his rank.

Assuming that Yohzin likely performed his covert functions without his family's knowledge, Federov focused his initial search on the general's sanctum. Like the rest of the apartment, it was sparely appointed. The central feature was a beautiful antique oak *secretaire*—a roll-top writing desk—that looked like it dated back to the Tsarist era. Besides the ornate desk, which was likely an heirloom, there were two large bookcases that were filled with hundreds of engineering references and science books. Scanning the titles, which were about half German and half Russian, Federov frowned; the abundance of books didn't bode well, since it was likely that his technicians would have to painstakingly remove and examine every single book to check them for secret compartments. That would be a time-consuming effort, and time wasn't a luxury available to them.

After the entire building was evacuated, Federov's technical search team—four men attired in gray coveralls and also lugging tool boxes—rushed up the stairs and entered the apartment. To dispel potential suspicion by casual observers, similarly dressed men entered the other units in the building.

Federov met his men at the door and guided them to the study. He decided to start with the desk. At his direction, a technician methodically brushed a thin layer of fingerprint dust on the various surfaces of the desk. The delicate procedure was not intended to determine who had been in the room, since that was effectively already established, but to identify the drawers and compartments used most often.

The fingerprint expert whistled quietly to gain Federov's attention, and then gestured at one of the lower drawers. Faint smudges in the fine dust indicated that the drawer had been opened and closed often, and probably recently. Federov smiled; no matter how well they might have been trained, clandestine operatives seldom take the time and effort to adequately tidy up after conducting secret activities in their own homes.

Federov motioned for another agent to open the drawer. Since clandestine operatives sometimes go to great lengths to safeguard their secrets, the drawer-opening process was not simple. He knew of instances where operatives had laid elaborate traps with explosives and even poisonous vipers, and since he was painfully aware that Yohzin had ready access to volatile chemicals, he wasn't taking any chances. To reduce the team's exposure, one agent—the "inside man"—would carefully rig a mechanism using string and a series of pulleys to remotely open the drawer. Federov and the others would temporarily retreat to the stairwell and wait either for the inside man to call them in or for an abrupt indication—screams, perhaps, or a violent explosion—that Yohzin had indeed covered his tracks.

As the team waited on the stair landing, one of the agents absentmindedly lit a cigarette. "Put that out, moron," snarled Federov, glowering at the offender. "Can you not remember that we're here to investigate a gas leak?"

After a few minutes, Federov heard a low whistle from within the apartment and waved the other men back inside. By the time he reentered, the inside man had not only opened the drawer, but had also employed a dental mirror and other special tools to locate a secret compartment concealed inside the roof of the drawer. He had already opened the compartment to extricate a wooden box, which he placed on a metal tray for Federov's inspection, and a small shortwave radio receiver.

Federov donned rubber gloves to examine the contents of the secret box. It was obviously a kit for writing secret messages. Since he had seen several similar kits in the past, this one held pretty much what he expected: two vials of tiny gray capsules, a third vial which was empty, several thin strips of message paper, and several larger sheets of graph paper for composing messages. The gray capsules were a new twist; he assumed that they were some type of concealment device, since the strips of paper were specially trimmed to fit inside them.

At the bottom of the box, underneath the other contents, he found the specific item that he sought. Approximately the size of a small pocket notebook, it was a bundle of "one-time" encryption sheets. Each sheet contained a unique transcription key; once a message was composed and encrypted with the code, the sheet—which was probably made of chemically treated "flash paper"—was destroyed. The resultant message could be decrypted by a receiver with the matching code, but it was virtually unbreakable otherwise.

Federov signaled for one of the technicians to photograph each remaining page of the code book, and for the sake of redundancy, he directed another technician to repeat the process with another camera. He smiled; only a few hours ago, he was ready to dismiss the allegations against Yohzin as frivolous, but now he was well on his way to completing the puzzle of how the messages were conveyed. He had not yet determined exactly how Yohzin executed the drop, but he was confident that he would possess that information after watching more film this afternoon.

Most importantly, he was entirely certain that the message capsules were unguarded for at least eighteen hours—the interval between the Yohzins' evening stroll and the housekeeper's lunch—during the transmission process. Consequently, it was vulnerable to interception. Now, armed with Yohzin's one-time codes, Federov could surreptitiously interdict and read each message with virtually no risk of detection. Provided that he received approval from his superiors at the Aquarium, he could allow that scoundrel Yohzin to continue his subversive activities indefinitely, oblivious to the fact that he was being monitored. That capability could eventually yield a coup of unprecedented magnitude, probably resulting in the compromise of a major Western espionage network and the capture of scores of operatives.

R-7 Launch Facility, *Burya* Test Complex, Kapustin Yar Cosmodrome
2:45 p.m., Monday, October 23, 1972

Soaked to the skin and shivering, Yohzin slowly inched his way along a treacherous catwalk, pretending to verify pressure levels on various valves. It was a miserably dreary day; a fast moving thunderstorm had swept through the cosmodrome earlier. Even though the deluge and deadly threat of lightning had passed, a persistent drizzle spattered the slick metal decking. His cotton coveralls were drenched, and he didn't have a dry set to change into.

Even though he had not yet revealed the problem to the pad crew, Yohzin had long ago diagnosed the flaw that threatened to delay Gogol's launch this evening. In his left breast pocket, he carried a three-centimeter scrap of copper wire that he intended to jam into a critical pressure relief valve in the R-7's first stage. The wire would prevent the valve from fully opening, which would cause an overpressure in an oxidizer line moments after the booster left the pad. All evidence of the wire fragment would be vaporized as the oxidizer line ruptured and the rocket exploded, so there was little chance that his sabotage would be detected in the investigation that was sure to ensue. It was still roughly four hours until launch; to ensure that it was not inadvertently discovered, he wanted to wait as long as possible until he planted it.

Yohzin looked down. At the base of the R-7 rocket's gantry, the launch pad crew had jury-rigged a rubber-coated tarp to fashion a shelter. Crouching under the makeshift tent, a gaggle of technicians studied an electronic schematic. At least Magnus had the sense to escape the lingering rain; licking his front paws, the Alsatian huddled under an equipment console at the bottom of the tower. Yohzin watched two vehicles—a GAZ-66 utility truck and an unfamiliar *Lada* sedan—pull up at the site.

A few minutes later, as he continued on his rounds, he heard the faint squeak of rubber-soled shoes on the steel decking behind him, and slowly pivoted to see a heavy-set man in a dark blue topcoat and straw fedora. Livid, he bellowed, "You idiot! Who gave you permission to wander around up here? We're launching this rocket today. No one but launch personnel are authorized on this gantry. You need to climb down immediately!"

"Comrade General Yohzin?" asked the stranger. He tilted his hat, exposing wet red hair matted to his crown. "I'm sorry for the disturbance, but you are Major General Yohzin, are you not?"

"*Da*," replied Yohzin. "That's me. Look, I'm very busy and we're behind schedule. What's this about?"

"You must come with me immediately," said the stranger.

"*Bosh!*" exclaimed Yohzin. "Look, I'm very busy and we're behind schedule."

The mysterious man frowned and then glanced from side to side, obviously to see that no one was watching. He stepped forward and subtly grabbed Yohzin's right shoulder in a grip that would have rivaled a heavy machine vise. Simultaneously, the hulking stranger nudged Yohzin's upper body backwards just a few centimeters, leaning him over a flimsy railing and pitching him off-balance.

Wobbling, Yohzin gasped silently; he had spent virtually his entire adult life clambering up and around the rickety scaffolding used for servicing rockets, so he was as sure-footed as a six-toed cat and absolutely comfortable working at perilous heights that would reduce most men to blubbering idiots. As a result, the sensation was doubly disconcerting; not only was he stricken by the terrifying knowledge that he would fall to his death if the stranger simply let go, but he was also startled by the man's casual ease in applying precisely enough physical force to cause him to lose both his physical and psychological equilibrium. His ire was swiftly dissipated, replaced swiftly by abject fear.

The interloper quietly declared, "*Listen* to me, Comrade General. I'm GRU. I know exactly who you are and I am not confusing you with anyone else. You *will* come with me immediately. If you insist on making a spectacle of this, then I will accommodate you, but the consequences will be much less than pleasant. Do you understand?"

Yohzin nodded, and the GRU officer yanked him forward and released his grip.

Overcome with a sense of impending dread, Yohzin leaned over the rail and shouted down to the pad supervisor, "Oleg, General Abdirov has called me back to headquarters. I will return as quickly as possible."

As they clambered down the ladders to the concrete base of the tower, Yohzin saw a man in a padded suit slipping a rope noose over Magnus's neck. Magnus bared his fangs and snarled, but it was to no avail. The noose was fixed to the end of a long metal pole, which the man used to guide the subdued dog into the canvas-covered bed of the GAZ-66.

Yohzin started to protest, but the GRU officer shoved him toward the black Lada. He swung open the flimsy door, brusquely jammed Yohzin into the back of the tiny car, and then climbed in afterwards. The driver was a burly, fearsome man with a face as coarse and impassive as a millstone. The sedan's interior was filled with an almost overwhelming stench of wet wool and body odor.

"You're GRU?" declared Yohzin, striving to regain his composure and at least some pretense of the authority due his rank. "I don't recognize you. I *demand* that you identify yourself. Show me your papers."

"Gladly, Comrade General." The red-haired GRU officer flipped open his credentials. "Colonel Felix Federov, *Glavnoye Razvedyvatel'noye Upravleniye.*" He quickly frisked Yohzin, confiscating his German-made wristwatch, his pocket-sized slide rule, a cigarette lighter, two pencils, a clasp knife, an engraved cigarette case, a soggy pocket notebook, and the tiny strand of copper wire.

Federov gestured at the driver. The car lurched forward, pulling away from the launch site. Yohzin glanced at the rear view mirror and watched the R-7 rocket fading in the distance behind them. Trying to avoid eye contact with Federov, he shifted his gaze down to the sedan's plywood floorboards.

The three men rode in silence. A few kilometers away from the pad, Federov directed the driver to stop at an abandoned test site, and then ordered Yohzin out of the car.

The secluded location resembled a ghost town in a Western cowboy movie. A pair of wary vultures gnawed at a jackal's decaying carcass while more of the scavengers circled overhead, impatiently waiting their turn. An old telemetry antenna swayed against slackening guy wires. A rusting rocket gantry creaked in the wind. Parked at the base of the structure, a derelict armored car—with scorched paint and melted tires—was evidence of an explosion and subsequent fire during a refueling accident three years ago. The foul air hinted of various spilled chemicals that still permeated the barren soil.

"What's this about?" demanded Yohzin. He suspected that Federov was investigating the leak that Abdirov had described and was trying to intimidate him into confessing. But since he had nothing to do with *that* leak, he knew that Federov had absolutely nothing on him.

Federov reached into the pocket of his blue wool coat and extracted a tiny gray plastic pellet, which he held gingerly before Yohzin's eyes. "I think you *know* what this is about."

Yohzin swallowed deeply. "Oh," he mumbled, staring at the blue-stained ground. His hands trembled, his lower lip quivered, and tears trickled from his eyes.

"So, this is yours then?"

"*Da*," replied Yohzin in a faint and faltering voice.

"We know that you've been sending messages in a dead drop," stated Federov. "I'll admit that it took me a while to deduce how the messages were being passed. It was difficult, after all, because no one would have ever conceived that the messages were literally being *passed*. Ingenious, I must admit. Very cunning. Your own idea?"

Overwhelmed with fear, Yohzin shuddered as he nodded solemnly.

"I suppose that your dog must share in the credit," added Federov sarcastically. "After all, he did contribute his bowels to the effort. I can just picture him now, stuffed and mounted in a glass display case in the counter-espionage exhibit at the Aquarium."

Cringing at the image of his beloved Alsatian rendered by a taxidermist, Yohzin pleaded, "You wouldn't harm him, would you? He's just an animal."

"*Nyet*," replied Federov, shaking his head. "No need for you to fret over that, not for the moment, anyway. I suspect that your mutt might be holding another transmission in the hopper, so to speak. To be honest, I had considered slicing him open to extract it, but then I came to my senses and realized that it was vital to leave him intact, at least for the time being, just in case we have you continue your activities."

"My activities?"

"*Da*. Passing messages. I'm awaiting guidance from my bosses at the Aquarium, but I am recommending that you stay here, under my supervision, so that you will continue to service your dead drops. Of course, henceforth, we will furnish the script. So, can I count on you to be cooperative?"

Staring out at the bleak landscape, Yohzin nodded his concession.

Federov continued his droning commentary. "I assume that you're intelligent enough to know that you're destined for some serious inter-rogation at the Aquarium. Quite frankly, it's more likely that you'll be sent there, post haste, possibly as early as this evening."

Yohzin had overcome some of the initial shock of his arrest and remembered that, despite his absence from the proceedings, Gogol's rocket was still scheduled to depart in a few hours. Since there was no way to know what his immediate future held, Yohzin realized that this might be the opportune moment to reveal General Abdirov's plan for the Egg.

"Look, Colonel," he blurted. "I know that I will suffer for my trans-gressions, but there's something I need to describe to you, a crime much greater than what I am guilty of."

Federov laughed. "If only I had a kopek for every time I've *heard* that one," he said. "Everyone seems so willing to conceal terrible crimes and treasonous activities, right up until the moment they get rolled up, and then suddenly they're so stricken with guilt that they feel obligated to sell out their contemporaries."

'It's not like that!" muttered Yohzin.

"Really?"

"Are you familiar with *Perimetr*?" asked Yohzin.

"I am. The *Dead Hand*. What of it?"

"Do you know what I was working with here?" asked Yohzin.

"Vaguely."

"We sent up a Proton rocket over a month ago. It carried a two-man space station called a *Krepost*. It is armed with a nuclear weapon, a big one. We sent two men up to watch over it, but one has since returned."

"This all seems rather far-fetched, Comrade General."

"Wait. There is a *Perimetr* device called the interlock installed on the *Krepost*, to prevent the nuclear warhead from being deployed without authorization from the General Staff of the High Command.

Federov lit a cigarette and drew deeply. Frowning, he puffed on it as he listened to Yohzin talk.

"There's a secret code that can disable the interlock, that only the *Perimetr* people are supposed to know," said Yohzin urgently. "I stole it from them and brought it to Lieutenant General Abdirov. He gave it to the cosmonaut who is being launched today. The two of them are conspiring to drop the bomb without authorization. You *must* pass this information on to the High Command, so they can halt the launch and arrest General Abdirov."

"Oh, *that's* all?" asked Federov. He dropped his smoldering cigarette onto the blue-shaded mud, crushed it beneath his boot, and smiled. "I'll admit that I'm as ambitious as the next officer, but I'm not foolhardy enough to haul in an RSVN lieutenant general to be investigated for a seditious act, with no tangible evidence, based solely on the testimony of someone who was caught dead to rights, neck deep in espionage with sworn enemies of the State."

"But…"

"Very imaginative ploy, Comrade General, but I'm not falling for it. Your time is nigh, so let's go."

Federov shoved him back inside the Lada and yanked a canvas bag over his head. Yohzin realized that he obviously wasn't the first one to wear the coarse veil; the damp sack reeked of grime, sweat and someone else's vomit.

Like virtually every officer in the Soviet armed forces, particularly those entrusted with high level secrets, he had endured a multitude of grueling practice interrogations, so he was well aware of his tolerance for physical pain and deprivation. Consequently, he had no misconceptions about how he might bear up under an actual interrogation, particularly if the intense inquisition was heavily seasoned with brutal torture, as the GRU are like to do.

As they resumed their journey, Yohzin thought of Luba and the boys. Probably the only way he could ensure his family's escape was to be dead, so that he was incapable of answering questions. To that end, Yohzin yearned for a bit of cyanide to nibble on.

Ironically, Smith had provided him with a black capsule during his hurried training at a Moscow hotel, but he had misplaced it months ago. Then, amused at Smith's melodramatic instructions and the absolute absurdity of committing suicide, he hadn't bothered to request a replacement. Now, faced with agonizing reality and an uncertain fate, he desperately wished that he still possessed the means for a prompt exit.

GRU Internal Security Office, Kapustin Yar Cosmodrome
4:30 p.m.

Federov tugged the canvas hood from Yohzin's head as the Lada creaked to a stop. Blinking, Yohzin struggled to orient himself to his surroundings. The rain had subsided completely. A light wind blew from the north as the sun periodically peeked through breaking clouds. A burly GRU sergeant exited a small brick building, which Yohzin recognized as the GRU's field office; he saluted Federov before handing him a note. The sergeant swung open the door, pulled Yohzin out of the sedan, and then escorted him into the building.

Federov perused the dispatch, and then said, "Wonderful! Your fate is now more clear, Comrade General. The Aquarium beckons for your presence. You'll be detained here until an aircraft is dispatched to fly you to Khodinka."

Picturing what was in store, Yohzin swallowed deeply.

"But while we're waiting," said Federov. "I want you to watch a movie with me."

"A movie?"

"*Da*," answered Federov, directing Yohzin to a chair facing a movie screen. "Since you had been so fastidious in servicing your message drop, I was curious whether you knew who was picking up your messages and how they left here. So, did you know who was plucking your delicate little pellets out of your dog's crap?"

Yohzin glumly shook his head.

Federov chuckled. "There's no need to lie. At this point, it makes no difference, since the people who assisted you with passing your messages will not be harmed."

"You don't intend to arrest them?" asked Yohzin incredulously.

"*Nyet*. Well, truthfully, we won't arrest them until you're of no further value to us," said Federov, threading film into a projector. "And that day, when you are of no further value to us, is swiftly approaching. When that time comes, and it will, we'll snatch up those people forthwith, and you can rest assured that they will answer mightily for their sins. So, I will ask you again, Comrade General, do you have any notion who it was that passed your messages from here?"

Even though he honestly did not know, Yohzin felt as if the GRU officer was trying to lure him into a trap. "Honestly, I have no idea," he answered. "I felt that it was prudent that I not be too curious."

"Words to live by," noted Federov. "In any event, it doesn't matter, but I do want to show you the kind of people who will suffer as the result of your shenanigans. I can assure you that they had no earthly idea of what they were passing, not that it matters."

Federov turned to Yohzin, grinned, switched off the lights, started the projector, and said, "Watch this. Here's your James Bond, but she's probably not the double-naught secret agent you might have imagined. I doubt that she has ever been granted a license to kill, but she obviously has been issued a license to scrub."

The old projector clattered as the movie played. The film depicted the commons area between the apartment buildings in the complex where his family lived.

Yohzin's heart sank as he recognized the kindly old woman who cleaned apartments of the higher ranking officers. She entered the commons from the east, took a seat on a concrete bench for several minutes, pitched crumbs to doves, then stood up, smoothed out her threadbare woolen coat, and walked exactly twenty-three steps to the south. She paused for a moment and then knelt down as she pretended to extract a miniscule pebble from her shoe. Then she slowly stood up and departed the commons to the southwest, apparently toward the next apartment on her route.

"And that, Comrade General," gloated Federov, switching on the lights, "is how we came to find your little gray capsules.

"Comrade General, I've been observing you for quite some time. Just so you're aware, we interdicted five of your gray capsules between the time your dog passed them and the old maid picked them up. We took the capsules apart, removed and copied your messages, reassembled the capsules, and then put them back in the dead drop.

"Now, I must confess that we have yet to crack the transcription code for the messages, but it's just a brief matter of time until we do, and then we'll know exactly what you've told the Americans over the course of the past five weeks. With that said, I can assure you that the remainder of your life will be considerably less miserable if you're forthcoming with the information that we will eventually discover, whether you choose to cooperate or not."

Although he was nearly exhausted, Yohzin had a flash of insight. He was so weary and disoriented that he almost missed it. Federov just made a significant blunder. In so doing, the otherwise wily GRU officer had inadvertently given him a reference point in time. Yohzin had only started using the *gray* capsules this past Thursday, after exhausting his stock of *brown* capsules. *The supposedly all-knowing Federov was bluffing!* Most importantly, although Thursday's message requested that the escape plan be executed, it did not specify the details of the plan, but just a code name, along with a code word and recommended timeframe. So, even if they were capable of cracking the transcription code, the location and details of his family's escape would still be unknown to them.

"Tell me about your family," said Federov. "I actually met your wife briefly this morning, but was not aware that she was making ready to go on a journey."

"They are," said Yohzin. "She is taking my sons to visit her parents. My father-in-law is frail from his war injuries and is not expected to live very much longer, so she wanted to take my sons to see him before he passed. She has the appropriate travel documents, signed by General Abdirov and approved by the KGB."

"I know that she has the correct papers," said Federov. He patted his breast pocket. "I have copies as well. Your family is supposed to return in about two weeks' time?"

"*Da.*"

"Just as well, I suppose," said Federov. "I've dispatched some of my people to fetch them, so they can accompany you to the Aquarium. Of course, in the time it takes my men to muddle through the KGB's paperwork, it's highly likely that your family will have returned on their own."

"Perhaps," said Yohzin. He was very cognizant of the poor relationship between the GRU and the KGB; the two bureaucracies routinely fought pitched battles over perceived turf. "The boys do have to go back to school. Luba will not let them miss that."

"Well, I do hope that they make it back to school," said Federov menacingly. "But that depends upon you, Comrade General, and your willingness to cooperate with us."

Yohzin swallowed deeply. "Please don't harm my family," he mumbled, staring at the concrete floor. His hands trembled, his lower lip quivered, and tears trickled from his eyes. "They had *nothing* to do with this. They know absolutely nothing, I swear."

"I don't doubt your sincerity, but you're not exactly in a position to negotiate, Comrade General," replied Federov. "With that said, let me assure you: what happens to your wife and children is *entirely* up to you. If you cooperate with us, then they will not be harmed. We might even let you say goodbye to them before it's all over. But if you *refuse* to cooperate, then I won't have the slightest qualm with sawing them to pieces right before your eyes, if that's the motivation it takes to alleviate your reluctance. Understood?"

"Da," said Yohzin meekly.

"Excellent," noted Federov. "Now, I think we have a much better understanding."

6:40 p.m.

Yohzin heard a resounding roar and felt his cell's concrete floor tremble. He instinctively glanced for his wristwatch, but then realized that Federov had taken it from him. It had to be Gogol's departure. He closed his eyes and wished to hear an explosion, the sound of the R-7 detonating before it even left the pad, but heard instead the unmistakable sounds of the rocket climbing in the air. Almost deafening at first, the roar gradually diminished as the rocket accelerated and climbed to the northeast. In time, the sound was gone altogether.

And there goes Gogol with his code, he thought. *Now it's just a matter of time before the world ends.* He hoped that Smith would follow through on the plan to extricate his family to safety, but now he realized that there soon might not be any safe place left on this planet.

If only he had some way to convince Federov of Abdirov's plan, and then maybe he could halt this nightmarish scenario. He looked up; his eyes met Federov's. The GRU colonel sat at his desk, leaning back in his chair, gazing into the cell like an explorer studying a prized predatory big cat captured out on the African veldt. Yohzin was certain that Federov looked forward to triumphantly parading him into the Aquarium, leading him to the basement, and delivering him to his waiting inquisitors.

The eight digits of Bogrov's secret code kept appearing in his thoughts, as if they were the magic key to unlocking this dilemma, in much the same manner that they disabled the interlock. And then it came to him: Bogrov's code *was* the key to stopping Gogol and Abdirov.

"Colonel Federov," he said. "I have something important to tell you."

"Important?" asked Federov. "If I can make a suggestion, Comrade General, you might concentrate on getting some rest now, while you can, because you will have plenty of important things to share with your interrogators."

"No, honestly, I need to talk with you. It's about General Abdirov and that secret code."

"The *secret code?*" sneered Federov. "You're still pursuing that angle, huh?"

"I can prove the existence of the code," declared Yohzin.

Federov stood up from his chair, stretched, and then walked over to Yohzin's cell. "How?" he asked.

"Write this down," said Yohzin. "Seven, Six, Eight, One, Zero, Seven, Two, Three."

Federov took out his pocket notebook and jotted down the eight numbers. "So what?" he asked. "What am I supposed to do with this? Confront General Abdirov with it? Do you think that this piece of information will somehow compel him to fall to his knees and confess?"

"Listen," said Yohzin. "Those eight numbers are the secret code for *Perimetr.* As I told you, that code is used to bypass the *Perimetr* interlock, so it is available only to the *Perimetr* engineers."

"I still think you're trying to lull me into a wild goose chase."

"It's real," replied Yohzin. "Go find a *Perimetr* engineer named Bogrov. I stole that code from him." Yohzin related how he had employed a powerful laxative to gain access to the secret code written in Bogrov's notebook.

Grinning, Federov shook his head. "It's a shame you missed your true calling, Comrade General," he said. "If all that's valid, you could have worked for us."

"It's true," said Yohzin. "Find Bogrov and you'll find the truth. I gave the code to General Abdirov, and he gave it to Gogol. If Gogol makes it aboard the *Krepost,* then he'll use it to drop the Egg. Simple as that."

Federov tore the page from his notebook, folded the paper in half, and tucked it into his breast pocket. "I'll check this out," he said. "But hear me, Comrade General, if this is all a ploy meant to delay your ultimate fate, then you will soon regret it. I swear that I will skin your sons alive while you watch."

Federov turned and left with one of the sergeants. Surprisingly, Yohzin felt as if a tremendous weight had been lifted from his shoulders. He spread out on his uncomfortable cot, pulled the blanket over him, and promptly fell asleep.

9:02 p.m., Monday, October 23, 1972

Awakened by the sound of keys clinking together, Yohzin looked up to see Federov unlocking his cell. "I found Bogrov," he said solemnly, removing his straw fedora. "And he corroborated your story about the code. To be honest, I knew immediately that you had told me the truth, because I have never seen anyone so frightened as when I read the numbers to him. He shook like a newborn kitten lying on the snow."

"And General Abdirov?" asked Yohzin.

"I placed him under arrest. Presently, he is in his quarters under guard and will go to the Aquarium tomorrow."

"Listen, I take no joy in this situation," said Yohzin.

"Of that, I'm certain," replied Federov. "Get up."

"The plane has arrived?" To take me to Khodinka?"

"*Nyet*," answered Federov. "No need for you to be so hasty. You're still destined to go there, but not now. Don't think that you've received a reprieve, but there's been an interesting turn of events."

"How so?"

"As it turns out, the High Command is not anxious to lose their investment in this damned *Krepost* of yours. Since it seems that you have such extensive knowledge of it, as well as the *Perimetr* systems, they've come to believe that you can salvage it. As I said, considering the fact that you're under arrest for treason and other high crimes, this is a rather unusual turn of events."

"This is a joke?" asked Yohzin. "You're pulling my leg?"

"*Nyet.* Get up and get dressed. I'll have the sergeant fix you some tea, and then we'll head to your office."

Yohzin threw his blanket aside and stood up, then realized that his coveralls were still damp.

"I'll loan you an overcoat," said Federov. "But I will expect it back before you go to the Aquarium."

14

EXPEDITED

Krepost **Station, On Orbit**
20:46 p.m. GMT, Monday, October 23, 1972
GET: 39 Days 20 Hours 1 Minutes, REV # 637

Vasilyev studied the luminous second hand marking time on his
wristwatch and then switched the radios on for another contact
window, this time with Control.

He had been busy with the radios. He had been talking with the
Americans for several hours now, but was confident that Control was
not aware of his extracurricular communications. In his weakened state,
he had to remain very cautious to remember who was on the other
side of the line. Even the slightest slip of the tongue could yield drastic
consequences.

He was amazed that he had survived this long, given the horrendous
conditions. After all, it was really just a short time ago that he was so
distraught that he was ready to give up altogether. But all that was due
to change; Gogol was on his way and should arrive in less than an hour.
The only question that remained was how Vasilyev should handle the

awkward situation, because as much he feared Gogol, and was petri-fied of what would ensue once the repulsive cosmonaut was aboard, the long-awaited Soyuz was Vasilyev's ride home.

Running the air scrubbers had helped his circumstances immensely, since he no longer constantly felt on the verge of suffocating, but cleans-ing the air of carbon dioxide also came with a price. Even though the solar panels were operating at capacity, electrical power was still very much at a premium. As much as he yearned to switch on the heaters, if but for just a few minutes, he had to be cautious to reserve plenty of battery power for the docking mechanism.

He heard General Yohzin's voice again: "*Krepost*, this is Control. I have some specific instructions for you, but I will keep this as brief as possible to conserve your battery power.

"Soyuz *Kochevnik* will rendezvous with you at roughly 2100 GMT." Confused at first, Vasilyev remembered that Kochevnik—"Nomad"—was Gogol's operational codename.

"*Krepost*, your fundamental mission has not changed," stated Yohzin. "You will occupy the Soyuz and wait for a freighter to arrive. Once it does, we will instruct you on how to scuttle the damaged freighter. When another crew arrives, you will assist them in restoring the *Krepost* to full operation, and then you will return home. Do you understand?"

"I do."

"There have been some very disturbing developments. You should be aware that the *Perimetr* interlock has been compromised, and that we have learned that Gogol intends to exploit this compromise."

That explained Gogol's insistence that they would drop the Egg.

"In order to restore *Perimetr*, the system must be shut down and recycled. We will explain how to accomplish this later, but until the interlock is recycled, you should not operate or even touch its controls. Understood?"

"I understand that I am not to operate or touch the *Perimetr* inter-lock," replied Vasilyev.

"Because he knows how to bypass the interlock, Gogol is extremely dangerous. We want his Soyuz to dock so that you'll have a way to return home in an emergency, but under no circumstances do we want him to physically enter the *Krepost*. Is that clear?"

"You don't want him to come aboard the *Krepost*?" asked Vasilyev. "I'm not sure how I'm supposed to discourage him from doing that. Do you want me to *incapacitate* him?"

"*Incapacitate* is a nice word, but it is also a little vague," replied Yohzin. "I don't want you to *incapacitate* Gogol. I want you to *kill* him. Is that clear enough?"

"Understood," answered Vasilyev. "I will not allow him to enter the *Krepost*."

"Good," said Yohzin. "Gogol will have a tool that resembles a screwdriver. It has a yellow handle and a special tip that can loosen the fasteners on the Egg's control station. After you take care of Gogol, you need to find that tool. During the next contact window, the *Perimetr* engineers will explain how to gain access to the Egg's controls, so you can reset the power breakers for the *Perimetr* gear.

"I know that some of this matter may sound drastic," said Yohzin, "but as I said, your basic mission will remain effectively the same. You'll occupy the Soyuz until we send the next freighter up, and then you'll receive your relief crew about two weeks later. Understood?"

"I understand and will comply," replied Vasilyev. Seconds later, he heard a warbling sound in his headset, and knew that he was out of radio range.

He removed the headset, pulled his sleeping bag over his head, and massaged his throbbing temples. The updates from Control were largely what he expected: with the minor exception of being ordered to commit homicide, his mission requirements had barely changed. But thankfully, the nuances of Control's plans were now starting to align with his own, since he had already contemplated the potential difficulties with subduing Gogol. But although General Yohzin and Control now seemed more inclined to let him peek behind the curtain, they obviously weren't ready to reveal why he had been fed a steady diet of falsified intelligence reports.

His body shook in a spasm of uncontrollable shivering. As he shook, he struggled to convince himself that it was just a matter of hours before he would be warm again, hopefully for the rest of his life.

Krepost Station, On Orbit
21:50 p.m. GMT, Monday, October 23, 1972
GET: 39 Days 21 Hours 5 Minutes, REV # 638

Shivering in his sleeping bag, Vasilyev waited. Gogol had docked almost twenty minutes ago. After enduring a steady stream of Gogol's lascivious comments over the intercom, the moment of truth was upon him: Gogol had worked his way forward to the Orbital Module and was making ready to open the nose hatch of the Soyuz.

Vasilyev felt ready. He had mentally rehearsed his actions, time and again, and had arrayed his arsenal to subdue Gogol. Ironically, he had acquired most of his weapons from the senior cosmonaut's own survival kit. For his initial attack, he had selected Gogol's survival hatchet, which he had wrapped with strips of cloth salvaged from a set of coveralls.

Vasilyev twisted open a valve to equalize the pressure in the access tunnel and then opened the hatch from the docking hub side. He listened for the muted clicks that indicated that Gogol was manually actuating the back-up latching mechanism that ensured the spacecraft would remain sealed to the docking port.

Knowing that it would likely be less than a minute before Gogol cracked the hatch, Vasilyev climbed out of his makeshift sleeping bag, wadded it into a ball, and then crammed it behind the docking hub hatch. Incredibly strong and wiry, able to adapt almost immediately to weightless conditions, Gogol was extraordinarily dangerous under any circumstances. Vasilyev was conscious that he was in even greater danger than usual, since he was physically debilitated by nearly forty days in weightlessness, compounded by near-starvation and constant exposure to brutal cold. He switched off his flashlight and waited in the dark. His body shook uncontrollably, half from the frigid temperatures and half from abject fear.

Grunting, Gogol swung the Soyuz's nose hatch open. "At last, darling, I'm home!" he exclaimed. As he peered into the darkened docking hub, dimly backlit from the lights in the Soyuz, his shocked expression revealed that he was obviously taken aback by how cold it was. Vasilyev was pleasantly surprised by the warm air that rushed from the hatchway.

"What, kitten, no bread or salt to welcome me?" asked Gogol.

Poised for action, Vasilyev grabbed the lip of the hatchway with his left hand for leverage, and then swung the hatchet with all his might to clout Gogol on the side of his head. The cosmonaut's head thumped against the hatch's rim. Losing his grip, Vasilyev tumbled backwards through the air in slow motion, rebounding off the aft bulkhead of the docking hub.

Gogol had apparently not locked the nose hatch open; the spring-loaded portal half-closed, gripping Gogol like a crocodile's jaws. The hatchet's impact split open Gogol's closely shorn scalp. Globs of blood spurted from the gash. Stunned and disoriented, grunting in pain, Gogol flailed with his arms in a futile effort to defend himself.

Vasilyev made his way back up the access tunnel and gave Gogol another swing for good measure, although with not nearly as much force. The second blow did the trick.

Gogol stopped moving. Switching on his flashlight, Vasilyev realized that Gogol was bundled up in a heavily insulated dark blue snowsuit, of the sort worn by ice fishermen or dog sledders. He felt at Gogol's neck to check his pulse; although unconscious and momentarily harmless, he was still very much alive. He considered finishing the job, as prescribed by General Yohzin, but as much as he despised Gogol, he just could not bring himself to kill the repulsive ogre. Gasping for air, he was amazed by how much energy he exerted.

He unzipped his coveralls to find the plastic box of hypodermic syringes, pre-loaded with powerful sedatives, that he had stashed close under his shirt to keep thawed. He opened the box, fumbled for a syringe with fingers numb from cold, and then jammed the needle into Gogol's massive bicep. Just for good measure, he repeated the process, plunging a second needle into Gogol's arm, hoping to make sure he would be unconscious for a long time.

He shoved the hatch open and locked it in place, and then pushed by Gogol's body to enter the Orbital Module. He was awestruck at the massive volume of food and other supplies packed into the orbital module; there was just enough empty space to gain access to critical equipment, as well as a passageway barely wide enough to crawl through to the Descent Module. He paused to bask in the luxurious warmth of the Soyuz. He spotted the galley station, stuck the water dispenser directly in his mouth, and drank a copious volume. He rummaged in the galley cabinets and found a chocolate bar; barely pausing long enough to unwrap it, he chomped into the confection, delighted to finally sink his teeth into something that wasn't half-frozen. He quickly made a pouch of hot tea, and kneaded the crystals in its pouch as he made his way into the Descent Module.

He was excited to find a set of insulated coveralls, matching Gogol's, in his couch. In an officious touch, his rank and name were stitched above the breast pocket of the heavily padded garment. After donning the coveralls, he spent several minutes looking for the special tool that Yohzin had described, but could not locate it.

He snaked back through the Orbital Module and frisked Gogol's pockets. He found the tool and then headed back to the control area. Following the instructions of the *Perimetr* engineers, he used the tool to carefully remove the fasteners from the specific panel they had designated. Having no intent to replace the panel, he allowed the fasteners to float. He picked one out of the air, examined it closely, and realized that it was intentionally designed to shear apart if someone tried to force them without using the special tool.

Vasilyev aimed his flashlight's beam into the guts of the Egg's controls and identified the circuit breaker that the engineers had described. He laughed. Control had wanted him to recycle the Egg's power, but now it was time to deviate from their plan. He opened the code safe, referred to the targeting book, and slowly punched a set of latitude and longitude coordinates into the targeting computer, verified them against the book, and then pushed the green Arm button.

Almost immediately, he heard maneuvering thrusters firing as the station made a few minor adjustments, and then heard explosive bolts fire, signifying that the Egg was now free of the *Krepost*. Entirely self-contained, the Egg's on-board computer would wait until it reached the correct position in orbit, then align itself for reentry and fire its braking rockets. Slightly more than thirty minutes later, after a scorching descent through the earth's atmosphere, the Egg would arrive at the place that Vasilyev had selected.

With that noxious chore out of the way, Vasilyev methodically went through the next steps of his checklist. He switched on the radios to establish contact with his new American friends below, to let them know that he would be returning to his beloved Earth shortly, in accordance with the plan they had coordinated. Afterwards, he switched off the circuit breakers for the Egg and disabled the power for the *Krepost's* radios.

He moved forward to the docking hub, grabbed his makeshift sleeping bag as a souvenir, shoved Gogol's inert body through the Soyuz's nose hatch, and then entered the Orbital Module. He closed and secured the nose hatch and disabled the safety latches for the docking mechanism. He didn't look forward to eventually wrestling the unconscious Gogol into his specially fitted couch, but there would be plenty of time for that later. Right now, he focused on undocking the Soyuz so that he could finally be free of his miserable outpost in space.

Gemini-I, On Orbit
2:40 a.m. GMT, Tuesday, October 24, 1972
GET: 39 Days 21 Hours 5 Minutes, REV # 33

The nuclear warhead had physically separated from the *Krepost* over an hour ago, and was orbiting on its own, several miles away from the station. Unaware of the drama that had taken place since the Soyuz had docked with the station, Carson and Ourecky focused their attention on the warhead, trailing it at a distance of roughly a mile as the right-seater literally kept his finger on the Disruptor's Destruct button. Just as he had predicted, they had been able to slip the Disruptor onto the warhead without alerting Vasilyev.

Tew had reluctantly conceded to their scheme, but only after Ourecky had already fixed the jammed Disruptor boom. At present, they were acting as the final backstop to the overall plan; if it became obvious that the warhead was not going to reenter to a safe location, Ourecky would fire the Disruptor's main charge to destroy it. Granted, the explosion would yield a cloud of radioactive debris that would eventually return to Earth, but even that was still a more favorable outcome than a nuclear detonation over a major American city.

"Time," announced Carson, shifting his gaze from the warhead to the mission clock on the instrument panel. They were just entering the light period of an orbit; the sun popped over the horizon as the Gemini-I passed over Australia. "We're in the window right now."

Tense moments passed as they watched for the warhead's retro rockets to fire. And then it happened, almost to the precise second that Vasilyev had stated. They witnessed the solid rocket motors flare brightly to life, and then the warhead slowed almost immediately, quickly passing below them and outside their field of view.

"Whew," mumbled Carson. "Now I can breathe again."

"Me too," said Ourecky, thumbing a toggle switch. "Safe on."

"Confirm Safe on," observed Carson. "Well, that's it. Our work here is done. It's finally time to go home and put all of this behind us."

"That it is," replied Ourecky, immediately picturing Bea in his thoughts. *Maybe they could finally move on with their lives, now that all this was behind them.* "Hey, Drew."

"Yeah, brother?"

"Do you have any more of those Fig Newtons left? I'm famished."

Mission Control Facility, Aerospace Support Project
1:45 a.m., Tuesday, October 24, 1972

As the mission swiftly drew to a close, Wolcott and Tew emerged from their glass-enclosed office to take seats at the otherwise deserted last row of consoles. The past few hours had been packed with momentous events. Vasilyev's Soyuz had safely landed in a remote wilderness area roughly fifty miles north of Gallup, New Mexico. A veritable armada of rescue aircraft had been dispatched in advance to search the area, and within an hour of landing, an Air Force SAR helicopter had spotted the bullet-shaped Soyuz capsule. Vasilyev had already built a roaring fire to warm himself and his unconscious companion, and was busy making a pot of tea when the chopper arrived.

The Egg had plunged into the Indian Ocean roughly midway between Madagascar and Australia; not detonating, just as Vasilyev had programmed it, it sank safely into the depths, far from land and human habitation. The threat of nuclear weapons in space was over, at least for the time being.

Just a few minutes ago, Carson and Ourecky returned safely to Earth, landing at Minot Air Force, North Dakota, under the cover of darkness. Blue Gemini's twelfth and last mission was complete.

Weary from the strain of the past few days, Wolcott breathed a sigh of relief. While the flight had certainly been a tightly orchestrated team effort, the Gemini-I definitely turned out to be the little engine that could and did. Now it was time to make good on the promises that had been made to its stalwart crew.

Further down on the floor, a noisy party was underway. Champagne gushed, confetti flew, flags were waved, and the room was soon awash in a dense pall of cigar smoke. Wolcott looked to his left, studying Tew for a negative sign, but the general was silent, almost unnaturally so.

Tew usually frowned on any celebrations in the control room. While he understood how high emotions ran and the controllers' need to vent off steam after completing a stressful mission, Tew had a good point. Voicing valid concerns that things could readily get out of hand and that sensitive equipment could be damaged, he normally insisted that any festivities be moved to a more appropriate venue.

Maybe he was silent now because he knew that this chamber would soon be obsolescent. Soon, the consoles would be stripped out and the room gutted. And although they were jubilant now, most of Gunter Heydrich's civilian workers would be unemployed in just a matter of days. They would be returning to the job market just as the Apollo-fueled aerospace boom was swiftly drawing to a close, emerging from the shadows with an inexplicable five-year void on their otherwise impeccable resumes.

"Virgil," said Tew in a wavering voice. "There's something I want you to do for me."

"*Anything*, Mark," replied Wolcott, rolling his seat closer. "Your wish is my command."

Lightly patting his chest, Tew leaned toward Wolcott and whispered in his ear.

Wolcott grinned, nodded and said, "You're right. He deserves that. He's certainly earned it. I'll do my best to make it happen, pard. I promise."

With his trembling hands on the desktop, Tew smiled weakly. He exhaled softly, closed his eyes and slouched forward until his ashen face rested on his hands. He made a faint gurgling sound, and then was silent.

Puffing on a huge cigar, grinning widely, Heydrich walked up. His hair and clothes were drenched with champagne. "Care for a Macanudo, Virgil?" he asked, offering a brown torpedo.

Wolcott shook his head and gently placed his hand on Tew's shoulder. "No, thanks, Gunter, but I do have a mighty hankerin' for a coffin nail, if you might have an extra one handy."

Heydrich pulled out a pack of Winstons, tapped out a cigarette, and handed it to Wolcott. Looking at Tew, he said, "I guess all this excitement just finally wore Mark out, huh?"

"You could say that," replied Wolcott. He snapped the filter off the cigarette, casually flicked open his Zippo, and lit it.

"Virgil, is he okay?" asked Heydrich, nudging Tew. He had obviously been so caught up in the exuberance that he failed to recognize that anything could be wrong. "Should I call the hospital for an ambulance?"

"No, Gunter," replied Wolcott calmly, reaching into Tew's jacket to retrieve a white envelope. He tucked the envelope into his pocket, leaned back in his chair, drew deeply on the cigarette, and then puffed a smoke ring into the air. "That won't be necessary."

Artsyz'kyi District, Odessa Oblast, Ukrainian Soviet Socialist Republic
1:32 a.m., Thursday, October 26, 1972

It was the middle of the night, but Luba Yohzin was wide awake, lying under a veritable mountain of quilts and blankets. Coming here to visit her parents was always a bittersweet experience, because she gradually watched as her father lost touch with reality. She knew that like so many others who went to fight in that terrible war, her father had died in battle, but his living body remained.

She and her sons slept in a small outbuilding behind her parents' cabin. Her father used the shed to store winter fodder for goats and sheep. The cabin itself was just barely large enough to accommodate her parents, so there was certainly not enough room to accommodate the five of them. Besides, her father woke up frequently during the night, screaming and crying, which frightened her sons, so Luba transformed the rustic shed into a temporary bedroom. The structure was so rickety that it rattled and swayed with every strong gust of wind.

The boys loved it. They both enjoyed camping expeditions with the Young Pioneers, so sleeping in the shed was a huge adventure. Nestled in a pile of hay, sleeping soundly, she listened to the rhythm of their snoring.

She was startled by a persistent buzzing noise. Curious, she shoved her covers aside and tugged on her winter coat over her housecoat. As she walked toward the door, she noticed the shotgun her father kept for wolves; as an afterthought, she took it with her.

She stood outside in the cold night, looking up into the brilliant mantle of stars. In a short while, the buzzing sound faded, replaced by a faint swishing noise. Shielding her eyes against the glare of the full moon, she saw the shadowy silhouette of a man descending under a parachute. He landed in the distance, perhaps a few hundred meters away, and she walked toward him, shotgun at the ready.

As she drew near, padding quietly in her stocking feet, she saw that he was stuffing his rolled parachute into a large satchel.

"Halt," she ordered quietly, summoning up her lessons from para-military training.

He looked up and raised his hands into the air.

"Explain yourself," she ordered, menacingly pointing the shotgun at the stranger. "And make it quick."

"What is the fastest way to travel to Tiraspol?" he asked in a soft voice.

What? Incredulous, Luba said, "You fall out of the sky, and then you ask…"

"What is the fastest way to travel to Tiraspol?" he repeated quietly.

And then she recalled what Gregor had told her. "You can travel by train," she said, lowering the shotgun. "But an airplane would be faster."

"I'm Smith," said the stranger quietly.

"Smith?"

"Listen to me," he said. "In a few minutes, a plane will land here to take us away. We don't have much time, so you need to gather your sons and your belongings. Quickly."

"A plane will land *here?* Where are we going?" asked Luba. "And when will we come back?"

"Obviously, your husband has not told you very much. If you board this plane when it lands, you will leave the Soviet Union and never come back. That's what your husband wants for you and your sons."

"And what if I don't want to go?" she asked.

"That's your choice, but I assure you that your husband wants you to go with me."

Gregor had urged her to trust the stranger without question, but he certainly had not mentioned that they might meet under these condi-tions. She thought to wake her parents, but quickly reconsidered. The less that they knew, the better. She heard the strange buzzing sound again and glanced up at the sky.

"Go now," said Smith. "Get your boys and your belongings and walk down the trail. I'll collect you and make sure that you get aboard."

2:02 a.m.

In minutes, Luba had awakened her sons, gathered their belongings, and then rushed down the trail. She found Smith lighting red flares, like the sort used on railroads, to mark the corners of an expedient landing strip.

"The plane will be here in just a few minutes," said Smith, escorting Luba to a place beside the road. "The pilot won't shut off the engine while he's on the ground, so it's going to be very loud. Most importantly, stay right here until I come to get you."

The boys were excited with the nocturnal adventure, but what teen-aged boys would not be delighted with the idea of a nighttime ride in a mysterious airplane?

As Smith stood to talk on a small radio, Luba crouched down with the boys as she heard the buzzing noise again, this time from the south. In seconds, a dark silhouette swooped out of the sky and landed on the dirt track. If Luba didn't know any better, and if she didn't see with her own eyes that the plane was still intact, she would have been sure that it had crashed.

"Remember!" yelled Smith, standing up to walk toward the plane. "Stay right here!"

Smith approached the plane and slid open a large side door, much like the sliding door on a cargo van. He yanked out a pair of blocks and quickly chocked the wheels of the airplane.

In the moonlight, Luba could see that the cargo area was almost completely filled by three large metal drums. The pilot climbed down from the airplane and immediately started disconnecting hoses from the drums. He and Smith wrestled the containers out of the airplane and rolled them into the low scrub growing beside the road, not too far from where Luba and the boys waited. Residual fuel sloshed onto the ground. Once they had concealed the drums, Smith returned to the airplane, pulled out some objects, and returned to Luba.

"This is an inflatable life vest," he explained, holding an object toward her. "You pull this toggle to inflate it. Put this one on, and put the other two on your sons."

"Is this really necessary?" she asked, pulling the vest over her head.

"It is," replied Smith. "We're going to be over the Black Sea about fifteen minutes after we take off, and will be flying very low over the water for most of the trip. You will be glad you have that in case we're forced to ditch."

Luba noticed that the pilot appeared to be limping. "Is he injured?" she asked.

"He nearly died in a bad crash," answered Smith. "Now he has a wooden foot."

"Oh."

"Come with me," said Smith, throwing his canvas parachute bag onto his shoulders.

"Climb in," yelled Smith, heaving his bag into the cargo compartment. "We'll be in the air for a few hours, so you will probably be more comfortable sitting on your suitcases instead of that bare metal."

Once Luba and the boys were embarked, Smith retrieved the chocks, clambered in, and slid the side door closed. The pilot revved the engine, the plane lurched forward, and in seconds, roared into the air.

While still a teenager, Luba had taken flying lessons at a sport aviation club, so she knew the mechanics of what she was witnessing. She was fairly certain that the plane was a Swiss-made single-engine aircraft built especially for STOL—Short Take-Off and Landing—operations in high mountain terrain. The pilot had to be exceptionally competent and extremely brave to land his plane in the dark, on unfamiliar terrain, on a stretch of road barely as long as a soccer pitch.

So even if she didn't know who these people were or where they were headed, she trusted her husband and knew that she was in good hands.

**Headquarters of the *Glavnoye Razvedyvatel'noye Upravleniye (GRU)*
Khodinka Airfield, Moscow, USSR
5:15 p.m., Friday, November 3, 1972**

Yohzin's day of reckoning had arrived. After enduring five excruciating days of brutal interrogation at the Aquarium, his inquisitors had declared it was time for him to accept his final judgment.

His impending death sentence had endowed him with incredible clarity: absolutely shattered by his interrogation, he was more than ready to die. His eyelids twitched, and he was delirious from hunger and exhaustion. Blood-laced saliva dribbled from his lower lip.

If nothing else, the interrogators' methods were barbaric but extremely efficient. Yohzin was painfully sure that his tormentors had successfully encouraged him to recall the contents—including even the most mundane of technical trivia—of every single message that he had passed to the Americans in the past three years.

Probably the oddest aspect of their interrogation was that they believed that he was somehow responsible for the fate of the *Krepost*. Vasilyev had stopped communicating with Control shortly after Gogol had docked. For a while, before Yohzin was dragged away from Kapustin Yar last week, the prevailing theory was that both men had somehow perished in a confrontation, but the GRU now suspected that Yohzin had somehow sabotaged the mission when he was allowed to return to Control. His interrogators spent an almost inordinate amount of time in an attempt to force him to admit his complicity.

He was confident that Smith had managed to gather Luba and the boys, as he promised, simply because Federov had not brought his family here to the Aquarium. Certainly his interrogators would have used his family as leverage against him, but they obviously had been deprived of that option.

Federov, wearing full GRU dress uniform, entered his cell and announced, "It's time."

"Listen, Comrade Colonel, before I die," mumbled Yohzin, "I want to...apologize to...General Abdirov. I know that you sent him here."

Federov shook his head, and replied, "I'm sorry, Comrade General, but that's not possible."

"But is he not...in a cell down the hall?" stammered Yohzin. "I've heard his voice. I know he is close."

"He *was* in a cell down the hall, but no longer. He died yesterday of a heart attack. The doctor tried to revive him to finish his session, but it was too late."

Yohzin groaned in anguish.

"It's a pity you didn't have a chance to say farewell," said Federov, clicking his tongue.

"If I cannot see General Abdirov before I go, then can I at least say farewell to my family?"

"*Nyet*," replied Federov. "They are not here."

"They're not?"

Obviously embarrassed, Federov leaned close, and quietly said, "I suspect that you know where they are, even if I don't. They never returned to Kapustin Yar as you claimed they would."

Yohzin smiled to himself.

"Although your family had been spared from harm, I'm afraid to report that your precious dog was not so fortunate," declared Federov. The GRU colonel dipped his hand into a pocket and held out a photograph. With his abdomen slashed open and his pink guts stacked to the side, Yohzin's beloved Magnus was splayed open on a stainless steel table.

"In a week or so, after the taxidermists have finished stuffing him, your dog will be on permanent display in the Aquarium's museum," noted Federov. "I'm sorry that you won't be able to witness the exhibit."

"*Poshol nahuj,*" hissed Yohzin. His curse was barely audible.

"Stand up and come here!" ordered Federov angrily, beckoning him toward the opposite wall. Once nearly pristine at the beginning of the week, the wall was spattered with dried spots of Yohzin's blood.

Famished and thirsty, weak from a lack of sleep, Yohzin could scarcely push himself out of the chair he had occupied for the past five days. As he finally managed to stand up, he came to recognize the importance of having toes; now lacking most of the digits on both feet, he found it almost impossible to maintain his balance. Wobbling, struggling to stand even somewhat erect, he half-waddled and half-staggered toward the broad-shouldered GRU colonel.

"Major General Gregor Mikhailovich Yohzin, do you understand that you must die for your crimes against the Motherland?" asked Federov, drawing a diminutive Makarov automatic from a highly polished belt holster.

Grimacing with what remained of his face, Yohzin croaked, "I do."

"Turn to the wall," ordered Federov, sliding back the pistol's slide to chamber a round.

Yohzin pivoted clumsily to face the wall.

"Comrade General, was it worth it?" asked Federov.

Was it worth it? Yohzin had lost his dearest friend, but stopped the madman he had become, and in the process may have saved most of the world from destruction.

Was it worth it? He started to answer, but heard the flat crack of the Makarov firing behind him. His head snapped forward with the bullet's impact and his face rebounded against the concrete slab. His knees buckled as his limp body sagged downward. Sprawled on the floor, he was surprised that he didn't die immediately.

As he felt his limbs grow cold as his life drained away, he saw the faces of his beloved Luba and his sons. Even if he did not, they would have the opportunity to bask in the glow of freedom. *Was it worth it?* he thought, recalling Federov's last question. *Yes, it was.*

15

OPPORTUNITY KNOCKS

Arlington National Cemetery
10:55 a.m., Monday, November 6, 1972

"We need to chat, gentlemen," said Wolcott, slipping off his white gloves. In honor of Mark Tew, he wore his dress uniform; it was the first time he had donned the blues since retiring more than four years ago.

"Here?" asked Carson. He looked back toward the still-open grave. The official guests had departed over an hour ago, and the throng of mourners was just now dispersing; only a handful of well-wishers remained to express their condolences to Tew's widow and grown children. A pair of civilian caretakers slowly collected the huge array of wreaths and flower arrangements. Silent evidence of the groundskeepers' next chore, two shovels leaned against the side of their pick-up truck, which was parked next to a large pile of fresh dirt.

"Now?" asked Ourecky.

"Yup. *Here* and *now*. I need to chew the fat with both of you about some new developments." He led the two men toward the shade of a nearby oak tree.

"First things first," drawled Wolcott, stooping over to retrieve a pack of Winstons from his left sock. He stood up, shook out a cigarette and snapped off the filter. "Carson, provided you agree to abide by the deal that Admiral Tarbox has wangled for you, you're goin' to *Vietnam*."

"Vietnam?" asked Carson, removing his hat as he looked nervously towards Ourecky. "Vietnam? Uh, Virgil, we had talked about that as a possibility, but I thought it was contingent on..."

"No matter," answered Wolcott, waving his hand in dismissal. He lit his cigarette with an engraved Zippo lighter and then took in a deep draw. "If I'm nothin' else, pard, I'm a man of my word, as is Admiral Tarbox. I know that things didn't 'xactly work out as we agreed, but I feel like we promised you an opportunity to fly in Vietnam, so we're makin' good on it."

Grinning like he had just won an enormous jackpot in Las Vegas, Carson almost laughed at the absurdity of the situation. Since General Tew had always been so insistent that Carson would fly in Vietnam only over his dead body, Wolcott could not have picked a more appropriate place to break the news. But although he was thrilled with Wolcott's revelation, he realized that there had to be a knotty tangle of strings attached. "You said that I had to abide by some bargain that Tarbox has struck?" asked Carson.

"Yup." Wolcott looked toward the Pentagon, just a half-mile away in the east. "Right now, the admiral is over there workin' to drum up some additional funding, but he'll be back in Ohio tomorrow. We'll sit down for a powwow then. But let me forewarn you, son: some of his deal's provisions are pretty danged strict, and if you don't abide by them, the deal is *off*. Savvy?"

"Agreed," replied Carson. "But how about Scott? If I'm going overseas, is he going to MIT?"

Wolcott drew on his cigarette, blew out a pall of smoke, and shook his head. "Not immediately. He will remain assigned to the Project, at least until you get back."

Crestfallen, Ourecky muttered, "But General Tew promised…"

Gazing over the seemingly endless rows of headstones, as if making a tally of the deceased, Tarbox replied, "Here's a news flash, son: General Tew is *dead*. Admiral Tarbox is now our head honcho, and what he says goes. He wants you to stay here until Carson gets back. Period."

"But it's not fair," grumbled Carson. As the words left his lips, a formation of four F-4 fighters zoomed by overhead. As they passed over the massive cemetery, one aircraft split off, leaving the other three in the famous "missing man" formation. The same four aircraft had executed a similar salute for General Tew less than thirty minutes ago; this maneuver was obviously for a pilot recently killed overseas.

Wolcott waited for the noise to subside before speaking. "Fair? Son, it ain't about bein' fair or not," he snapped. "Carson, let me make this clear: Ourecky will stay at Wright-Patt until you get back from your sightseeing tour. And if *either* one of you insist on balkin' about it, then you can pretty much bet that any and all deals would come off the table. Do you understand, Carson? *Savvy?*"

Carson clenched his fists behind his back. "Yes, sir," he replied.

"And you, Major Ourecky, do *you* understand?" demanded Wolcott. "If you want to make some noise about the promises that Mark Tew made, then be my guest. But understand this, son, you might find a receptive ear and then zoom right off to MIT, but your buddy Carson would go no further west than Disneyland. Do you understand?"

Ourecky nodded. "Yes, sir," he croaked.

Wolcott stubbed out his cigarette against the bark of the oak tree and then "fieldstripped" the butt. Scattering the unburned tobacco, he softly said, "Look, boys, you have to trust me when I tell you that there is much more in the wind. Ourecky, I know how anxious you are to high-tail it out of here, but I have to tell you that Admiral Tarbox has some tremendous opportunities in store for you both. Really, I'm achin' to tell you about them, but I can't, at least not for the time bein'. All I'm askin' is that you be patient with me. Can you do that, son?"

Looking toward Carson, Ourecky swallowed deeply and replied, "Yes, Virgil, I can do that."

Auxiliary Field Ten, Eglin Air Force Base, Florida
2:35 p.m., Thursday, November 9, 1972

Glades had just emerged from the swamps, where he had been evaluating students during the last phase of the Ranger course, when he received an urgent message to report to Aux Field Ten. He had been in the field for the entire duration of the Ranger students' final twelve-day patrol, and hadn't even had the luxury to shower and change his uniform. His boots were caked with black muck, and his soaked OG-107 jungle fatigues stank from sweat and grime.

He parked his borrowed jeep in front of the headquarters for the 116th Aerospace Operations Support Wing at Aux Field Ten. As he strolled into the single-story, he had to maneuver around a string of airmen carrying out furniture, boxes, and filing cabinets.

Looking up from his packing chores, an Air Force master sergeant directed Glades into the office of Brigadier General Isaac Fels, the Wing commander. As Glades entered the office, he saw that Fels was using sheets of newspaper to wrap mementos and framed photographs.

As his uniform dripped water on the red carpet, Glades saluted and formally reported to the lanky, bald-headed Air Force general. "The commander at the Ranger Camp said that you wanted to see me, sir?"

"Nestor? Come on in and have a seat," declared Fels, cordially gesturing to a leather-upholstered chair in front of his desk. "I guess that you've heard that the Wing is being deactivated. The formal ceremony will be next week."

"I've heard you were shutting down, sir," replied Glades. "Seems like a shame to me. You've got a bunch of very well-trained troops here."

"No argument," said Fels, swathing an intricately carved wooden model of a CH-53 helicopter in newsprint. "But our mission is over. Anyway, Nestor, it's been quite a while since I've seen you. Back from Vietnam?"

"I rotated back in April, sir," replied Glades, dutifully occupying the plush chair that Fels offered.

"MACV-SOG again?"

Glades nodded. "Yes, sir, but I seriously doubt that I'll go back over again." With the South Vietnamese assuming responsibility for their in-country and cross-border clandestine missions, MACV-SOG had folded in May. It had been replaced by 'Strategic Technical Directorate Assistance Team 158,' a detachment that advised the Vietnamese special operations forces in their new role.

"Don't be too quick to assume that you'll never return," noted Fels.

"Do you know something that I don't, sir?" asked Glades. "As it stands, sir, I'm supposed to be here at Eglin through June, and then I report to Fort Bragg."

"Maybe." Fels placed a framed certificate in a cardboard box and then sat down in a chair next to Glades. "Nestor, I called you over here to ask a favor. I have a mission that's custom-made for you, if you'll accept it. It's strictly a volunteer assignment."

"A mission, sir?" asked Glades. "But didn't you just say that you folks were shutting down entirely?"

"This is different. *Very* different. You might consider it a personal favor. And as I said, it's entirely voluntary. If you're not comfortable with it, you can walk right out the door and I wouldn't think any less of you. Granted, it's your call whether you go or not, but I think I can dangle a little enticement to sweeten the deal."

"And what would that be, sir?"

"You won't be out of pocket any longer than five months, but once you're done, you can have your pick of any assignment or any post that you want. To be frank, Nestor, I've heard it through the grapevine that you're not too happy to be headed back to Fort Bragg, so this could be quite an opportunity for you."

Nodding, Glades replied, "You're right about Bragg, sir. I just never did blend in very well up there. The guys in the regular Special Forces groups play it a little too fast and too loose for me. But my choice of a follow-on assignment? I have to admit, that's a mighty powerful draw."

"I've already squared it with the personnel assignment branch at the Department of the Army. So, I have to assume that you're at least some-what curious about the task?"

"I am, sir."

"It's a stand-by assignment, sort of a preemptive contingency mission," explained Fels. "If you take the job, you'll go back over to Vietnam, where you'll put together a team to execute a search and rescue operation with limited or no notice."

Glades had been around long enough that he could read between the lines. Apparently, someone or something would be flying over Vietnam in the very near future, and it was obviously imperative that it—whoever or whatever *it* was—did not fall into enemy hands. He strongly suspected that the subject was the son or nephew of some high-ranking general or admiral trying to log some combat time before the curtains dropped completely. It was certainly logical; after all, John McCain, the son of a senior Navy admiral, the former Commander-in-Chief of the US Pacific Command, had been a prisoner of the North Vietnamese since 1967.

To avoid a similar situation, Fels was probably coordinating the expedient mission as a personal favor to someone high up in the food chain. Such arrangements weren't at all uncommon; the average American taxpayer would probably be amazed to know how many significant military activities—including major operations—took place only as the result of a friendly handshake between old acquaintances.

"Sir, am I to assume that this contingency is for someone in particular?" asked Glades.

Fels nodded. "It is, but you won't be exposed to that information unless there is an actual requirement to execute. To be honest, I seriously doubt that it will come to that, but we would like to have an ace in the hole just in case."

"So I'll be assembling a team? If that's the case, sir, I would like an opportunity to pitch this to a few guys. Most of them are stationed at Bragg, but a couple are assigned to the Ranger Department at Benning."

"So these are all Army people?" asked Fels.

Glades nodded.

"Nestor, write down the names of the men you want," replied Fels. "But I'll warn you: this is a very sensitive, close-hold mission that will rely almost exclusively on Air Force resources and personnel. Everything has to be kept strictly on the QT, so I seriously doubt that I will receive authorization to let you talk to anyone on the Army side. Sorry."

Glades pulled a notebook from his breast pocket, removed it from a thick plastic pouch, and used a pencil to jot down a list of names. He tore the page from the notebook and handed it to Fels. At this point, he strongly suspected that this mission was not entirely on the up and up.

"I'll see what I can do," said Fels, frowning as he examined the list. "But don't get your hopes up. But, on a positive note, you can have your pick of any of the men assigned here at the 116th Wing. Let me caveat that, though; a lot of our men have already departed Eglin, and they're scattered all across the planet."

Glades contemplated the offer. Most of the 116th's guys were at least as competent as the some of the men he had fought with in Vietnam, but he was still extremely leery of potentially staking his life on strangers. He had worked with a handful of Fels's men in Haiti just a couple of years ago and felt comfortable with working with them again, especially if this was only a contingency operation.

"Okay, sir. If that's the case, I have a few names for you. I'll definitely need a medic or possibly two. How about Steve Baker? He's a PJ."

Fels consulted some papers in a folder and then shook his head. "Baker *was* a PJ, but he's left the Air Force. He's at the University of Maryland, finishing his degree, with plans to go on to medical school."

"Ulf Finn?"

"Finn is still here. I'm sure he would be up for the challenge."

"Captain Lewis?" asked Glades, somewhat reluctantly.

"Died in a helicopter crash last year. Very unfortunate."

Glades scratched his head. "I have one more name. Matt Henson? I think he was one of your logistics support contractors."

Fels riffled through his paperwork for almost a minute and then replied, "Right after the mission in Haiti, Henson got bit hard by the PJ bug. He returned to active duty with the Air Force and went straight into the pararescue training pipeline. He's a fully qualified PJ, currently

assigned to the 40th Aerospace Rescue and Recovery Squadron at
Udorn Royal Thai Air Force Base in Thailand."

"So Henson's off limits, sir?"

Grinning, Fels replied, "Not necessarily. I'll make a few calls and see
if his services can be made available."

Milton, Florida
4:47 p.m., Thursday, November 9, 1972

Glades stood next to the washing machine, stripping off his filthy
fatigues, when Deirdre arrived home after picking up the kids from
their post-school activities. With his mud-caked web gear and rucksack
heaped in one corner, and his boots and uniforms piled in another, the
tiny laundry room was in shambles.

"You're home early," she said cheerfully, strolling in with a wicker
basket jammed with dirty clothes and football uniforms. "I didn't expect
you any earlier than tomorrow afternoon. Everything all right at the
camp?"

Standing naked before her, he gave her a quick kiss and nodded.
Turning around, he said, "Would you mind checking my back for ticks?"

"Oh, Nestor, darling, you're so romantic," she replied, nudging the
door closed with her elbow. "Check you for ticks, dear? You just sweep
me off my feet with such sweet talk."

Looking back over his shoulder, he grinned. "I'll make it up to you
after I grab a shower."

She laughed. "Well, it might have to wait until we put the kids to
bed. So, Ness, will you have to go back to work tomorrow? Is there any
chance you'll be home in time for the big game?"

"*Niceville Eagles?* I wouldn't miss that one. And to answer your question, no, I'm not going back to work in the morning, or this weekend,
either."

"Really?" she asked, lighting a kitchen match. She blew it out and
held the hot tip to a tick burrowed into the small of his back. In seconds,

the squirming tick dropped to the floor. "And why would you not be going to work this weekend, my love?"

"I have a mission. I fly out on Monday morning. Vietnam. I'll be gone about five months."

Deirdre gasped. "*Vietnam?*" she mumbled. "*No...*"

Although he didn't expect her to be overcome with joy, her reaction wasn't what he anticipated. Usually, she exhibited a seemingly nonchalant attitude about his comings and goings. After all, she was the daughter of a professional soldier in the toughest regiment in the British Army. She was descended from a long line of soldiers and soldiers' wives; the women in her family had been sending men off to battle for generations, all the way back to the Gaelic and Celtic eras. They had probably developed their detached demeanor as a defensive mechanism against the prolonged anguish of waiting for their warriors to come home, and the terrible grief when they didn't.

Glades turned to face her and was taken aback by the shocked look on her face. "You look like you've seen a ghost," he said.

"I had a dream last night," she replied, leaning against the dryer and slipping down to the mud-streaked linoleum floor. "Really, it was more of a nightmare. Nestor, it just really spooked me. I woke up crying."

Glades suspected that, given Deirdre's ancestral background and the fact that she had spent her childhood and most of her teen years watching her father's SAS troopers leave for dangerous places, and then watching as many did not return, she probably possessed something of a sixth sense about such matters.

"Was I killed in your dream?" he asked, sitting down beside her. He cupped her head in his hands and felt her warm tears on his bare shoulder.

"No," she stammered, trembling against his chest. "Much worse than that: you were captured. Nestor, we've talked about this, and as much as I could accept you being killed, the thought of you being captured just terrifies me. I just don't think I could bear it."

"Tell me: in your dream, did you see me being taken prisoner?"

"No. Actually, I never saw you at all."

"Then why did you think I was captured?"

Tears streamed down her face as she replied, "Because in my dream, I saw the inside of a cell in a POW camp, and your name was written on the wall."

16

WESTBOUND

Dayton, Ohio
3:30 p.m., Saturday, November 11, 1972

Grasping a bouquet of lilies and a bottle of white wine, Ourecky lightly rapped his knuckles on the front door. He hadn't seen Bea or Andy since returning from the final mission, and he was both anxious and apprehensive.

Bea opened the front door quietly. "*Shhh*," she whispered, pressing her finger to her lips. "Jill is asleep, and the little ones are also down for a nap."

"Okay," he replied, handing her the flowers and wine. "For you."

"Oh, thank you, honey," she said quietly. "Beaujolais? My favorite!"

He stooped over to slip off his shoes and then stood to hug her tightly. "I missed you, Bea."

"I miss you always, Scott," she answered. "Come on in."

In his stocking feet, he padded after her as she guided him to the kitchen.

"I'm sorry about General Tew," she said, taking down two wine glasses from the cupboard. "I wish that I could have gone to the funeral, but...."

"It's okay."

"How was his wife? It sounded like it was all so sudden. I wouldn't be surprised if she was just a wreck."

"Not as much as you might think," answered Ourecky, opening the bottle with a corkscrew. "I guess that you hadn't seen him in a while. He definitely wasn't in the best of health. As for his wife, she seems to be taking it in stride. I talked to her briefly after the funeral, and I think she was really surprised that he had lived as long as he had."

"Well, it's sad. He was such a sweet man."

Smiling, Ourecky said, "I don't know if I would necessarily describe him as *sweet*, but he was a good man, a good boss."

"So, does that mean Virgil Wolcott is in charge now?" she asked, holding out the stemmed glasses.

"No, they've brought in a Navy guy," he answered, pouring the wine. "Admiral Tarbox. I don't think you've met him."

"Navy? Interesting."

"It's part of a big consolidation," he explained. He sipped the warm Beaujolais and frowned; he wasn't nearly as fond of it as she was. "To save money, the Department of Defense is combining some programs. Mostly, it's to cut down on duplication of effort. Of course, with the Navy assuming control, there will be some big changes. A lot of folks will be laid off here at Wright-Patterson, and most of the...uh...test operations here will shift to California."

"California? Really? I suppose there are worse places to be."

"I suppose. I'm going out there next week."

"Permanently?" she asked, setting her glass on the counter. "I thought..."

"No, next week is just a visit. Just a couple of days."

"Well, how about the election?" she asked, gesturing toward a dog-eared McGovern flyer taped to the front of the refrigerator. "Are you going to be able to vote?"

"Yeah. I'm not supposed to leave until Thursday."

"So, since you finished that last big test, aren't you done here?" she asked, filling a glass vase with water from the tap. She took an aspirin from a bottle on the counter, crumbled it in a spoon, and poured the white powder in the water. "Are they going to make good on Mark Tew's promise to send you to MIT?"

"Not immediately," he replied. "I have to stay here for a few more months." He glanced at her face as she arranged the lilies in the vase; her expression clearly conveyed that she wasn't happy.

"I suppose they have to squeeze just a little bit out of you," she said. She drained her wine and then refilled both glasses. "Some things never change."

"Bea, I'm confident that I will eventually go to Cambridge," he said, changing the subject. "It will probably be around the end of March. I thought we might fly up to Boston some weekend soon and start looking around for an apartment and…"

She sighed, shrugged her shoulders. "Scott, how many times have I heard *that* story? And it always ends up the same: you're right on the verge of going, and then there's an urgent call from Virgil Wolcott, and you vanish yet again."

"But this time it will be the real thing."

"*Sure* it will be, Scott, but the Air Force has promised this same thing, over and over, but has yet to deliver. Why should it be any different this time?"

"It will be. I promise. Virgil says I'm going. We will just have to wait a little longer than I had expected."

"I can believe *you*, Scott, but I don't have much faith in Virgil's word. Until you're finally free of him, I don't think that anything will ever change."

Quickly changing the subject, he asked, "I did tell you that Drew's gone, right?"

"You did. When you called last week. He going overseas?" she asked. "Isn't that what you said? Vietnam?"

He nodded. "He's training in California now. He probably won't leave for Vietnam until next month at the earliest."

"But isn't the war all but over?" she asked, fingering the silver peace symbol dangling from a thin chain around her neck. "I just don't understand why the Air Force is making him go *now*. Aren't we bringing our guys home?"

He wasn't sure how he could possibly explain to her that Drew had been relentlessly lobbying for this opportunity as long as Ourecky had known him. And the *Air Force* definitely wasn't sending him. In fact, the Chief of Staff of the Air Force had personally decreed that Carson *not* go, so it was a safe bet that he would blow a four-star gasket if he ever discovered what was going on behind his back.

Holding the wine bottle in one hand and her glass in the other, she nodded toward the living room. He followed her, and joined her on the couch.

"If it's any consolation, Bea," he said, "I suppose that you can see that I won't be working with Drew anymore, so…"

"That's beside the fact, Scott. As much as I don't want you flying with Drew, I don't want him going overseas. It's not just Drew. I don't want *anyone* to go. I think too many guys have been killed and hurt over there as it is, and I don't think anyone has any real clue what it was all for. It all just seems so tragic."

They both turned as they heard a faint noise in the hallway. Rubbing sleep from his eyes, sucking his thumb, dragging a well-worn blanket in his wake, Andy waddled into the living room.

"Andy, look who's here," announced Bea.

Squealing with glee, Andy dropped his blue blanket, ran to Ourecky, and jumped headlong into his outstretched arms. "Daddy!" he cried.

"How is he?" asked Ourecky, gripping him tightly.

"He misses you. Otherwise, he's fine. Besides, he has a new best friend. He and Rebecca are thick as thieves."

"Rebecca?" he asked.

"Jill's little girl. She's a year older than Andy. I'm not looking forward to when we leave here and I have to tear them away from each other. That's going to be traumatic."

Ourecky smiled to himself. Her comment was a favorable sign; although she had not otherwise said anything about it, Bea was obviously thinking of the time when she would come home. Maybe there was some hope for the future. He smiled at Andy, kissed the top of his head, and held him closely.

A few minutes later, Rebecca joined them. Yawning, clutching a Raggedy Ann doll to her chest, she plopped down on the floor amidst the scattered toys. With shoulder-length raven hair, she looked like a younger version of Jill.

Andy wiggled out of Ourecky's lap to join his playmate. "Well, that sure didn't take much," he commented, watching them frolic. "I guess I'm all but forgotten now."

"Don't take it so personally," noted Bea. "It's not just you. Both of us might as well be invisible at this point."

"Boy, you weren't kidding. Those two sure get along well," he observed, watching as the two children noisily romped on the floor.

"That's putting it mildly," she replied, laughing. "They're inseparable, like two peas in a pod."

"So what happens to her when…"

"When Jill dies?" she asked matter-of-factly. "Jill's mother will keep her. Jill has already been to a lawyer to draw up the papers. It's in her will, also."

Hand in hand, Andy and Rebecca approached the coach. "Can we have a popsicle, Aunt Bea?" asked Rebecca, grinning. "We'll share it."

"It's too close to supper." Bea shook her head. "I don't want you two to spoil your appetite."

"What are we having?" asked Rebecca, smoothing her doll's crown of yarn hair.

"Spaghetti and meatballs." Bea gazed toward Ourecky and smiled. "Can you stay for dinner?" she asked, lightly touching his forearm.

"Spaghetti and meatballs? How could I *possibly* resist?" Ourecky grinned at her and then looked at the children. Suddenly, he started coughing as he noticed something he hadn't seen before. Rebecca's eyes were crystalline blue, almost unnaturally so, just like...

"Are you all right?" asked Bea. "You look like you've seen a ghost."

He gulped down the remainder of the water. "Those eyes," he gasped, staring into Rebecca's face. "Her eyes. They look just like…"

"That's right," noted Bea. "Scott, I love you, but for such an intelligent man, sometimes you're not very observant. We're pretty sure that Drew is her father. I can't believe that you're just now catching on. You've seen her before this."

"I guess I just never made the connection before," he muttered. "I can't *believe* that Jill could put herself in such a situation."

"*Jill?* You can't believe that *she* put herself in such a situation? Like she did it all by herself? Are you that quick to let your buddy Drew off the hook?" Have you ever heard that expression *'There but for the grace of God go I?'* Well, I've done plenty of stupid things myself, and I can tell you that there have been moments in my life when I could have been in exactly the same circumstances as Jill. I was just very fortunate."

"But she never said anything to Drew…"

"Jill doesn't want Drew Carson in her life, and she doesn't want him in Rebecca's life, either," said Bea. "She regrets ever meeting him, much less having his child. It was an accident, and she's tried to make the best of it."

Ourecky recalled their conversation in Idaho, when Carson described his concern about the prospects of dying by himself and not leaving anything or anyone behind. "Don't you think Drew has a right to know that he has a daughter?" he demanded. "Especially since Jill is dying? What if he wanted to raise her?"

Bea grimaced. "Do you *really* believe that Drew is even capable of raising a child? He has plenty of growing up to do himself. Besides, it's not an absolute certainty that Rebecca is *his* daughter."

"You don't think it's obvious?"

"Well, yeah, I'll admit it looks pretty certain, but there's no way to be really positive. They can do blood tests, but that doesn't prove anything absolutely. Even if everything matches up, it would only prove that he *could be* her father, not that he *is*."

Ourecky coughed and then replied, "I still think he has a right to know."

"Maybe, but that's Jill's decision to make, not ours. She has to decide what's best for her daughter, and she may not want Drew to be involved in her upbringing."

"Would you at least talk to her about it? He should at least have the opportunity to meet her, if nothing else."

Bea nodded. "I'll talk to Jill, but are you sure that you want to broach this issue with Drew *now*? He's going to Vietnam. I'm sure that he has enough on his mind and all this would do is distract him at the very time that he doesn't need to be distracted. If Jill agrees, then maybe we can talk to Drew when he comes back home. He's going to be gone about six months, right?"

"Right," answered Ourecky. He thought that Bea was probably right; going into combat, Carson needed to focus on the immediate matters at hand. If Jill agreed, then he and Rebecca would have a whole lifetime together, if he decided to bring her into his life.

She stood up. "Look, I have to start dinner. Your garlic meatballs aren't going to make themselves."

He stood beside her and held her forearms. "Bea, I want you to come home. I want our son to come home. Don't you understand that?"

"I *want* to come home, but it's not that simple. I need to know that you're going to be there."

"I will."

"Then when *that* time comes, we can all be together. You're always welcome here, but this is where I belong right now. I need to be here for Jill. She doesn't have long, and her mother just isn't handling it very well. I really need to be here. Can you understand that?"

"I guess so."

"Speaking of Jill, I have to help her with her shots," she said, looking at the wall clock. "Oh, I hate doing the shots. I'm so afraid that I'm going to hurt her."

"Shots? I can help you if you want," he said. "I know how to give shots."

"You do?" she asked, raising her eyebrows. "Well, Scott Ourecky, you just never cease to surprise me with the things you can do."

17

BIG CAN

Naval Space Operations Vehicle Assembly Facility
Vandenberg Air Force Base, California
10 a.m., Friday, November 17, 1972

Before he left Ohio yesterday, Wolcott had assured him that a big surprise lay in store, and he was absolutely right: it was a bombshell that Ourecky could have never anticipated. He joined Wolcott and Tarbox as they walked into a glass-enclosed observation room overlooking a massive assembly bay. Dressed in white coveralls, workers scurried about, preparing a huge yellow-painted cradle to receive a large cylindrical object suspended from a huge overhead crane. Sheathed in protective white film, the massive object looked like a giant Christmas present waiting to be opened.

"So, son, do you have any notion about what you're lookin' at?" asked Wolcott. "Care to speculate?"

Initially, Ourecky suspected that it was an intermediate stage for a Titan IIIC booster, since its dimensions and rough configuration looked approximately right, and then realized that it looked a little more familiar than he had anticipated.

"This is the *second* ocean surveillance MOL," declared Tarbox, confirming Ourecky's suspicions. "After the mission in August, we started to put all the MOL hardware in mothballs and shelve the entire project. Then we realized that the solar storm was an anomaly that no one could have predicted. Granted, if the comms gear had been working, we could have warned those boys and got them down safely. Anyway, except for the comms gear, all of the other systems—including the reactor—worked perfectly. We've resolved the issues with the comms equipment. Instead of shutting down, we intend to pursue operations."

"Savvy, Ourecky?" interjected Wolcott. "That danged rascal ain't goin' to be tucked into mothballs after all."

"We have tentative approval to launch next year," declared Tarbox. "We're awaiting on the final approval, at the highest levels of command authority, before it's a solid Go."

"But I don't understand, sir," said Ourecky. "Why is this relevant to me? Why are you showing me this?"

"We want you and Carson to fly it."

Staring at the MOL, Ourecky swallowed deeply. His mind reeled with the potential angles being worked by Tarbox; a consummate master of political intrigue, the admiral rarely did anything without some ulterior motive. Clearly, he had the President's ear, which explained why this particular revelation had to wait until after the Presidential election.

Although Ourecky knew that he couldn't slow the momentum of a speeding locomotive, and the Project would continue roaring down the tracks regardless of what he did or didn't do, he clearly knew that he had to disembark from this train, once and for all.

"Son, did you *not* hear me?" screeched Tarbox. "We want you and Carson to fly it."

"Sir, I'm *not* interested," blurted Ourecky.

"Before you up and do somethin' rash that you'll regret for the rest of your life, pard, you need to listen to the admiral's proposition," countered Wolcott. "After all, this ain't just about you, son, it's about Carson as well."

Tarbox cleared his throat and said, "Major, I understand your reluctance, but allow me to clarify a few things that might just change your mind. I'm sure that you're aware that the Air Force publically cancelled their MOL program in 1969, and that the Navy picked it up—in secret—afterwards. To make a long story short, I have convinced the National Command Authority to fly this next mission in public view, at least to a certain extent. The launch will be televised, the flight crew will be announced, and the public will be aware that we—the Navy and the Air Force, in unison—are conducting the mission. Beyond that, most of the flight will still remain classified; it will be billed as a reconnaissance mission, and we'll just leave it at that."

"So it will all be in public view?" asked Ourecky.

"I think that's what the admiral just said," said Wolcott, grinning ear to ear. "A public launch, for all the world to see!"

"But why?"

"Why did the NCA decide to throw back the veil?" asked Tarbox. "It's simple, Major. The last Apollo lunar mission will launch next month. Next up is Skylab, which should launch next year, and then there's a possibility of a US-Soviet joint flight in a couple of years, but that's it. Moreover, with Skylab, NASA is putting all their eggs in one basket. It's a complex machine, and if there are any major problems before the crew docks to occupy it, then the project could be over before it starts. In the meantime, the Soviets are rapidly moving forward with plans to send up a series of manned space stations..."

"Like the *Krepost*, sir?" asked Ourecky.

"No. Hopefully not, anyway. They plan to transition to a civilian station called the Salyut. So, I convinced the NCA that since we have the MOL hardware, this is an excellent opportunity to continue flying. We have three flight-rated MOLs ready to fly, so the United States can maintain a visible presence in space for two years or more, regardless of what happens with Skylab."

"And the crews would be publically announced?" asked Ourecky.

"Yup," replied Wolcott. "And we're very aware that your better half would have a tough time swallowing that one, but I'm sure that Bea will get past her misgivings when she sees your face on the cover of *Life* magazine!"

Yeah. Sure she would, thought Ourecky. *And Bea is plenty smart enough that she would connect the dots to figure out what I've been doing for the past three years, and then my marriage would really be over.*

Tarbox interjected, "This is not a done deal; we have tentative approval but won't get a final blessing until April of next year. But if this flies, and we're confident that it will, you'll be delayed a year going back to school, but we'll hold fast to Mark Tew's promise. You'll go to MIT in the fall of 1974."

Obviously sensing his hesitance, Wolcott nudged Ourecky's shoulder and urged, "But just think about it, son, you'll go to MIT as a bona fide, publically recognized *astronaut*. Granted, there won't ever be any public acknowledgement of your other flights, but what an opportunity this is!"

"With all the sacrifices you've made, we feel that we owe you this, Ourecky," said Tarbox. "You and Carson both."

Ourecky contemplated the opportunity. After nine flights jammed into the tiny cockpit of the Gemini-I, he was thrilled at the notion of spending unfettered days in the MOL's spacious cabin. But he was also painfully aware that the public MOL flight would almost certainly result in the dissolution of his marriage, particularly if Bea put the pieces together about the previous flights.

"I'm sorry, Admiral," said Ourecky in a pained voice. "As appealing as all this sounds, I'm still going to pass."

As Tarbox's face turned crimson, Wolcott cleared his throat and said, "Son, I figured out that you were a hardheaded danged cuss back when you turned down my offer to let you earn your wings, but this is entirely different. As far as me and the Admiral are concerned, you and Carson have earned this ride a thousand times over, but on a practical note, this offer ain't goin' to stand indefinitely."

"How so, Virgil?" asked Ourecky, looking out into the massive bay as the shrouded MOL was carefully mated with the storage cradle. Rotating yellow caution lights flashed and warning horns blared.

"This is a package deal," answered Wolcott. "It's a package exclusively for you *and* Carson. If you decide to jump on this horse, then we'll start preparin' you immediately. You'll spend some time out here in California, workin' with the hardware, but there will also be a lot of academic work back in Ohio. Once Carson gets back from his boondoggle, he'll fall right in with you, and you boys will probably fly about this same time next year."

"But if you're still reluctant and you decide *not* to commit," asserted Tarbox, regaining his composure. "Then we will have to immediately begin the process of preparing another crew. Timing is of the essence, particularly since we don't have any other crews with your expertise and level of flight experience."

"So, bub," interjected Wolcott. "As you ponder your reply, bear in mind that you're not speakin' just for yourself, but Carson as well."

Ourecky cringed as if he had been punched in the gut. Obviously, Carson would be absolutely elated if he were present to hear the Ancient Mariner's pitch. It was a no-brainer: even though Ourecky was content to remain in the shadows, the public flight would give Carson the recognition that he coveted and clearly deserved.

"Obviously, for some very practical reasons, we can't share any of this with Carson right now," said Wolcott. "And you should be forewarned, Ourecky, if *you* say anything to him before he gets back, all deals are off, to include Carson's cruise as well as your future academic dreams. Therefore, I urge you to keep your trap shut."

"Yes, sir," muttered Ourecky.

"Since Carson is not here, and since you speak for him in his absence, here's what I propose," said Tarbox. "Unless you want to decline this opportunity outright, today, let's just assume that Carson will be receptive to this plan when he returns. In the meantime, you'll begin your familiarization training with the MOL-specific systems and reactor operations. Do you concur?"

Ourecky felt trapped. "Yes, sir," he croaked quietly.

"What was that, son?" asked Tarbox. "I didn't hear you."

"Yes, sir. I concur, sir."

"And there's one other slight wrinkle," announced Wolcott. "Ourecky, you'll fill the *left* seat for this mission. I figure that's only fair, since you'll be busting your butt studyin' and trainin' while our friend Carson is gallivantin' overseas, getting his ticket punched."

Command pilot? Startled, Ourecky was speechless.

Grinning, turning to look towards the MOL, Wolcott drawled, "Shucks, pardner, I reckon that your stars have really lined up, haven't they?

Da Nang, Vietnam
8:45 a.m., Saturday, November 18, 1972

Walking out of the sandbagged plywood team house, Nestor Glades paused to soak in the familiar scene. A groaning bulldozer shoved a massive pile of red dirt to reinforce an earthen revetment. Manning a perimeter enclosed by multiple belts of concertina wire, bored South Vietnamese soldiers lounged behind machine guns, mortars and recoilless rifles. The still morning air was heavy with the noxious stench of human waste being burned in diesel fuel.

Glades had been levied for some odd assignments in the course of his career, but this was by far the strangest errand handed down to him. The mission was unusual in many ways. First, as best as he could determine, the mission focused on snatching one particular man out of harm's way, but neither Fels nor anyone else seemed predisposed to reveal the man's identity. Despite this, he was granted unprecedented latitude and resources to prepare for the standby task.

The pre-mission briefings were vague, except that it was almost certain that if the trigger was pulled, they would go into North Vietnam. And although there was virtually no likelihood that the mission would be executed, they had to be prepared to launch on an extremely short notice. The window for the potential mission would open in November and would likely last three to four months at most. So between now and then, they would go into seclusion, assuming a fireman's existence, training and preparing as they patiently waited for the bell to ring.

Although it would entail a ground incursion into North Vietnam, Glades still harbored lingering doubts that the mission wasn't officially sanctioned at the highest levels of the military. He personally knew most of the men who participated in the gutsy raid on the Son Tay POW camp—which turned out to be dry hole—so he was aware of the considerable preparations prior to that mission. It wasn't a casual endeavor by any means. In contrast, this job seemed to be planned on the basis of a wink and a nod, with Fels quietly cashing his wealth of personal chips with key leaders within the Air Force Rescue and Recovery Service and other organizations. Consequently, very few individuals were privy to the plans, and Glades was effectively fenced off from the Special Forces operators who were most proficient in executing this sort of mission.

Except for a handful of former MACV-SOG personnel who would coordinate logistics and planning, there were no Army personnel involved. Fels had made good on his offer to provide two Air Force sergeants—Matt Henson and Ulf Finn—whom he had worked with in Haiti. Finn had accompanied him from Eglin, and Henson just arrived this morning from Thailand. They were both a little rough around the edges, but he trusted them, and hopefully there would be adequate time to train them before they were called to action. As with the MACV-SOG recon teams he had led in the past, the team would be rounded out with indigenous soldiers familiar with the operational area.

But there was something different about this job, something that he couldn't pin down. He couldn't shake the unsettling feeling that his luck may have finally petered out, and that he might not make it out of Vietnam alive. It seemed as if Deirdre was of the same mind. Even after she had revealed her dream to him, and after he had convinced her that he would be safe, their departure embrace had been much more prolonged than usual. Unlike her usual lighthearted hug and peck on the cheek, it was if she was more than reluctant to let him go.

He shrugged, pushing gloomy thoughts from his mind, and focused on what had to be accomplished. Since he was assured that his new team would not go operational any earlier than mid-December, he proceeded at an almost leisurely pace. Accompanied by Finn, his first stop was the supply room, where a large consignment of mission equipment awaited. With his virtual carte blanche, he was granted anything he asked for, within reason.

He and Finn entered the dimly lit supply room. A Johnny Cash tune played from speakers wired to an Akai reel-to-reel tape deck. Seated behind a field desk, a ponderous supply sergeant—Ted Blair—perused the latest edition of *Playboy*. Blair's enormous gut flopped over his straining belt; it looked like he had a reserve parachute stashed under his faded green T-shirt. In any other circumstances, Glades would have been thoroughly disgusted, but he knew that Blair was a former "One-One" assistant team leader who had paid his dues and then some.

Once in peak physical condition, Blair was slowly recovering from a broken back sustained in a helicopter crash that killed most of his team over a year ago. Although thrown clear of the wreckage, he had run into the roaring flames to drag two of his teammates and the pilot to safety. While most MACV-SOG personnel had long since redeployed, he had drawn this cushy assignment because he was married to a Vietnamese nurse who was waiting for a work visa to accompany him to the States. Addicted to a mix of painkillers and powerful muscle relaxants, which he gulped down by the handful, Blair was marginally functional at best.

"All that stuff's yours. Everything you asked for. Just arrived this morning," said Blair, waving his hand toward several wooden loading pallets stacked with crates.

"Will I need to sign for it?" asked Glades, shivering. A window-mounted air conditioner ran at full blast, spewing out air frigid enough to raise goose bumps, but Blair still sweated like a fat missionary invited to a cannibal tribe's pot luck supper.

"Hell, Nestor, no one's signing for anything anymore," replied the bored-looking sergeant. "I have orders to hold this junk until you no longer have any use for it, and then I'm supposed to torch what's left over." He gestured at a crate of thermite incendiary grenades next to his desk.

Nodding, Glades tore brown paper wrapping from a brand new AK magazine. Examining the curved magazine, he recalled periods of scarcity in the past, when his teams had to scavenge, scrounge and barter for items as simple as radio batteries and blasting caps. After becoming so accustomed to deprivation, it was difficult for him to grasp the notion of almost inexhaustible plentitude. Of course, there was a very logical reason. He was aware that an entire warehouse was still crammed with off-the-books "non-attributable" gear and supplies; for various reasons, most of the goods would be destroyed instead of being handed over to the South Vietnamese. Most of it was captured material, but there was also a lot of ordnance and sundries from other countries. Some of the third-country armaments, such as ammunition from Warsaw Pact nations, had been obtained surreptitiously, while much had been sold or donated to MACV-SOG by countries not overly eager to advertise their relationship with the United States.

Circulating amongst the pallets, Glades and Finn surveyed the booty. One wooden crate was packed with spanking new AK-47 assault rifles, still coated in sticky brown cosmoline. Large burlap sacks swelled with NVA uniforms, pith helmets, rucksacks, and chest webbing. There was a collection of Soviet-built radios, as well as rocket-propelled grenades and their tube-shaped launchers. Grinning, Finn looked like a kid in a candy store, almost drooling over the abundance of equipment.

A stack of light blue metal containers, resembling oversized sardine cans, contained Israeli-manufactured AK ammunition earmarked specifically for the actual mission. The "sterile" Israeli rounds bore no case stampings or other identifying markings, and were produced to the highest standards so as to be absolutely reliable. Wooden crates with East German and Yugoslav labels contained the less trustworthy Warsaw Pact ammunition they would expend in training.

Not all of the allotment was foreign-made. Some of the stuff Glades had requested was proudly made in the good ol' *US of A*, as American as Ford, Chevrolet, Mom, and apple pie. Glades was a discriminating connoisseur of lethal goods, and there were certain instances where only the best would do and cheap substitutes just could not be accepted. Consequently, one pallet was filled with US-made C4 plastic explosives, Claymore directional mines, silenced Hi Standard pistols, and a pair of M-79 grenade launchers with an assortment of rounds.

Kneeling down, Glades picked over the pallet to ensure that it contained the M14 "toe-popper" mines he had specifically asked for. Intended to deter pursuers, the little mines were about the size and shape of half a soup can. They were easy to emplace but not designed to kill; rather, they contained just enough explosive force to mangle the foot of anyone unfortunate enough to step on them. And of course, besides materials intended to maim and murder, the pallet contained a large assortment of medical supplies requested by Henson.

Glades listened to a helicopter landing at the small helipad within the compound. He turned to Finn, pointed at three of the pallets, and said, "Go grab Henson and tote that stuff over to the team house. I'll be over as quick as I can to lend you a hand. We'll get the rest tomorrow."

As he headed toward the door, he heard Blair's voice behind him. "Let me know if you need anything else, Nestor. Official or otherwise."

"I'll do that," replied Glades. He thought for a moment about a problem which had been vexing him. "Are you still qualified as a parachute rigger?"

"I am," replied Blair. "Do you want me to scrounge up some parachutes for you guys? I have a wire on some HALO rigs if you're interested. Top quality stuff."

"No. I have something else in mind. Let me think about it, and I'll get back to you."

Stepping out into the bright sunlight, Glades squinted. He saw four men climbing out of a Huey helicopter on the helipad, and immediately recognized a South Vietnamese officer. The officer—Major Phan Lac Lahn—was a past member of the South Vietnamese LLDB Special Forces and had formerly been assigned to MACV-SOG as a liaison for intelligence. He apparently would be working with Glades and his team in the same capacity.

"*Glades?*" implored Lahn, extending his hand. He spoke perfect English with a Southern California accent; he could have easily passed as a used car salesman in Los Angeles. In the style so prevalent amongst Vietnamese officers, his camouflage uniform was tailored so snugly that it could have been a second skin. "This is *quite* a surprise. I didn't think you would be coming back here."

"Neither did I," answered Glades. "I understand you have some men for me, Major Lahn." He was never particularly trusting of anyone in the South Vietnamese Army, particularly Lahn.

"I do," replied Lahn, handing over three dossiers. He pointed toward the trio unloading duffle bags and footlockers from the Huey. "You requested former NVA soldiers. Here are the top three. I have several more if these don't work out."

Glades leafed through the documents. The former enemy soldiers, now members of the Vietnamese SMS—Special Mission Service— had each previously participated in at least one clandestine infiltration of North Vietnam. Although they smiled and seemed eager to please, Glades could tell that they were incredibly tough men, intensely hardened by years spent at war.

Lahn introduced Glades to the trio. Their captain—*Dai Uy*—was Cao Dihn Quan, a wiry but studious-looking former North Vietnamese field artillery officer. Sergeant—*Trung si*—Liu Xuhn Hieu was a former NVA sapper who had participated in several raids on major US military bases throughout the South. Half-Chinese, Hieu was huskier, taller and lighter-skinned than most ethnic Vietnamese. *Trung si* Trihn Van Dinh was a communications specialist who had served only a year in the infantry before being captured. With a pockmarked face that was a lasting souvenir of childhood smallpox, he was a small man, even by Vietnamese standards, but could clearly hold his own. As he shook their hands, Glades felt confident that he could build an effective team between the three SMS commandos and his two Americans.

9:55 a.m.

As the former NVA soldiers unpacked their gear and made themselves at home, Glades, Finn, and Henson collaborated with a mission planner who was responsible for refining the operational aspects of the rescue mission. An Army captain, Al Coleman, was a competent and meticulous planner, but bore an air of Walter Mitty about him. Constantly alluding to heroic—and almost certainly imaginary—exploits, Coleman obviously perceived of himself as a highly skilled clandestine operator.

"We've been burning the midnight oil. I think we've covered everything, except for one important piece." Coleman gestured at the map. "We're fairly certain we can extract you, unless it gets *really* hairy. We're just not sure how to insert you in there. We need your input. We're receptive to just about anything you might suggest, within reason."

"Well, I'm assuming we're coming out by rotary wing," drawled Glades, looking at the map while making mental calculations on distances and times. "Affirm?"

"Correct," answered Coleman. "If your mission launches, the Air Force will exclusively commit a heavy package, which will be on stand-by from the moment you depart this location. You can expect at least three CH-53s to be at your disposal, as well as plenty of fast mover close air support, and about all the Combat Air Patrol in theater to cover your extraction. When and if the time comes, you folks will be the tip-top priority in the theater."

Glades laughed silently to himself; he had heard that line plenty of times in the past. "Well, obviously the only practical way out of there is a helicopter ride, so rotary wing is off the table for the insertion," he said. "Too much noise, too much risk, and unless I don't have a choice, I don't like to show the same card twice. Do we have a Combat Talon available?" Glades was referring to the MC-130 "Combat Talon" variant of the venerable Lockheed Hercules. The turboprop transport, loaded to the gills with classified electronic warfare equipment, was expressly configured for clandestine infiltrations deep into enemy-controlled areas.

"Sure. All you have to do is ask," said Coleman. "Are you looking to jump in?"

"Maybe. Right now, I'm just figuring options. A HALO jump is definitely out, and our indig troops haven't jumped in months, so even a static line jump would be a stretch."

Coleman pushed his black-framed issue spectacles up on his sweaty nose. "I wouldn't rule it out. There's plenty of time before your window opens, so you shouldn't have any problem bringing your guys up to speed. I can coordinate any resources you need for training. And I wouldn't mind bustin' some clouds with you myself, if you're going to do any HALO jumping."

Glades shook his head. "I don't think so. I know it looks like we have an abundance of time, but we don't. I would rather focus our training time on what we'll be doing once we're on the ground. Besides, jumping just throws too many variables into the mix. Even if things go smoothly, it eats precious time to assemble after the jump. And when things don't go so smoothly, people get scattered, lost and hurt. I'm still interested in the Talon, though."

"Well, if not HALO or static line, what exactly do you have in mind?" asked Coleman.

"When I did an exchange tour with the Brits, I used to hop the tram down to the Imperial War Museum in London," explained Glades. "They had a bunch of German stuff on display from the war. Everyone knows that the Allies used to parachute operatives into France, but most folks don't realize the Germans were dropping agents into the UK at the same time."

Glades continued. "The Germans didn't want to waste time teaching their agents to jump, so they came up with something called a *PersonenAbwurfGeraet*, which meant 'Personnel Dropping Device.' I saw one on display at the museum. It works like this." He opened his notebook and pointed at a sketch he had been working on.

Enthralled, Coleman grinned as Glades finished articulating his concept. "Well, obviously not a carnival ride for the timid," he commented. "But it'll be your asses on the line. If that's the way you want to go, we'll make it happen."

Finn, on the other hand, turned white as a sheet. Grimacing, he implored, "Nestor, you're kidding, right? Please tell me that you're kidding."

"Oh, not in the least," replied Glades.

18

FOOLPROOF PLAN

Alameda Naval Air Station, California
3:35 p.m., Saturday, December 2, 1972

Carson had phoned last night to let Ourecky know that he would be leaving today for Southeast Asia. He and his RIO—Radar Intercept Officer—had just graduated from Top Gun at Miramar and would link up with maintenance holdovers and various stragglers from the fighter squadron. After launching from Alameda Naval Air Station, on the outskirts of San Francisco, they would rendezvous with their carrier about three hundred nautical miles offshore.

The timing was fortuitous; this week, Mike Sigler was slated to fly the last of their T-38s to Edwards Air Force Base to be handed over to the ARPS test pilot school. At Ourecky's urging, he flew a day ahead of schedule and dropped the engineer at Alameda.

An ensign at Flight Operations directed Ourecky to the flight line and urged him to hurry if he wanted to catch the departing flight. After frantically scrambling up and down the rows of aircraft, he literally found Carson at the last minute, in the process of boarding his F-4N

Phantom. His RIO was already strapped into the back seat, obviously anxious to depart. If Ourecky had been only a few minutes later, Carson would have already been in the air, zooming westbound over the Pacific.

Seeing his friend, Carson grinned and clambered back down the ladder. "Scott, you came all the way out here to see *me*?"

"You didn't think I would come to send you off for your big adventure?" asked Ourecky, bending forward at the waist as he caught his breath. "I sure wasn't going to let you just swoop off into the sunset without saying goodbye."

"It's really no big deal," declared Carson, using his hand to shield his eyes from the sun. "I'll be home in two shakes. As much as I relish the thought of mixing it up with some real bad guys, it doesn't look like that's going to come to pass. At least I'll get my ticket stamped, though."

"I'm so happy for you, Drew," said Ourecky in a sarcastic tone.

Carson smiled. "Hey, since it's going to be such a quiet cruise, are you *sure* you don't want to come along?" he asked, twisting the end of his now flamboyantly long moustache. "We probably won't do anything but hang out in the wardroom, gulp down sliders, and play acey-deucy. I could pull a few strings, work a few angles, and…"

"Thanks but no thanks. I hope you're not too offended, Drew, but I'm not going to miss flying with you."

"Ouch. That hurts, Ourecky," replied Carson, playfully grimacing as he pointed at his chest. "Right here, like Bea says. Speaking of Bea, I suppose the notion of going home to her is probably a lot more enticing than bunking with me and a few thousand other guys aboard a noisy, stinky ship."

Since Ourecky had not disclosed anything about their separation, Carson had no clue that he and Bea were not currently residing under the same roof. He glanced up at the cockpit and saw the name "LCDR DREW 'REAPER' SCOTT" stenciled in dark blue letters under Carson's canopy and "LT JOE 'BEANS' LEESMA" below the rear canopy. "So is this Phantom a loaner?"

Carson looked back over his shoulder toward the RIO, and quietly answered, "No, it's mine. Well, mine and the Navy's, I suppose. Anyway,

I think I told you that I can't fly under my own name overseas, so that's my *nom de guerre*. I hope you don't mind that I borrowed some of yours. It does have a nice ring to it, don't you think?"

"Drew Scott? I suppose," replied Ourecky, shaking his head. "I just think it's a little asinine that you have to go to war under an assumed identity."

"Assumed identity? I prefer to think of it as an alter ego, kind of like Batman or Superman. Hell, at this point you could call me Lois Lane, so long as it gets me into the fight."

"If you say so."

Adjusting the chin strap of his camouflage-painted flight helmet, Carson asked, "So was Virgil any more forthcoming about our future holds? Any idea of what's in store?"

Ourecky thought about his recent visit to Vandenberg Air Force Base and the impending MOL mission. "Uh...no. Nothing significant, anyway. I think he and Tarbox will have more to say once you get back. In the meantime, I'm having to cover the mission debriefings by myself. Thanks, slacker."

"Sorry," replied Carson, not relenting from his curiosity. "Scott, are you *sure* that you don't know anything? Didn't Virgil make a big deal about sending you out to Vandenberg last month? He didn't tell you *anything*?"

Ourecky swallowed; as much as he wanted to reveal the news about the MOL, he knew that he couldn't dare speak of it. "Not really. Mostly, the Navy guys gave me the nickel tour of their facilities. Very nice set-up."

"*Time*, Drew!" yelled the fidgety RIO, pumping his fist in a hurry-up gesture. "If we hold up the flight, you can bet Badger will torch your ass after we come aboard ship."

"In a minute, Beans. We're still on schedule."

"I guess this is goodbye, then," said Ourecky. He extended his hand.

The harsh wind coming off the Bay was bitterly cold, but Carson took off his flight gloves to shake his friend's hand. "Thanks for everything," he said, hugging Ourecky close to his chest.

"Take care of yourself, Drew. We'll see each other again, soon enough."

"Yeah, we will. Soon enough," replied Carson. He replaced his Nomex gloves and started back up the crew ladder.

"Hey, Drew, I…" Ourecky knew he could say nothing about the MOL mission, but he struggled over whether he should tell Carson about Rebecca. He decided that it would probably be better to share that news after Carson returned from overseas, just as Bea insisted. After all, even though it looked to be a quiet deployment, Carson didn't need anything to cloud his focus. He would be home in a few months, and after that he would have a lifetime to become acquainted with his daughter, if that's what he chose to do.

"What, buddy?" asked Carson, leaning over and stepping into the cockpit.

"Uh…be careful over there."

"*Always*," replied Carson. "Bye, Scott. I'm sorry how things turned out, especially your shot at MIT, but at least one of us got his wish granted. Now, you should get back home to Bea."

"Bye, Drew." *Yeah, at least one of us got his wish,* thought Ourecky as he watched the Plexiglas canopy whir closed over his friend. He backed away several yards and lingered, clasping his hands over his ears as the engines started. Carson went through the last stages of his pre-flight and then revved the engines to taxi. As the fighter rolled forward, he looked toward Ourecky, flashed a cheerful smile and threw a quick salute. Ourecky waved and Carson waved back. Minutes later, trailing twin plumes of dense black smoke, the sleek aircraft screamed off the runway, and Carson was gone.

Dayton, Ohio
2:20 p.m., Sunday, December 10, 1972

Sipping a cold Schlitz, relaxing in an old sweatsuit and faded suede slippers, Ourecky studied a ponderously thick Navy manual about the fundamentals of nuclear reactors and power generation. The television was on, and he looked up periodically to see if there were any updates about Apollo 17, NASA's last lunar flight. Launched three days ago, the astronauts should be going into lunar orbit anytime now, but there was nothing. Apparently, the television networks had lost interest in broadcasting news about the moon missions.

He had listened to the radio earlier, but most of the news focused on the United Airlines crash in a Chicago neighborhood, killing forty-five people, including an Illinois congressman and two persons on the ground. The other big story was the ongoing peace treaty talks in Paris. Rumors abounded that President Nixon was prepared to announce the end of the conflict at Christmas. If so, it would be a welcome gift for a nation weary of war and torn by internal dissent. Ourecky thought of Carson and knew that his friend was probably infuriated about being aboard an aircraft carrier making full steam toward a war that would likely be over before he arrived. At least he would be safe, and that was something to be thankful for. Maybe Carson would be allowed to return home early, and then the two of them could sit down to discuss Tarbox's offer of the MOL flight and public recognition as astronauts.

He heard a noise outside, stood up, and saw Bea's red Karmann Ghia pulling into the driveway. He frantically gathered the nuclear manuals and stashed them in his briefcase; none of the references were classified, but it was just something else that he didn't want to explain.

"I'm absolutely exhausted," confessed Bea, trudging through the front door, carrying two brown grocery sacks and trailed by Andy and Rebecca.

"Uh, I wasn't expecting you," said Ourecky, standing up. He was taken aback by her disheveled appearance. Covered with a floral print scarf, her blonde hair was unbrushed and unkempt. Her eyes were red, as if she had been crying, and she wasn't wearing makeup.

She slipped out of her Navy pea coat and hung it on the peg by the door. "I'm ready to come home, Scott."

Ourecky hugged her and asked, "Is Jill..."

Wiping a tear from her eye, Bea sniffled as she solemnly nodded. In a hushed voice, she answered, "This morning. The kids don't know. I don't think that they would understand anyway."

Ourecky glanced into the paper bags. One contained Andy's favorite blanket, some stuffed animals and a few toys. The other held a loaf of white bread, a jar of peanut butter, two boxes of breakfast cereal and a carton of milk. She knelt down and dug through the first bag before handing Rebecca her well-worn Raggedy Ann doll and Andy his blanket and a plastic car.

"I'm sorry, Scott, but I'm just spent," she said, standing up. She brought the groceries to the kitchen, put the milk in the refrigerator, and placed the sack on the counter. "Jill's mother is at the funeral home, making arrangements for next week."

She returned to the living room, plopped down on the couch, picked up his beer from the coffee table, and took a long swallow.

"I didn't think you liked beer," he observed, sitting down next to her.

"I don't mind it today." She took another sip and handed the bottle to him. "Have you heard from Drew?"

"Yeah. He called yesterday."

"Is he already in Vietnam?" asked Bea.

"He wasn't there yet," answered Ourecky, remembering that Carson had placed the call through the MARS—Military Auxiliary Radio System—network. His aircraft carrier had a "MARS shack," operated by a Navy radio technician, where sailors could place calls home through a network of volunteer amateur radio operators. When he called, the carrier had just pulled into Subic Bay Navy Base in the Philippines, where it would take on fuel and ordnance stores before sailing on toward Yankee Station in the Gulf of Tonkin.

"Well, I sure hope that he's safe," she replied. "So, have things changed between you and the Air Force?"

"No," he answered, shaking his head. "Not really. We're still tying up loose ends. I'll be here at Wright-Patterson at least until Drew gets back from Vietnam, maybe longer. Maybe *much* longer. Sorry."

"So you're not going back to school?"

"Eventually. So you said that want to come back home?"

"Yeah," she answered. "Very much so. I'm ready to sleep in my own bed. With you, Scott."

He grinned, but his smile was quickly replaced by a frown. "I wasn't kidding when I said that I might be staying here for a long time. To be honest, it's not going to be much different than before you left. Sorry. Are you sure that you're so anxious to come back?"

"I am," she replied, nodding. "We'll take it a day at a time, Scott. A day at a time."

"So Rebecca is going to stay with us for a while?" asked Ourecky, watching the children quietly play in the corner next to the television.

"For a while, if it's okay with you. At least until after Christmas, and then she will go to live with Jill's mother. Right now, I'm trying to give her a break. Not only is she grieving, but she's worn ragged, and there's still a lot left to be done."

"I can imagine."

"Are you still going to Nebraska at Christmas?" she asked, slipping off her shoes.

"I'm planning to. I put in my paperwork for twelve days of leave, and Virgil already signed off on it. I'll drive over the day before Christmas Eve and come back the day after New Year's."

"That sounds so good. Do you think your parents would mind if we brought Rebecca?"

"Not at all," he replied. "How is she?"

"Rebecca is all right now, but at some point it's going to sink in that her mother is gone forever, and that's going to be very difficult to deal with. If nothing else, a change of scenery might help. Besides, I would really like to see her have something resembling a happy Christmas, so I can't think of a better place to be than at your parents' house."

"I'll call later and let them know," he said. "I'm happy that you're home, Bea."

"And I'm happy to *be* home."

6:35 p.m., Wednesday, December 13, 1972

Reclining on the floor, Ourecky played with the kids as Bea fixed dinner in the kitchen. The phone rang as he was helping Andy stack a collection of colored wooden blocks.

Bea rinsed off her hands, dried them on a kitchen towel, and answered the phone. After listening for a moment, she said, "Over? Over what? What? This is so confusing. Let me see if my husband knows anything about this. No, no, I won't hang up."

"What's up?" asked Ourecky, carefully balancing a red block on a teetering tower on the verge of falling over.

She covered the phone's mouthpiece and used her free hand to spin a finger over her head. "Some crazy guy, asking if we'll accept a call from Mars. I'll wait a minute, then tell him you're not here and hang up."

Mars? Initially, Ourecky laughed at the absurdity of it, and then suddenly realized what Mars meant. He jumped up and raced towards her. "Carson! It's Drew Carson! Don't hang up!"

He took the phone from her. Curious, she stood close by, listening in on the conversation as she arranged rolls on a baking pan.

Ourecky listened to the operator's instructions, and then replied, "Yes, operator, we'll accept the call. Yes, I do know the procedures. Over." He felt self-consciously awkward saying "Over" on the phone, especially within earshot of Bea and the kids, but it was just part of the goofy rules associated with the MARS calls.

After a moment of silence, he heard Carson's voice, slightly scratchy and distorted. "Hey, buddy, I only have a five-minute block, so we have to make this quick. Over."

"Okay. Good to hear your voice. Over."

"Hey, I wish that I had brought my wool scarf. It's a lot cooler than I expected. Over."

Wool scarf? Ourecky breathed a sigh of relief; with a phrase pre-arranged between them before his departure, Carson was telling him that he had arrived at Yankee Station, off the coast of North Vietnam.

After years of pestering Wolcott and Tew, Carson had finally achieved his goal: he was in Southeast Asia.

"So, have you had a chance to drive around and look at the neighborhood?" asked Ourecky. "Over."

"Yeah," answered Carson. "Very quiet. Not much traffic. Kind of disappointing. Over."

Quiet was good, thought Ourecky; quiet was *safe*. Maybe the peace process would come to quick fruition, and the war would be over by Christmas.

"So how's our old gang back there in Ohio?" asked Carson. "Any new developments that you can tell me about? Over."

Ourecky looked at a yellowed water stain on the ceiling. There was so much that he wished that he could tall Carson. He wanted to tell him about Rebecca, the daughter he didn't know about, and the news about the MOL. At this point, with the reality that his friend was now flying in a combat zone, more than anything else, he just wanted him to be *safe*. Certainly, he didn't want to distract Carson or diminish his focus, but he wanted to say something, *anything*, to motivate him to be safe, so he could make it home in one piece. If he could somehow make Carson aware of the MOL mission, then maybe the fearless pilot might not be so prone to take unnecessary risks.

He thought of a potential tack. He glanced at Bea, smiled, and then took a stab at it. "Hey, Drew, you know how you used to talk about applying to become a NASA astronaut? Over."

Bea raised her eyebrows as she brushed melted butter on the dinner rolls and then slid the pan into the preheated oven.

Other than a mild buzz of static, there was a prolonged silence on the phone line. Finally, Carson replied, "Uh...NASA astronaut? What about it? Over."

"Well, you might have a shot at it when you get back," said Ourecky tentatively. "Wouldn't that be cool, Drew? Wouldn't you like to see your face on the front cover of *Life*, just like all the other astronauts? Over."

"*Life* magazine? Yeah, that'll be the day, brother. Over."

"You just *never* know, Drew. You really, really should think about it. I'll even help you with the paperwork when you get back. And you know that the admiral has a lot of connections in high places. With his signature, you might really have a good shot. Think about it, Drew: *Life* magazine! Over."

"I'll think about it. Over." They chatted for a couple of minutes, mostly idle talk about what was going on with the Paris peace talks and war protests in the States.

"Hey, the operator just gave me the one minute warning," declared Carson. "We need to wrap it up. Over."

Ourecky glanced at the second hand on his watch, and asked, "When will I hear from you again? Over."

"I'll sign up for a MARS block on Christmas Eve. My squadron has the daylight shift, from dawn until dusk, so I'll have to call late at night. I'm thirteen hours ahead of you, so the call should probably come through around noon your time."

Before he left Pensacola, Carson had explained to Ourecky that three aircraft carriers were typically stationed at Yankee Station at any given time. To ensure uniform coverage and continuity, one carrier's air wing conducted flight operations from noon to midnight, the second carrier covered midnight to noon, and the third carrier—the assignment that Carson had apparently drawn—launched and recovered aircraft from dawn to dusk.

Remembering that he and Bea weren't going to be at this number on Christmas Eve, Ourecky blurted, "Hey, Drew, we'll be at my parents' house, in Nebraska when you call. Let me give you that number. Stand by."

Ourecky slowly recited the phone number and then waited as Carson read it back to him. "Good copy. Over," he confirmed.

"Christmas Eve, around noon. Over," said Carson.

Bea nudged him, grinned and pointed at the phone. "Bea wants to say hi. Over."

"Can't," replied Carson. "Sorry. My time block is over and there's a line of guys waiting. Give her a hug from me, and tell her I'll talk to her on Christmas Eve. Over."

Pointing at a clock on the wall, Ourecky shook his head at Bea. "Roger. Be safe. Goodbye. Out."

"Bye. Out."

Ourecky hung up the phone and went to the refrigerator to retrieve a beer. "Sorry," he said, snapping the cap off with a can opener. "He ran out of time on the call. He's calling back on Christmas Eve, though. You can talk to him then."

"I just wanted to tell him to be safe," she said, stirring a pot of steaming succotash. "I just want him to come home."

"He'll be safe. And he'll come home."

"I know, but that doesn't make it any easier for me," she said. "Anyway, that was very enlightening. After all, it's not every day that we get a call from Mars. And Drew Carson wants to be an *astronaut?*"

Ourecky sipped the beer and nodded. "Yeah. That's really why he went to test pilot school in the first place."

"Didn't you also want to be an astronaut?" asked Bea. "Back when you were in high school?"

"Yeah, but I got over it," answered Ourecky. He wiped a bead of condensation from his Schlitz bottle and added, "I'm entirely content to keep my feet on the ground now."

"Well, you *never* know," she said. "I've read that NASA is accepting scientists now, and they don't even have to be pilots. You know, I heard that one of those astronauts on the last mission was a geologist."

"Schmitt," he said. "Harrison Schmitt. So, are you saying that you wouldn't have a problem with me applying to be an astronaut? Granted, it would never happen, but you were so dead set against me being a pilot, I couldn't possibly imagine you wanting me to be an astronaut."

"It's certainly not what I *want*, Scott Ourecky," she replied, sliding the lid back onto the pan. "More than anything, I want us to be happy. I want *you* to be happy."

"I am happy," he replied, stepping toward her and wrapping his arms around her waist. He looked at Andy, playing with his blocks, and Rebecca, thumbing through *Green Eggs and Ham*, and added, "I'm *very* happy now that we're all back together."

"Maybe you're satisfied now," she said, playfully smearing a streak of flour on his nose before kissing him. "But I don't want you to ever look back and regret not pursuing something that you could have done or should have done. Face it, Scott, you're a smart guy, so much so that the Air Force wants to send you to MIT for your doctorate. You have a lot going for you, and that admiral's signature should carry as much weight on your application as anyone else's."

As he held her close, Ourecky struggled hard not to laugh at the irony of their conversation. Trembling, he managed to keep his composure until her final comment caused him to burst out laughing. "You know, Scott, as handsome as Drew is, and he is a very handsome man, that could be *your* face on the cover of *Life* magazine."

19

THE PEACE TRAIN LEAVES
THE TRACKS

Yankee Station, Gulf of Tonkin
6:35 p.m., Sunday, December 17, 1972

The squadron's wardroom was jammed with men passing the monotonous hours as they waited for a call to action. A table full of pilots and RIOs raucously sang along with the squadron's unofficial theme song, "Bad Moon Rising" by Credence Clearwater Revival, as it blared from a cassette player. Other aviators—equally bored but far quieter—played cribbage, cards, checkers or chess. Two men busied themselves setting up a movie projector in preparation for the evening's feature attraction: The Pearl Harbor war drama *Tora! Tora! Tora!* The men had seen the film so many times that most knew the dialogue—including the Japanese parts—by heart, and loudly recited each line as it was spoken on screen.

Nibbling on popcorn, Carson flipped through a recent issue of *Life* magazine as he waited on the movie to start. He had long admired

Life's coverage of the NASA astronauts, and had read each such article—many written by the astronauts themselves—with a fervor that bordered on obsession. He couldn't believe that the illustrious news magazine would cease publication at the end of the year, joining several other similar magazines—*Look, Saturday Evening Post*—that had folded in recent years. Supposedly, the advent of color television had hastened their collective demise. It made sense, since the average citizen seemed more inclined to spend his leisure time watching *Laugh In* or *M*A*S*H* than reading a book or magazine.

He grinned, thinking about Ourecky's recent comment about someday seeing his face on the cover of *Life*; obviously, like so many other things that might have been, that time had passed. Finishing an article about Joe Namath, Carson set aside the glossy magazine and reflected on last week's conversation with Ourecky. It had been a very odd and restrained exchange; Ourecky had diligently tried to convey some insight about something happening back at Wright-Patterson, but was unable to say outright whatever it was that he wanted to say. Carson didn't know if his friend's reluctance was due to communicating over open phone lines and the MARS network, or whether it was simply because he couldn't freely talk in front of Bea. In any event, it was immensely aggravating, since Carson yearned to know more about what lay in store once he returned to Ohio.

Because Ourecky had seemed so fixated on NASA's astronaut selection program, Carson suspected that Virgil Wolcott and Admiral Tarbox were probably working some sort of angle to allow him to expeditiously transfer to NASA. Perhaps the move might be in conjunction with NASA's next call for astronaut candidates, whenever that might be. With Tarbox's considerable political pull, Carson could likely be slipped into the mix as a surefire ringer.

Considering how long he had aspired to be an astronaut—more specifically, a legitimately *recognized* astronaut and not some guy who had just anonymously snuck into orbit nine times—the idea just didn't appeal to Carson as it once had. After all, NASA already possessed an abundance of astronauts on their roster, and even when the program was in full swing, there was a constant competition for the few available seats aboard spacecraft.

Now, with the Apollo lunar program literally ending the day after tomorrow—with the planned splashdown of Apollo 17 in the Pacific—flight slots would become even more scarce. The only scheduled mission after Apollo was Skylab, and only nine men—three three-man crews—were slated to occupy that makeshift space station cobbled together from Apollo hardware. Sure, the Space Shuttle would eventually fly—maybe—but how long was Carson willing to wait for *that* opportunity? Regardless of his flight experience, which he could obviously never speak of when and if he was assigned to NASA, or any sort of influence that Tarbox was able to exert, he would be at the lowliest bottom of the pecking order, and it was extremely likely that he would be too old to fly before he even had a chance to catch a ride on the Shuttle.

So, even if he was handed a golden ticket by Tarbox, was it even worth it to make the transition to NASA? Maybe it was just as well to leave things as they were and to never go to orbit again; by the grace of God or sheer damned luck, he and Ourecky had escaped the law of averages, so was it really worth it to take that chance yet again? With this cruise, Tarbox had granted him a unique opportunity to get his military career back on track. Maybe he wouldn't have the opportunity to engage in the sort of hot air-to-air combat that he so craved, but at least his records wouldn't reflect that he had just sat it out on the sidelines, either.

Carson thought about his future. Maybe the time was right for him to move on with his life. Although he was sure that Wolcott and Tarbox probably had some grandiose scheme in store, perhaps his best bet was to ask Virgil for the squadron command that Tew had offered. Maybe it was time for him to finally grow up, abandon his playboy lifestyle, and build a future for himself. Carson really envied Ourecky for what he had with Bea; with a little bit more stability in his life, maybe he could finally meet someone worthwhile, get married, have kids and settle down. But as appealing as those things were, he still had to finish this cruise.

To this point, his much anticipated combat tour had been much less than eventful. He had flown exactly nine times since arriving at Yankee

Station. Eight sorties were orientation and training flights, but on the ninth mission, he and Badger had escorted an RF-8 Crusader on a photo recce run near *Cam Pha*. Although the mission had proceeded without incident, except for dodging a SA-2 surface-to-air missile on the way out, at least Carson had gone "feet dry" by actually crossing the coast and flying in North Vietnam's airspace. In contrast, several of his squadron mates had yet to do even that. But although he had theoretically flown in combat, he wanted *more*. He felt as if he was being denied the experience that would indelibly define his generation of airmen. But maybe there was a chance: the current scuttlebutt was that the Paris peace talks were quickly unraveling, and that the Air Force had already dispatched a few B-52s to bomb targets north of the 20th Parallel. As anxious as he was to believe the rumors, Carson forced them from his mind.

A hush fell over the wardroom as the squadron's executive officer hustled in and switched off the cassette player. "No movie tonight, gentlemen," he announced tersely. "Badger wants everyone in the ready room in five minutes. Full squadron muster."

"Everyone?" asked "Beans" Leesma, Carson's RIO.

"*Everyone*," reiterated the Exec, stretching out the syllables. "All hands. That includes *you*, Beans."

As he navigated his way to the next deck, Carson heard an excited chorus of whoops and yells coming from the attack squadron's ready room, which was adjacent to theirs. The Intruder drivers were usually a lot more subdued, so if the attack pukes were this wound up, then something momentous was obviously afoot. Moreover, Carson was thrilled because he would likely be smack in the middle of it, since A-6 strike packages were habitually shepherded by F-4 fighters.

Usually more rambunctious, the twelve F-4N flight crews quietly entered the ready room, quickly took their seats and sat in rapt anticipation of Badger's announcement.

Carrying a large briefing binder, Badger strode purposefully to the metal podium. Holding the black book over his head, he declared, "This is Operation Linebacker II, gentlemen. The Paris peace talks have broken down, and the President is apparently anxious to motivate the

North Vietnamese back to the table, so he has authorized the resumption of *aggressive* bombing operations above the 20th Parallel, beginning tomorrow."

As the squadron's aviators clapped and yelled, Beans pounded Carson's back. *The scuttlebutt was true*, he thought. Maybe there really was some hope.

Badger shushed them and then launched into an hour-long overview of the operation, as well as the general missions assigned to the squadron. As Carson had suspected, the majority of their squadron's sorties would entail escorting A-6 daylight strike packages on targets in the vicinity of Haiphong.

After informing the crews that they would receive specific mission assignments tomorrow morning, Badger concluded the briefing by asserting, "The gloves are coming *off*, gents. It's official: the MIG killing lamp is now lit. Get a good night's sleep tonight, because it might be your last until this war is over."

Several minutes passed as the men gradually filtered out of the ready room to return to the wardroom or their cabins. Every few minutes, the ready room reverberated with the resounding sounds of the steam catapults on the flight deck overhead; launch operations were in progress, probably for night strike packages on their way to targets inland.

"Drew," said Badger, pointing toward the front row of chairs. "Hang back after everyone leaves."

As the last men departed, Carson took a seat beside Badger. "What's up, boss?"

"I guess you know that Admiral Tarbox gave me some very specific directives about how to handle a situation like this," said Badger quietly. "His instructions were to bench you if things started getting hot over here."

Carson's heart sank in his chest.

"Hey, I have absolutely no idea what you did to grab this ticket," said Badger. "But obviously whatever it was, it must have been something truly significant, because Leon Tarbox doesn't dole out favors lightly, and he's also not one to bend the rules or allow others to bend them.

As far as I'm concerned, Drew, you must have done something of an extraordinary magnitude. So, whatever that was, I will never ask, but you obviously earned this ride. Moreover, you complied with every measure that Tarbox prescribed, and you've done your utmost to assimilate yourself into this squadron, so not only did you earn your ticket from the Ancient Mariner, but you've earned your ticket with me as well."

"Thanks," replied Carson. "But..."

Badger yawned, held up his hand and said, "You're welcome to sit this one out, particularly since that's precisely what Tarbox ordered. I'll talk to the flight surgeon, and I'm sure that he can come up with some phony baloney excuse to ground you, so there won't be any sort of stigma or shame attached. We'll shoot you off the boat on the next COD Greyhound, and then zoom you back to the States."

Grinning, Badger continued. "But even though Tarbox clearly told me to bench you, he's not here and I'm commanding this squadron. This is *war*, my brother. I need all hands on deck, and I'm damned sure not too enthusiastic about leaving one of my hottest hands back on the boat when we go feet dry tomorrow. So, Drew Carson—I mean, Drew *Scott*—do you want to *fly* or ride wood?"

"*Fly*," answered Carson excitedly.

"Good. Pre-mission brief at Zero-Four and cats at dawn. Go get some sleep."

Flight Crew Office, Aerospace Support Project
Wright-Patterson Air Force Base, Ohio
4:04 p.m., Monday, December 18, 1972

Ourecky threaded a strip of 16-mm film into a Bell and Howell reel-to-reel projector, locked the door, adjusted the screen, switched off the lights, and sat down to peruse yet another classified Navy training movie. This particular reel described the fundamentals of synthetic aperture radar technology, but most of the movies were part of a lengthy series on nuclear propulsion systems. In the past week, he had watched enough Navy films about nuclear reactors on submarines that he felt sufficiently qualified to set sail on the USS Nautilus for a submerged voyage under the North Pole. He had to endure the training films as part of his theoretical training for the upcoming MOL mission. While he actually enjoyed the technical information, the voice-over narrations—typically a lecturer droning on in a tedious monotone—could readily be used as a cure for insomnia. Struggling to remain awake, he sipped from a cup of lukewarm coffee and periodically splashed his face with cold water.

He was fifteen minutes into the somnolent feature when the phone rang; Admiral Tarbox's administrative aide, an Air Force captain, tersely stated that the admiral demanded Ourecky's immediate presence.

Concerned that it might be bad news about Carson, he shut off the projector and rushed down the stairs. As he entered the external office and was waved on by the aide, he heard Tarbox screaming at the top of his lungs.

"*Authority?!*" screeched Tarbox. "Authority?! Who the hell are you to tell me that I don't have authority in this situation? He's *my* officer, and I can order him back any time I damned well please! Damn it, Commander, count yourself fortunate that this isn't the *old* Navy, or I'd boil your head and serve it at evening mess with black pudding and toasted cheese." Tarbox's hand trembled as he slammed the receiver down on the telephone.

Ourecky was taken aback by Tarbox's behavior. He had seen the Ancient Mariner angry, but never this angry. In stark contrast to his closely cropped shock of white hair, Tarbox's face was almost crimson red. Veins visibly throbbed in his neck and temples. His eyes bulged out so far that Ourecky was concerned that the orbs might pop out and land on the floor.

After taking a series of deep breaths, Tarbox asked, "Major Ourecky, have you had *any* contact with Major Carson, official or otherwise?"

"Affirmative, sir. I talked to him last Sunday, right after his carrier arrived at Yankee Station. He called on MARS."

"MARS?" growled Tarbox. "Is there any possibility that you might converse with him again in the near future?"

Ourecky was extremely reluctant to reveal that he and Carson already had plans to talk on Christmas Eve. "Possibly, sir. Am I to assume that you want me to convey some sort of message to him, Admiral?"

"Yes, Major. If you do talk to him, then you need to make it abundantly clear that he is to return here by the most expeditious means available. In lieu of that, if he is not able to return here quickly, for operational or other reasons, then he is to *physically* remain aboard that aircraft carrier until I expressly authorize him to depart it. Is that clear, Major?"

"Yes, sir, but didn't you authorize..."

Tarbox glowered menacingly, as if he were looking for a convenient puppy to strangle. "Tell Carson to get back here *immediately*. You are dismissed, Major."

Aerospace Support Project
Wright-Patterson Air Force Base, Ohio
9:15 a.m., Tuesday, December 19, 1972

The projector's take-up reel clattered as the film ran out. Ourecky reached out to switch it off, stood up and stretched. He was distressed about Carson, so much so that he could barely concentrate on the subject matter in the training films. The news coming out of Vietnam was scant and contradictory; the Air Force had admitted to some aircraft being shot down, but the North Vietnamese made it sound as if planes—especially the big B-52s—were raining out of the sky over Hanoi. Ourecky craved more detailed information, and especially wanted to talk to Carson so that he could pass on Tarbox's directive to come home.

He went downstairs to the communications room to see if there was any up-to-date news coming through on the classified Teletype, but the big machine was silent. He cursed under his breath; he wanted to know that Carson was safe.

He decided to take a break before watching yet another Navy film, so he headed across the parking lot to see Gunter Heydrich in the simulator hangar. The lot held very few cars; compared to Blue Gemini's heyday only a couple of years ago, the asphalt expanse was effectively deserted. Most of Project's Wright-Patt operations were closing, with only a small liaison office slated to remain. Ourecky would remain here until Carson returned, and then the two of them were likely to be transferred to California. This was Gunter Heydrich's last week; like most of the other civilian workers, he was being laid off.

Ourecky strolled into the hangar and watched several technicians disassembling the simulator systems. The infamous "Box"—the upright procedures simulator—in which he had spent many hours and days in seemingly endless agony, was destined for permanent storage in Arizona. He wasn't surprised; the simulator's technology was dated when he had first started in it, and now it was downright obsolete. On the other hand, the Paraglider Landing Simulator was being dismantled and crated so that it could be moved to Vandenberg Air Force Base.

As if he were locked in a time machine stuck in 1968, Heydrich still dressed in a white button-down dress shirt, black tie, and black dress slacks. His black hair was still greasy and largely unkempt, and he wore the same black-framed eyeglasses made popular by Clark Kent.

"*Guten Tag*, Gunter," said Ourecky, walking up and extending his hand.

"*Und Guten Tag* to you," replied Heydrich, standing up and taking Ourecky's outstretched hand. He sat back down and consumed the last two bites of a glazed doughnut. "Are you all right, Scott? You don't look very well."

"I'm a little concerned about Drew," replied Ourecky.

"*Ja.* I've been watching the news. I thought the war was supposed to be over by now. Who could have possibly foreseen this turn of events? I can see why you would be alarmed, but are you really sure that he's even flying in that operation?"

"Am I sure that he's flying missions? Gunter, he's *Carson.* Need I say more?"

"Really, Scott, you needn't be so fretful. I have it on good authority that Admiral Tarbox set some provisions on Carson's deployment to preclude him from engaging in any serious combat actions. Of course, Carson didn't know about it, but..."

Heydrich's words were little solace to Ourecky. "But I'm sure that he found some way to skirt around any traps that Tarbox may have laid," he interjected. "After all, he's Carson."

"He'll be fine. Regardless of what's he's doing, he's an excellent pilot."

"Yeah. Hey, I understand that you're leaving soon, Gunter," said Ourecky, changing the subject. "Any plans?"

"*Ja*. In fact, we'll probably see each other someday soon. I interviewed last week with the Instrumentation Labs at MIT, and they offered me a position. We're still discussing salary, but I intend to take it. Aerospace jobs are scarce now; I would be a *verdammt* fool to pass this offer by. After all, MIT is a very prestigious institution."

"Agreed. Anyway, I wanted to tell you that it was truly an honor to work with you, Gunter. I don't think we could have done it without you."

Heydrich laughed. "If you say so, Scott, but the honor is mine, and I *know* that we couldn't have accomplished what we had without *you*."

"Thanks," replied Ourecky, taking a seat. "That's very kind of you."

Heydrich glanced from side to side, apparently making sure that no workers were within earshot, and quietly confided, "Virgil told me about the MOL mission, Scott. Congratulations. I'm glad that you and Carson will finally get the recognition that you deserve."

Surprised that Heydrich knew about the pending MOL flight, Ourecky said, "I wish that I was as enthused as you are. To be honest, I'm a lot less than thrilled about it, but I'm still obligated to crack the books until Carson comes home. At this point, I have resolved myself to abide with his decision, whatever that might be. If he wants to go, then we'll go together. As for myself, I could certainly live the rest of my life without the recognition, but it's very important to Carson, so I will let him decide."

"Do you really think there's even the most remote possibility he might turn it down?" asked Heydrich.

Ourecky shook his head. "No. It's a no-brainer. Carson will leap on this opportunity like a glory-seeking Marine jumping on a hand grenade.

But to be frank, Gunter, I'm a little perplexed why Tarbox would offer this mission to us. Even though I'm leery about strapping on another rocket, I have to hand it to the Ancient Mariner for even coming up with an idea like this. It's certainly changed my impression of him."

Leaning back in his chair, Heydrich chuckled and said, "Scott, do you really believe that someone as manipulative as Tarbox would ever make such a magnanimous gesture without some sort of ulterior motive? Do you *really* think that he would send you and Carson on this MOL mission, with public exposure and media attention, instead of one of his Navy crews? After all, his Navy guys have been training for this mission for the past three years, so he's going to willingly give it to you gentlemen just because he says that you deserve it?"

In the process of removing one of the hatches from the Gemini-I mock-up, a worker clumsily dropped a socket wrench, which clattered from the raised platform to the concrete floor. Startled, Ourecky and Heydrich both jumped out of their chairs.

"But what does Admiral Tarbox stand to gain, Gunter?" asked Ourecky, sitting down as he regained his composure.

"*Everything,*" answered Heydrich, slipping a Dietzgen slide rule into cordovan leather case. "More specifically, he stands to gain *everything* if he sends you and Carson. Conversely, he's subject to lose everything if he doesn't send you."

Heydrich continued. "The joint Air Force-Navy merger has always been tenuous. Tarbox knows that in order for the joint program to go forward, he has to be in good graces with the Air Force's senior leadership. I don't know if he bothered to inform you, but his MOL program was on the verge of being cancelled altogether after the incident in August. The Navy was still supportive of it, mainly because they have such a pressing need for the ocean surveillance system, but the Air Force leadership was dead set against it, particularly because they thought it was too risky to put a nuclear reactor in orbit. To keep the program going, Tarbox had to convince the Air Force of the MOL's viability. Additionally, he had to provide them with plenty of incentive to put their chips on the table to back it."

"But what does that have to do with me and Carson?" asked Ourecky.

"Flying you two is *the* incentive to keep the Air Force on board," stated Heydrich. "Scott, you have to understand that there are very senior leaders in the Air Force who think that you and Carson deserve legitimate recognition for what you've accomplished, especially for the MOL rescue mission. That's why Tarbox came up with his scheme to fly it in public view, with you and Carson as the first crew. As I've heard, it was fairly easy to sell the Air Force. So the Air Force's senior leadership is willing to support continuation of the MOL ocean surveillance program, contingent on you and Carson flying the first—or rather, second—mission."

Thinking about yesterday's encounter with Tarbox, Ourecky now had a better understanding why the admiral was so unsettled by Carson's apparent reluctance to return home from Vietnam. Not only did his vaunted MOL program rest in the balance, but if Carson was harmed during his deployment, then Tarbox would certainly be compelled to explain to the Air Force's senior leaders—who would be far less than amused—why he could have placed their star astronaut at such tremendous risk.

"But if the MOL is so contingent on Carson and me," asked Ourecky, "then why would Tarbox ever allow Carson to fly overseas?"

"Two reasons," answered Heydrich. "First, that deal was well underway long before the public MOL mission was even tentatively approved, so it made sense to let Carson finish his training and deploy. Second, Carson wasn't ever supposed to be in real danger. If you recall back to October, when Carson's deal was done, Henry Kissinger publicly stated that he believed that peace was at hand in Vietnam. Obviously, Admiral Tarbox never anticipated that the war would go hot again, especially this quickly, but to his credit, he did put some provisions in the plan, just in case."

"Right," answered Ourecky. "And like I said before, I'm confident that Drew has skirted those restrictions."

"He'll be back in no time. There's no need for you to worry."

"I suppose you're right. I guess I need to focus on getting ready for this mission, since it's almost a certainty that Drew will want to go back up."

"Probably, but what do *you* want, Scott?" asked Heydrich. "Do you really want to go back up?"

Ourecky shook his head. "No. Gunter, what I really want is to get on with my life. I don't have any desire for publicity. More than that, this Project has stretched my marriage to the breaking point. Bea is a smart lady, so if this MOL mission happens, she's going to figure out that the Air Force didn't just arbitrarily pick my name out of a hat to suddenly become an astronaut. I don't know how she'll take it, but I doubt that it will be good."

Dayton, Ohio
1:30 a.m., Friday, December 22, 1972

The exterior of the teardrop-shaped window was smeared with a filmy layer of grime, so Ourecky loosened his shoulder restraints, leaned forward and pressed his face against the glass so he could see more clearly. He was in a forty-five-minute period of orbital night; the only thing he could see outside was a faint blob of light floating against a background of absolute blackness. As his eyes gradually focused, he recognized that the blob was actually a round viewport, even though the rest of the spacecraft was shrouded in darkness. He saw an indistinct face in the center of the viewport; it was a man, apparently staring out at the stars. Even in the poor light, the features looked vaguely familiar; mesmerized by the visage, he struggled to recognize who it might be.

Orbital dawn was fast approaching. As sunlight dispelled the gloom, the vague form before him progressively took shape. What he saw surprised him; he expected to see the derelict *Krepost*, still spewing debris, but the object wasn't the Soviet space station, but the MOL. The cylindrical fuselage, emblazoned with big block letters that spelled out "UNITED STATES NAVY," was unmistakable. As the light improved, the face in the viewport became clearer. Struggling to see through the obscured glass, he squinted, and then finally recognized the face: it was Carson! His expression was despondent, like he was hopelessly lost and looking to the stars to find his way home. Startled, Ourecky yanked his face away from the glass and looked to his left, and realized that he was alone in the Gemini-I.

Gasping, Ourecky shoved the heavy quilt aside and sat up in the bed. His heart beat so hard that it felt like it would thrash out of his chest. He was drenched in sweat and fought for breath.

"What's wrong, honey?" asked Bea sleepily. Yawning, she slipped her arm around his shoulders. "Bad dream?"

"Yeah," he replied. "*Really* bad dream."

6:12 p.m., Friday, December 22, 1972

Wringing his hands, Ourecky sat watching the evening news. None of it was good. Operation Linebacker II, a full court press intended to force the peace process in Vietnam, was in its fifth day, with no end was in sight. The toll was heavy: two B-52's and two fighter-bombers had been shot down today, with sixteen airmen missing. In total, according to the news, the United States had lost eight B-52s, four fighter-bombers and forty-three airmen.

Despite the Pentagon's claims to the contrary, the North Vietnamese alleged that bombs were falling on POW camps, wounding American prisoners and civilian hospitals in Hanoi. The North Vietnamese claimed to have shot down thirty-four planes.

The stakes were incredibly high, and everything was in play. Even longstanding allies were being uncooperative. General Alexander Haig, an assistant to Secretary of State Kissinger, had been dispatched to Saigon with an ultimatum from President Nixon to President Thieu, threatening to cut off aid to South Vietnam if Thieu refused to sign the peace treaty the US wanted. The apparent sticking point was that Thieu insisted on a version of the treaty that recognized the sovereignty of South Vietnam.

Ourecky cut off the television as the station cut to a commercial. He walked down the hall to Andy's bedroom, where he found Bea reading to Andy and Rebecca from a children's book. She couldn't bear to watch the news anymore, at least since the raids had started, and left the room as soon as he turned on the TV.

"Hey, we have to talk," he said quietly.

"I don't want to hear about it, Scott," she replied, slowly closing the book. "It makes me sick to even think about it. I don't want to know what's going on over there. We need to finish packing the suitcases, head out early tomorrow morning, and leave all of that behind us."

He helped her to her feet and led her to the living room.

She held his hands close and examined his fingers. "You've been chewing your nails," she observed.

"Yeah," he replied self-consciously. "I'm really worried about Drew."

"You think I'm not?" she asked. "It's just killing me, especially since there's nothing that we can do about it."

"I think we can do something. I've spent enough time around Drew and probably know him as well as anybody can know him. Right now, as far as he knows, he has nothing to lose, so he's probably taking a lot more risks than he has to. That's just the way he is; he has a naturally aggressive nature."

"Granted. But what could we possibly do to change that?"

"Let him know that he has something to live for."

"And exactly what would that be? A fancy watch and his new Corvette? A never-ending string of bimbo girlfriends?"

"No. Once, he told me that he was terrified of dying alone. I think he really wants to settle down and start a family, but he just hasn't been able to. If we told him about Rebecca..."

"*No*," countered Bea. "First off, Scott, we're not absolutely sure that Rebecca is even his daughter."

"I'm sure that she is," replied Ourecky. "And I think you know she is, also. If we told him that he had a daughter, then maybe he would be more careful over there."

"Maybe," she answered. "Maybe you're right. When are you two supposed to talk again?"

"Christmas Eve. Drew said that he would call around midnight his time, which would be noon our time."

She nodded. "Then we have some time to talk about it, but whatever we do, I want us to be very cautious about how we handle it with Rebecca. Fair enough?"

"Plenty fair," he answered.

20

LINEBACKER II

Over North Vietnam
9:35 a.m., Sunday, December 24, 1972

"Checkpoint Three. Feet dry," announced Badger over the radio. "Eyes up for gomers, gents. Reaper, stay tight on me."

Carson keyed his mike and replied, "Roger, Badger. I'm on you." Just for an instant, he looked down to his left to observe the sudden transition from blue ocean to green landscape, and grinned. Just two weeks ago, as he arrived at Yankee Station, he had gloomily speculated whether he would even have a single opportunity to fly over Vietnam. Now, just three hours after downing a hearty breakfast aboard ship, he was well into his first mission of the day. According to the flight schedule, he would fly two more missions before the sun went down over the South China Sea.

He still couldn't believe his good fortune. Initially, it looked as if his cruise would be relatively uneventful, but then the peace train had veered entirely off the tracks. Weary of stalled talks, the President ordered a new round of bombing, dubbed Operation Linebacker II. Combining the powerful resources of the US Seventh Air Force and the Navy's Task Force 77, the massive aerial onslaught was intended to coerce the recalcitrant North Vietnamese back to the peace table. Although hampered by bad weather, over a hundred American aircraft took to the skies daily to pound targets that were previously off limits.

Instead of twiddling his thumbs back in the wardroom, Carson was in the thick of the fray, flying CAP—Combat Air Patrol—in support of Alpha Strike operations in the Haiphong region. Since the air offensive had kicked off six days ago, he had already flown fourteen combat missions. Tomorrow, he and his fellow aviators would remain aboard ship to celebrate Christmas, and the CAP escort sorties would resume the following day. He reminded himself that he had a five-minute block scheduled for midnight at the MARS shack; he looked forward to talking to Ourecky again, but wished that he could be more forthcoming about his missions over North Vietnam.

For Carson, it would be the MIG-hunting safari of a lifetime, if only the pilots of the Vietnamese People's Air Force—VPAF—were slightly more cooperative. Of course, the wily Vietnamese certainly had their own agenda, which likely didn't include lining up in Carson's sights to be killed.

The circumstances had caused the VPAF to swiftly develop a marked preference for nocturnal hunting. Instead of flying in the daytime, where they were unlikely to achieve dominance in the air, they preferred to rise up after the sun fell, to gang up against the fattest and slowest targets, the B-52s. Besides being big and slow, the bombers operated under some restrictive constraints that rendered them into even easier prey.

In accordance with orders issued from SAC headquarters in Nebraska, the B-52's flew in three-plane "cell" formations along set routes, executing very predictable post-target turns after dropping their ordnance. The turns were intended to prevent the big planes from running into each other, but the prescribed maneuver made them easy fodder for SAMs fired in salvo. The consequences of the repetitive tactics were costly and themselves predictable; on the third night of the raids, the North Vietnamese destroyed six B-52s after downing three on the first night and severely damaging another on the second night.

When they did emerge during daylight, the VPAF pilots didn't seem too enthusiastic about engaging Navy aircraft. Consequently, Badger and his men attempted every trick in the catalog to lure the VPAF pilots up for a fight. This morning, it looked like they were in luck. Six MIG-21 "Fishbed" fighters scrambled from Kien An Airbase west of Haiphong, apparently tempted to jump the flight of eight A-4 Skyhawks protected by the Phantoms.

By the time the MIGs had cleared the runway, the A-4s had completed their attack runs and were turning back for the carrier. Watching the MIG-21s climb, Carson was almost giddy in his excitement, but then was chagrined as the less-than-valiant VPAF fighters scattered. Unwilling to risk a confrontation with the Navy Phantoms, the MIG-21s departed for safer airspace.

"Damn it," swore Carson over the intercom. "I was sure we would grab some of that action."

"Wait...wait...I'm still tracking one," replied his RIO confidently. "Two o'clock low."

Carson looked to the right and spotted a bogie. Apparently one of the stubby-winged MIG-21s had elected to fight instead of hightailing it for home.

"We're engaging," announced Badger. "Reaper, stay with me." In mere seconds, they closed the gap. Badger's flight of four F-4s was dispersed in two-plane mutually supportive "Loose Deuce" teams. Flying as Badger's wingman, Carson appreciated the flexibility of the Loose Deuce tactic; instead of complying with rigid rules about which of the two fighters would engage the enemy, the aircraft in the most advantageous position assumed the lead.

"Reaper, you're in," announced Badger over the radio. "Take him."

"Tally ho," confirmed Carson enthusiastically, tugging the stick as he maneuvered. "Reaper is engaging."

Even after his exhaustive training at Top Gun, it was an incredibly close match. Carson realized that his shrewd adversary probably had years of combat experience. Except for the rare instances where they might travel to China or the Soviet Union for advanced training, VPAF pilots typically flew in combat every single day. Moreover, they didn't have the option of rotating home, as their American opponents did; they *were* home, and the air war took place in the skies overhead. Consequently, most VPAF pilots flew until they were dead.

Carson didn't anticipate a long fight. Designed as an interceptor, the MIG-21 was smaller, lighter, faster and more agile than his F-4N. Despite these advantages, it lacked endurance as a consequence of its limited fuel capacity. But even though he was aware that the clock was on his side, Carson struggled to keep up. The tenacious VPAF pilot obviously wasn't throwing in the towel without a pitched brawl. Instead of a churning fur ball sprawling across the perfect skies, the duel was more like a knife fight in cramped quarters .

Carson knew he was much too close to engage with one of his AIM-7E Sparrow missiles; since his F-4N lacked an internal cannon and wasn't fitted with an external gun pod, he would have to rely on his short-range AIM-9 Sidewinder missiles. For several minutes, they were trapped in a stalemate; neither man could gain the advantage long enough to trigger a snap shot to close the deal.

Finally, after several minutes of gut-churning maneuvers, twisting and turning across the sky, the VPAF pilot made a significant error. He was about a quarter-second late in nosing over, which allowed Carson to momentarily seize the advantage.

Carson aggressively slid in close at the opportune moment, framed the delta-winged MIG in his heads-up pipper, and heard the distinctive lock-up tone growling in his earphones. "Reaper, *Fox Two...Fox Two*," he blurted, toggling the thumb switch to launch an Sidewinder. The heat-seeking missile surged off its rail, flew cleanly up the MIG's tailpipe and detonated.

Carson snap-rolled right to dodge a flurry of debris spewing from the rapidly disintegrating MIG. Morbidly curious, he whipped into a tight left-hand bank and looked over his shoulder to watch the bandit's demise. He was almost relieved as he saw the pilot eject. Tucking the F-4 into a descending spiral, he continued to watch the falling pilot.

"Splash One," noted Badger. "*Bravo Zulu*, Reaper. Good kill."

Carson's exuberance was swiftly supplanted by sheer horror. Just as it appeared that the VPAF pilot would escape death, his circumstances began to unravel. The ejection seat appeared to function as it was supposed to, but for some reason, the pilot didn't cleanly separate from the seat pan as his parachute automatically opened. As best as Carson could tell, something—perhaps a foot restraint or something as simple as a snarled bootlace—refused to break free, so the pilot tumbled madly through the air, still physically attached to his seat.

As he fell, squirming and thrashing in a frantic attempt to extricate his trapped foot, the parachute continued to deploy. The streaming chute furled tightly around the whirling seat, entangling the pilot like a spider's spinneret wrapping up a fly. Ignited by residual heat from the ejection seat's rocket motor, the chute caught fire. Engulfed in a molten cocoon of burning nylon, the hapless pilot plummeted to earth, trailed by a plume of orange flames and black smoke.

"Reaper, break contact and follow me," ordered Badger.

Carson rolled out of the spiral, climbed, and slipped into formation with Badger and the others. He had mixed emotions about the shoot down; it was the third time in his life he had witnessed a death as the result of a deliberate act, except this time it was by his own hand.

"SAM launch," stated one of the other pilots. "SAMs in the air! I have visual. One o'clock, looks like about five miles. I see three coming up."

Carson heard a warbling sound in his earphones. It was a warning tone indicating that enemy radar was tracking their aircraft. The NVA air defense crews had apparently determined that the Americans focused their SEAD—Suppression of Enemy Air Defenses—and electronic warfare countermeasures to protect the ingress phase of any major operation.

The NVA gunners knew that they assumed a huge risk when they lit up the big radars associated with the SAMs; they had learned to be patient, wait until the specialized anti-SAM electronic warfare aircraft like the EA-6B Prowlers had already returned to the carriers, and then ambush the fighters and attack aircraft on their way back to the safety of the sea.

Another pilot calmly chimed in, "Make that *four*."

"Jink Three-Zero on four-count," ordered Badger, like a quarterback calling a play. "Four…Three…Two…One…Break."

To evade the SAMs, the four F-4Ns broke left in unison and changed altitude. Intent on downing some American planes, the NVA SAM batteries continued to fire. Carson's aircraft shuddered as an enormous missile came up fast and detonated a few hundred feet underneath his right wing. Despite the force of the explosion, he maintained control as he assessed the aircraft's performance. The controls lagged slightly but were still functional. The aircraft had picked up a considerable vibration; Carson suspected that they sustained at least some airframe damage.

"Ouch," noted Carson. "That was too close for comfort. Beans, you okay back there?"

"I'm copacetic," replied his RIO. "Shaken but not stirred."

"Reaper, that was *mighty* close," reported Badger. "You lost some sheet metal from the underside of your starboard wing. Damage report?"

"We're fine, Badger," replied Carson. "Controls are sluggish but responsive."

"Okay. Everyone, break right in three and come back on heading. Resume Blue Two-Three Egress Route, Angels Eight. Mark."

Carson did a three-count, tugged the stick to the right, and applied right rudder to smooth out the turn. There was an unsettling wobbly quality to the controls, and the airframe was rattling much more so than before. "Badger, I'm experiencing a slight control problem here."

Seconds later, he heard a shrill tone in his right ear. He looked at the upper right of his control panel; the Master Caution light was lit. "Master Caution alarm," he announced.

"*Master Caution*," acknowledged Beans from the back seat.

What Carson saw next was definitely not a welcome sight. Looking to the right side of his panel, he checked his tele-light display. The

CHK HYD GAGES tele-light was illuminated, a cue for him to focus his attention on hydraulic indicator gauges on his pedestal panel. The needles on the three gauges—PC-1, PC-2 and Utility—were gradually creeping down.

The three hydraulic systems were largely redundant; he could safely fly the plane if one hydraulic system failed, but their emergency procedures dictated that they discontinue the mission and land as soon as possible. The concurrent failure of all three systems was catastrophic; ejection was their only option. Just as disconcerting, noise and vibration had increased exponentially in the past few seconds. Obviously, damaged sheet metal on the underside of the aircraft was being shredded by the slipstream; the clattering din sounded like a chicken house being ripped apart by a tornado.

Looking to his right, he could see that they were flying roughly parallel to the Gulf of Tonkin. The coast was tantalizingly close, so much so that he imagined he could see sparkling waves and smell salt-laced air. The tranquil sea was *safety*; ejecting over water astronomically increased the odds that they would be rescued in short order.

He spoke over the intercom: "Beans, our hydraulics are bleeding out. I'm going to trim up before we lose them altogether."

"Are we punching out?" asked Beans excitedly.

With luck, they could make it out to sea, but they were ejecting, whether Carson liked it or not. Lacking adequate hydraulics, putting the aircraft down on the carrier was not a viable option. "Yeah," he replied, scaling back power to diminish their airspeed. "There's no possible way we're going to plant this thing back on the deck, so we'll get feet wet and eject."

Carson felt the stick pushing against his palms. He knew that once the F-4N started losing hydraulics, its stabilator had an inherent tendency to depress, pushing the aircraft into a nose-high attitude. If he didn't quickly attend to the problem, the stick would eventually lock full aft and the aircraft would try to stand on its tail, at least briefly, until it stalled out and fell from the sky. To counter the fault, he fought the stick to push the fighter into a slight dive, with the goal of at least neutralizing the stabilator's orientation before the hydraulics failed altogether.

He didn't have to wait long for the hydraulics to give up the ghost. The fighter's two big General Electric J-79 engines roared right along as if nothing had happened, but the F-4 had effectively become an unguided missile. Provided nothing else failed or they weren't shot down by a MIG or SAM, they would remain in the air until their fuel was exhausted.

He fought to maintain level flight as he assessed the situation. The fighter's controls were all but unresponsive; Carson would probably have better luck steering a rail-bound locomotive. He quickly found that by tweaking the thrust of the engines, he could carve very slight turns, but every degree of turn cost them a significant amount of altitude. He determined his location and checked his compass; they were currently about twenty miles northeast from Haiphong, pointed northeast.

Try as he might, he could not force the crippled plane far enough to make it "feet wet." And that wasn't the only problem they faced. By his reckoning, at their current airspeed, they had less than four minutes before they encroached into Chinese territory. The Chinese didn't look favorably on incursions into their airspace, intentional or otherwise, and had established a precedent of shooting down US pilots who had strayed over from North Vietnam. The Chinese shot down two Navy A-6 Intruders in 1967, with one pilot killed and the other held as a prisoner.

Their options were quickly dissipating. Making it out to sea was no longer a viable alternative, so it came down to punching out over North Vietnam, where they could hopefully evade until rescued, or China, where rescue would probably not be forthcoming.

"Badger, we're punching out," he declared, pushing the Eject button on his left.

"We're with you," avowed Badger. "Good luck, guys."

"Beans, I'm squared up," stated Carson, throttling down the engines to further reduce their airspeed. "Ready to eject?"

"Ready," replied the RIO nervously.

Ensuring that his spine was straight and his legs tucked in, Carson sucked in a deep breath and grunted, "Eject, eject, eject!" The F-4N was equipped with Martin Baker seats that fired in a dual sequence. Beans, the RIO, would eject first; Carson's seat would spontaneously fire after Beans was away. Carson had punched out twice before, but never from a two-place aircraft.

He yanked down the face curtain handle, which initiated the rapid sequence of events. As the seat's restraints locked him into position, he thought of the VPAF pilot's tragic fate. He felt a sharp blast of wind as the canopies jettisoned, and then was surprised by the wash of explosive heat that engulfed him as his RIO's ejection seat blasted out of the cockpit. He tightly held his position as he felt the violent blast that blew him clear of the aircraft.

The next few seconds were a blur, but the seat and parachute functioned as advertised. The welcome canopy blossomed overhead. Suspended in his harness, with his one-man life raft dangling below him on a tether, he took stock of his surroundings. He was descending into a wooded area that was mostly uninhabited.

He observed a network of footpaths, car-sized outcroppings of limestone, and some partially cut clearings. Orienting himself, he spotted two rivers that converged near the coast; a small town was located at their confluence. Looking over his right shoulder, he briefly glimpsed another parachute and realized it was Beans; the planned delay in the ejection sequence had separated them by about a mile.

Except for the muted roar of jets in the distance, he was amazed with how quiet it was. Back aboard ship, he had listened to other pilots recount their own ejection ordeals, and several related that they were fired at while still in the air. Almost all claimed that they were on the lam, running full tilt with hostile locals hot on their heels, immediately after hitting the ground. As the trees came up, he steeled himself for the race that was sure to ensue.

Locking his feet together and guarding his face, he crashed through a tree, slamming into thick branches on the way down. The crescendo of blows felt like he was being pummeled by a gauntlet of heavyweights.

He crashed into the ground. He swung open his oxygen mask, caught his breath, climbed out of his parachute harness, and took off his helmet. His chute hung overhead, draped over the drooping branches of a low tree. He briefly tried to tug the chute down to conceal it, but realized that it wasn't coming down. He knew that the chute was like an enormous flag distinctively marking the place that he had come to earth; if he expected to remain a free man, it was imperative that he get far away from it, as quickly as possible.

Bashing his way through the lush vegetation, he made his way downhill for several minutes. Locating a dense stand of shrubs and vines, he stopped briefly to dig out his PRC-90 survival radio and pistol. At Badger's recommendation, he had purchased a Browning 9mm High-Power automatic at a gun store in Pensacola. While he attended Top Gun in California, Badger had smuggled it aboard the carrier for him. The Browning's double-stack magazine held thirteen bullets, considerably more firepower than the pipsqueak .38 caliber revolver that was the standard issue sidearm for Naval aviators.

He pulled back the pistol's slide and quietly jacked a cartridge into the chamber. He flinched as he heard the earsplitting roar of a jet passing immediately overheard, and guessed that his squadron mates were watching over him. He activated the radio, pressed it tightly to his ear, and listened for a moment. Crouching low in the dense undergrowth, he placed his mouth close to the microphone and furtively whispered, "This is Reaper. Anyone up there?"

The reply was immediate: "Reaper, this is Badger. I see your parachute. Are you injured?"

"Badger, I got a little banged up coming through the trees, but otherwise I'm okay."

"Good copy. Beans is about two klicks northwest of you, in some pretty rough terrain. Reaper, bad news: Beans broke a leg coming through the trees. Compound fracture. He's not going anywhere."

"Got it. I'll head that way. Badger, what do I have around me?"

"I'll orient you to the surrounding terrain," answered Badger. "You're about six klicks northwest of a ville called *Duong Hoa*."

"*Duong Hoa*. Got it," replied Carson quietly, committing the name to memory. Once he found a place to hole up temporarily, he would consult his waterproof evasion chart to more completely familiarize himself with the area's landmarks.

Like a bus driver narrating a rushed tour, Badger continued his concise travelogue: "*Duong Hoa's* right on the coast. Right now, you're between two rivers. The *Duong Hoa* is about a half-klick to your north. The *Song Lai Pan* is approximately three klicks to your south."

Badger continued to orient him: "There's a hardball road about five klicks to your southwest. It runs southwest to northeast and crosses both rivers. I don't see any vehicles on it. I think we've dropped both bridges. There's a dirt road a klick to your northwest. It also runs southwest to northeast. I've spotted several military vehicles on it, mostly medium trucks. Beans is on the far side of that road."

"I'll start moving that way," replied Carson. He pulled a metal tube of camouflage greasepaint from his survival vest and quickly smeared his face to cover his light skin.

"*Negative*," answered Badger. "It's too hairy over there. We're watching enemy troops converging on Beans. They'll be on him before you can make it. SAR is launching in less than five. If you're secure where you are, I'll ask them to focus their efforts on Beans. How copy?"

"I copy that SAR will concentrate on Beans," replied Carson. "That's a good plan, Badger. I'm fine down here, at least for now. I don't see or hear any bad guys in my vicinity. I'll find a place to hole up and wait for SAR. Get Beans out of there."

"Reaper, if you're pressed, I advise you to evade toward the south, toward the Gulf. Have your mirror and day-night flare handy."

"Will do," said Carson, feeling for the items.

"Reaper, I'm almost bingo fuel. I'm headed out over the water to plug a tanker, and then I'll come back to you. Someone will hang over you until there's sufficient air cover so the SAR helicopters can safely work their way in. We won't leave you and Beans hanging out there."

The sound of Badger's engines gradually faded to the south. Carson was surprised at how much the short run had taxed him. His heart thumped and he was soaked with sweat. Perspiration gushed from every pore. Then he realized that he was still wearing his "poopie suit." Worn under his flight suit, the impermeable anti-exposure coveralls were designed to protect him from hypothermia if he came down in the drink. Obviously, that was no longer an issue. He knew that he had to cover ground, and that if he didn't come out of the poopie suit, it would overheat and exhaust him in short order.

Although he was leery of a risky wardrobe change at this point, he shed his G-suit "speed jeans," which were also now superfluous, left his boots on, and pulled his flight suit down to his ankles. He yanked out his saw-backed survival knife and sliced through the poopie suit's rubberized fabric and the thin thermal underwear worn underneath. He stuffed the discarded garments under some thick vegetation and concealed them with a layer of dead leaves. He examined his bare leg; a souvenir of his descent through the trees, a large purple contusion covered most of the outside of his left thigh. Grimacing, he donned his flight suit and the torso harness that contained his SV-2A survival kit.

Tightly grasping the Browning's square butt in his right hand, Carson cupped his left hand to his ear and listened intently. Other than the rustle of a winter breeze blowing through the trees, he heard dogs barking in the distance and the muted sounds of trucks in the distance. He didn't hear gunshots, voices, or irate villagers crashing through the undergrowth. Perhaps all was not lost, if only the SAR guys would get here in time.

21

SIX MEN IN A BOX

Da Nang, Vietnam
12:15 p.m., Sunday, December 24, 1972

"They're serving turkey and dressing at the American compound tomorrow night," announced Finn cheerfully. "*Turkey and dressing*, Nestor. Cranberry sauce, shrimp cocktail, mashed potatoes, sweet potato casserole, pumpkin pie, apple pie, the works."

Spooning up a clump of rice with heavily spiced chicken, Glades shook his head. They had been subsisting exclusively on local cuisine since they had arrived. A day's supply of the chow was packed in insulated "Mermite" containers at one of the Vietnamese mess halls on the base and delivered every morning.

Glades was a firm believer that the food Americans ate caused them to exude different scents than their Vietnamese enemies. He suspected that more than one recon team had been compromised by their distinctive American odors. Along with other preparations for the mission, they were literally altering their body chemistry to blend in. Their rucksacks and field gear, already packed for the mission, contained only "indig" rations—mostly dried rice, shrimp and fish—specially stocked by the clandestine logistics office—"CISO"—of MACV-SOG. Besides the stringent menu, Glades placed other restrictions on them as well: no beer, liquor, candy, chewing tobacco or cigarettes.

"Nestor, you didn't hear me?" said Finn, glowering at being ignored. *"Turkey and dressing.* Just this once? C'mon, man, we've been eating this Vietnamese slop for weeks. I'm *sick* of it. Can't you give us a break for one day? Hell, it's Christmas! Surely our guy won't be flying on Christmas. Man, have a heart and let us wolf down some American chow for a change."

Their three Vietnamese SMS soldiers squatted on their haunches in a corner, chattering in their native tongue as they eagerly dined on boiled fish and rice. They might as well have been beamed up into the Buddhist equivalent of Heaven. Finn groaned as one squirted *nouc man* sauce on his food. The pungent liquid was distilled from fermented fish. Since the entire team lived together and partook of most of their meals here, the team house reeked of *nouc man*.

Henson sat cross-legged on the floor with the SMS soldiers, sharing their meal as they taught him bits and pieces of their language. Glades was thrilled to have the black PJ on the team; not only was Henson an exceptionally proficient medic, but he had also established an excellent rapport with their SMS counterparts.

There was a tap at the door. Coleman and Major Lahn entered, bearing two rubber-coated waterproof bags and a thick sheaf of maps and aerial photographs. Coleman wore a garish Hawaiian shirt over a ridiculous-looking pair of plaid Bermuda shorts. Lahn was customarily attired in his snugly tailored tiger-striped camouflage uniform, replete with brightly colored patches that loudly declared his various qualifications and affiliations.

"I hate to interrupt your Christmas Eve," announced Coleman, handing the two waterproof bags to Glades. "But we may finally have a task for your team. We've just received word that your guy is on the ground. His Navy F-4 Phantom was on a MIGCAP near Haiphong. It was apparently knocked down by a SAM missile. The Navy has planes overhead and they've been talking to the pilot and the back-seater. The back-seater was severely injured after he ejected."

"So is our guy the pilot or the back-seater?" asked Henson, setting down a bowl of rice and standing up.

"Honestly, we don't know," replied Coleman. He opened a pair of dossiers and examined the pertinent information. "The pilot is named Scott and the Radar Intercept Officer's handle is Leesma. Sound familiar?"

"No," replied Glades. "Not that it matters. We'll grab whoever they tell us to grab."

Coleman smiled. "That's a good attitude to have, Nestor."

"So we're launching?" asked Finn.

"Maybe, but not likely," replied Coleman. "A Navy SAR package is on the scene right now, making a play for Leesma. They're building up a heavy air cover over him. He had a flock of bad guys swarming in his proximity, but the Navy has pushed them away, at least for now. Once they have a good curtain hung, their helicopters should be able to break in there and extricate them quickly. The Navy is confident that their SAR guys will snag Leesma in short order, if they don't already have him."

"And the pilot?" asked Finn, sipping water from a metal canteen. "Scott?"

"Since he's not injured and in a better position, he's gone to ground temporarily," explained Coleman. He gestured at one of the maps. "He's in this box here, bounded to the north and south by these two rivers, and to the east and west by these roads. Since Leesma is the immediate priority, Scott will remain under cover until SAR sends the next bus. That probably won't take long, either. The Navy has another package making ready, and the Air Force is launching one of their SAR packages as well.

"If the SAR effort is unsuccessful and he's not picked up by nightfall, Scott has already been instructed to proceed to a link-up site, here," explained Coleman, gesturing at a spot on the map.

Coleman started to light a Marlboro, but quickly extinguished it when Glades glared at him. He continued. "We want you folks ready to launch tonight, to meet him at the link-up site. The chances are good that this whole thing will blow over shortly, and we'll be able to stand down this mission and send you all back to the States."

"Well, maybe we'll eat Christmas dinner after all," snorted Finn. "Obviously, one way or another, we should be closing out this mission very shortly."

"Don't go staking your life on it," noted Glades, examining the map and making notes on an index card. "How about the Combat Talon I requested?"

"It will arrive here right after nightfall," answered Coleman. He pointed at the two waterproof bags that he and Lahn had brought with them. The parcels were marked with strips of masking tape bearing the aviators' names in big block letters. "By the way, Nestor, you'll haul in an evasion kit for each guy. If Leesma is yanked out today, you'll only need to tote the kit for Scott."

Glades nodded as he hefted one of the bags. Approximately the size of a pillow case and filled to capacity with sundry items, it was considerably heavier than he would have imagined. He opened the bag and examined the contents. When and if they went into North Vietnam, Glades and his team would be wearing OG-107 jungle fatigues specially dyed to match the coloration of NVA field uniforms. They would also wear the distinctive green NVA pith helmets and canvas chest bandoliers. Accordingly, each waterproof bag contained a uniform that matched theirs, along with a set of black canvas "Bata" rubber-soled boots. Glades assumed that the clothing boots were pre-sized for Leesma and Scott. Additionally, each bag contained an extensive survival kit, including a PRC-90 survival radio and Colt .45 pistol.

Glades was somewhat surprised to see that the bulging bags also contained a variety of bartering items, including a leather pouch that contained ten coins minted of solid gold, that had previously been standard issue for MACV-SOG recon teams operating in denied areas. The gold coins obviously accounted for why the bags were so heavy.

Rounding out the collection was a tightly folded American flag sealed in a plastic bag. The reason for the flag was simple. Glades and his team carried them as well; in a pinch, since they would be dressed as the enemy, they could display the flags to helicopters or other aircraft, ideally to keep fratricide to a bare minimum.

Coleman dropped the two dossiers, maps, and other documents on a table. "After you've had some time to look that stuff over, give us a yell. Major Lahn and I will come back to brief you on the operational area and current enemy situation."

Current enemy situation? thought Glades. That was amusing. He didn't need to gaze into a crystal ball to divine the *current* situation. He knew that the NVA were *currently* piling into the area with everything they owned in an effort to grab the two Americans. Moreover, they knew that scores of men would be dispatched to rescue them, and they clearly wouldn't waste an opportunity to kill or capture as many of them as they possibly could.

Henson flipped open one of the two dossiers and began reading. His eyebrows arched as an alarmed expression passed over his face. Softly clearing his throat, he closed the folder and slid it across the table toward Glades. "Check the blood type," he commented. "I'll need to tote in a few cross-matched units, just in case. We have no matches on the team, so a field transfusion would be out of the question."

"Good catch, Henson," noted Coleman. "The doc over at the advisory detachment already has some stuff in his refrigerator for you. Cross-matched blood, serum albumen, etc."

Glades was slightly puzzled; he couldn't understand why Henson would call such an arcane medical issue to his attention, so he must have seen something more pressing in the folder. He glanced at Lahn, and then studied the map a few minutes longer before nonchalantly opening the dossier folder marked "SCOTT." The first page was a basic information summary on the subject; as soon as he saw it, he immediately realized why Henson was concerned.

Looking at the photographs stapled to the top of the sheet, Glades immediately recognized the man. The dossier identified the pilot as "Scott, Andrew C. LCDR USN," but since he had met "Scott" previously in Haiti, he knew that he wasn't a Navy pilot nor was his name Andrew Scott.

And at this point, name games were the least of his concerns. Recalling their mission in Haiti, he realized that only he and Henson had been in contact with Carson, the misidentified man in the photographs. Given enough clues, Finn could probably put the pieces together; electing to limit access to the potentially dangerous knowledge as much as possible, Glades decided not to raise the issue with him. He swallowed quietly, looked toward Coleman, and asked, "Any idea who these guys are? I mean, why is all this special effort so necessary?"

"Not a clue," answered Coleman. "Nestor, I subscribe to your theory that one of these guys is related to someone high up in the food chain, and they're doing their damnedest to make sure that he doesn't fall into enemy hands. I don't know if it's Scott or Leesma who's the golden child, but the fact that the Navy has placed priority on bringing out Leesma should be a clue."

Glades nodded solemnly. He looked toward Lahn. If the Vietnamese officer knew anything was unusual, his inscrutable face revealed no clues.

Coleman broke the awkward silence. "Look, we're headed to the chow hall to grab some lunch. We'll be back in about forty-five minutes. That'll give you folks plenty of time to give this stuff a good once-over. To the best of our ability, we'll answer your questions when we return."

Glades waited until their visitors had left before standing up. "Hey, Matt," he asked. "You mind taking a look at my left eye? I think I've got something in it. Finn, you can start building our terrain model."

"Sure, Nestor," replied Finn, intently comparing the details of a topographic map and some recent aerial photographs. The Vietnamese soldiers looked over his shoulder, talking excitedly amongst themselves. Glades surmised that at least one of them had previously been in this region, probably on a POW snatch mission, and was intimately familiar with the lay of the land.

Carrying a flashlight and his medical bag, Henson joined Glades at his bunk at the far end of the long room. "So you have something in your eye?" he asked quietly, switching on the flashlight as he tugged down Glades' lower eyelid. "I suspect that it's the same thing I have in my eye."

Glades nodded and said in a hushed voice, "We need to keep this little detail to ourselves. Don't discuss it with Finn. I don't want him contaminated. And one more thing…"

Wielding a cotton swab as he pretended to remove a fleck of dirt, Henson replied, "Yes?"

"If we go on this trip, we *cannot* be captured, regardless of what happens. Do you understand?"

"Nestor, I understand completely."

"Good." For the first time in his life, Glades felt nervous butterflies in his stomach. This situation was a lot spookier than anything he had previously dealt with. Obviously, his earlier concerns about not making it home were well grounded, even if he didn't know why at the time.

Over the Gulf of Tonkin, twelve miles east of Haiphong
North Vietnam
11:48 p.m., December 24, 1972

Earlier in the day, a combined Navy and Air Force SAR operation successfully extracted the F-4N's back-seater—Leesma—from harm's way. Despite their early triumph, later efforts to pull out the pilot—"Lieutenant Commander Drew Scott"—had ended in failure. NVA ground forces had poured into the area, establishing a cordon, just as Glades had predicted. While they initially suffered extensive losses from the F-8 Crusader jets providing close air support for the SAR mission, they fought tenaciously to prevent rescue helicopters from penetrating far enough inland to reach the downed pilot.

Tonight, after they infiltrated, the team would proceed to a link-up site in the densely vegetated terrain just south of the Duong Hoa River. They would remain there at a contact site, patiently waiting for the pilot to come to them tomorrow morning; after they had him, they would escort him south to be picked up by a boat from the coastline as early as tomorrow night.

In the dim red lighting of the MC-130 Combat Talon's cargo area, wearing a bulbous flight helmet, the Air Force loadmaster resembled an alien creature. He operated a series of controls that hydraulically lowered the transport plane's rear ramp. "You know you guys are crazy, right?" he shouted over the intercom as he flicked open an orange-handled switchblade knife.

Nodding in affirmation, Glades smiled. Looking forward, he saw the black canvas curtains that divided the cargo bay in half; the area forward of the tightly drawn drapes was the exclusive domain of electronic warfare specialists who operated massive racks of highly classified equipment.

He hoped that the EW wizards and their radar-spoofing gear would confuse the NVA's air defenses in this most vulnerable phase of the infiltration flight. Evidently aware that a potentially valuable commodity was within their immediate grasp, the North Vietnamese had aggressively repositioned SAM missile sites and other anti-aircraft weapons into the area, effectively isolating the terrain where the pilot was hiding.

Any close air support of the mission would be dicey, and slipping a helicopter in for an emergency extraction would be virtually impossible. Once they got their boots on the ground—*if* they survived the insertion—the six men would be largely on their own. Of course, they wouldn't be nearly as alone as the pilot currently eluding capture.

Glades felt the plane roll into a tight left bank as the pilot executed the last turn before they crossed overland from the Gulf of Tonkin. "Final run-in. Two minutes," declared the loadmaster, pressing his microphone close to his lips to muffle the roaring wind noise. He crouched down to slice through a pair of cotton safety cords. "Good luck. Merry Christmas to you and your little elves. I sure hope this sleigh works for you."

Glades removed his headset and handed it to the loadmaster. He donned his leather "bunny" helmet, strapped it into place, and then bent forward vigorously as he interlocked his right arm with Henson's. The six men were snugly jammed into a box roughly the size of two chest freezers, constructed of sturdy plywood sheets lashed together with nylon rigging straps.

They sat on a spongy layer of inflated air mattresses; the padding was intended to cushion the landing impact. Their heavy rucksacks were strapped down underneath the air mattresses. There were three men on each side of the box; a nylon cargo strap was hitched across their laps, effectively serving as a communal safety belt. The top of the box was partially enclosed by a plywood lid. Two G-12 cargo parachutes were positioned atop the lid; one would be automatically opened by a static line as the crate fell away from the aircraft. If the first parachute failed to open, Glades would pop open a pilot chute to immediately deploy the back-up G-12 chute. While it was a risky and untried endeavor, the unconventional technique would keep the team intact instead of scattering them across the landscape.

At this point, as they traversed the coastline of North Vietnam, only a short length of nylon parachute cord held the box in position. A green light flashed on, and the loadmaster slashed the final restraint. He nudged the box with his shoulder, and it slipped down a track of conveyor rollers and out into the darkness.

In his left hand, Glades tightly gripped the D-ring that would activate the reserve cargo parachute. He counted off the seconds, hoping that he had not sentenced the men to their deaths, as they fell. He felt a sharp jolt and breathed a sigh of relief as the giant cargo parachute, sixty-four feet in diameter, opened above them. Relieved, the six men braced themselves for impact.

Moments later, the box thudded to a bone-jarring landing. As a gentle breeze tugged at the cargo chute, the box promptly rolled on its side, pitching the men into an untidy heap. Wriggling, they extricated themselves from the pile and quietly climbed out to execute their post-landing tasks, just as they had rehearsed.

As the first man to emerge, Glades quickly activated quick release snaps that freed two of the four riser straps that connected the parachute to the crate; this action spilled the air from the billowing chute, preventing it from dragging the box and perhaps drawing unwanted attention.

Other than leaves rustled by the slight wind, the night was still and quiet. Thousands of brilliant stars adorned the firmament. The men silently loaded their AK-47s before collecting their rucksacks and other equipment. While Glades and Quan pulled security, the other four men loosened the cargo straps that held the box together, and then stacked the pieces.

They swiftly stretched out the cargo chute, rolled it into a bundle, and then heaped it onto the neatly piled plywood. Less than ten minutes after they had touched down, they moved out toward the north, with four of the men carrying the disassembled crate like an oversized litter.

Moving quietly in the darkness, Glades assessed the terrain. As best as he could tell, the Combat Talon crew had dropped them precisely where they had planned, in a clearing deep within the forest. From here, they would find a suitable place to conceal the dismantled crate and cargo parachutes and then proceed to the next phase of the mission,

22

WAITING IN NEBRASKA

Ourecky Homestead, Wilber, Nebraska
10:25 a.m., Sunday, December 24, 1972

Sipping black coffee at his parents' kitchen table, Ourecky was frazzled. He had barely gotten any worthwhile rest last night, even after an exhausting day of driving from Ohio. His MOL nightmare had become a recurring dream, haunting his sleep for the past week, so he had spent most of the night tossing and turning, afraid to lapse into unconsciousness. The dream was the same every time: he was alone in the Gemini-I, hovering close to the MOL, and saw Carson's face in the viewport. And in every instance of the dream, Carson always looked the same, like he was desperately trapped or lost. But more troubling than anything else, Ourecky knew that there was absolutely nothing he could do to help his friend. He hadn't shared his dream with Bea, although he was fairly sure that she knew that he wasn't sleeping well.

He studied the clock, anxiously waiting for the phone to ring. As she helped Mama Ourecky prepare the big carp that would be the centerpiece of this evening's meal, Bea looked at him and smiled reassuringly.

In accordance with family tradition, Papa Ourecky was concealed behind a makeshift curtain of bed sheets, hanging decorations—hand-made over the course of generations dating back to their forbearers in Czechoslovakia—on the fir he had chopped down early this morning. Striving to distract their attention, Ourecky tried to read a children's book to Andy and Rebecca, but his efforts were all but futile, since the *Cat in the Hat* was not nearly as intriguing as Papa's clandestine activities.

He was still distraught over the news from Southeast Asia, but reconciled himself to the notion that Carson must be safe, at least for the time being. Carson's fighter squadron flew a dawn-to-dusk mission cycle, and since it was nearly midnight in the Gulf of Tonkin, surely he was on his carrier, safe and sound. Moreover, last night's news broadcast had announced that there would be a twenty-four hour cessation of the bombing missions, so that the Navy and Air Force crews could celebrate Christmas. So, Ourecky convinced himself that Carson *had* to be safe right now, which would soon be confirmed when his MARS call rang through.

On the long drive over from Dayton, as the kids dozed in the back seat, he and Bea had agreed upon a plan about Rebecca. When the MARS call came, Ourecky would introduce her to Carson; immediately afterwards, Bea would take the little girl out of earshot, so that Ourecky could confide to his friend that Rebecca was likely his daughter. While he could just tell Carson that he might have a child, he decided that also hearing Rebecca's voice would carry a lot more psychological impact, perhaps enough to convince Carson not to take any undue risks.

He felt as if the weight of the world was on his shoulders. As worried as he was about Carson, he was very aware that in just a few weeks, his friend would probably be back from Vietnam and the Navy MOL mission would become public knowledge. He had no earthly idea of how he was going to break that news to Bea when the time came. It was ironic; here he was, spending Christmas in the house where he'd grown up hoping and planning to be an astronaut, and now the *last* thing that he wanted to do was to fly into space—yet again—but he knew that he would likely have little choice in the matter. Even though it would likely destroy his marriage, he just couldn't deprive Carson of the opportunity that he so rightly deserved. Ourecky closed his eyes and clenched his fists; he was incredibly frustrated, but right now more than anything, he just wanted to talk to his friend and know that he was safe.

"Daddy, can we go outside?" asked Andy, scampering out of his high chair and rushing to the window. His socks were mismatched, a consequence of his insistence to dress himself this morning. "Can we go outside?"

"Yeah. Can we go play outside?" begged Rebecca, tugging at the sleeve of Ourecky's red flannel shirt. "It's *so* pretty! Me and Andy want to make a snowman."

"No," snapped Ourecky. "Uncle Drew is supposed to call soon, and I want both of you to wish him a Merry Christmas. He's far from home, so it's very important." He glanced out the condensation-beaded window. The yard and pastures were covered with a thick blanket of absolutely pristine snow, so much different than the gray stuff they were used to in Ohio. Big downy flakes continued to fall. In the distance, past a grove of bare trees, a faint wisp of smoke rose from the chimney of his cousin's farmhouse. It was beautiful outside, a veritable winter wonderland. He felt sorry for the kids, but he wasn't willing to risk not having Rebecca immediately at hand to speak to Carson.

Bea sighed. "They'll be *fine*, Scott. I can't comprehend why you have to make such a fuss about it. I'll bundle them up and go out with them. We'll be right outside, in the back yard. They'll be back inside in plenty of time when Drew calls, I promise."

"We can't take that chance, Bea," replied Ourecky, shaking his head. He poured his coffee in the sink and ran some water to rinse out his mug. "Scott has a five-minute block on MARS, but that also includes the time it takes to make the connection, so by the time he finally gets through, we may only be able to talk for a minute or two."

"But these kids are getting cabin fever," urged Bea, rinsing flour from her hands. "It's so pretty outside. I'm sure that it will be fine."

"Please, Bea," he replied, staring at the old phone on the wall, as if he could will it to ring. "Please just give me this today. That's all I ask."

3:45 p.m.

Bea was worried about her melancholy husband. Despite a few anxious false alarms, the MARS call had not come through, and after hours of waiting by the phone, Ourecky had all but given up. While she knew that he was distressed about Carson in Vietnam, she sensed that something else was troubling him, but whatever it was, he wasn't forthcoming. She knew that he was having nightmares, and his sleep was suffering as a result.

His extended family was due to begin arriving in an hour or so, to sit down for the Oureckys' traditional Czech Christmas Eve feast. All of the decorating was done, the tables were carefully set, and most of the meal—with the exception of the dishes that would be contributed by other family members—was either in the oven or ready to be served. In any event, as much as Bea wanted to help Mama Ourecky, most of the remaining cooking chores were far outside her area of culinary expertise. Now, all that remained was to pass the time until the others came.

Yawning broadly, Ourecky peered out a window at the carrot-nosed snowman that Bea and the children had built earlier, after he finally conceded to let them go out to play. "I can't just *sit*," he confided to her. "This waiting is just driving me crazy."

"He'll call sometime, if he doesn't call today," said Bea, gripping his hand as she tried to console him. "There was probably some technical problem earlier. You have to know that there are thousands of guys trying to call home right now, and maybe the system was so overloaded that it just couldn't handle them all at once."

"Hey, Papa," said Ourecky, turning toward his father as he pulled his old field jacket from a wooden peg next to the door. "Didn't you say you wanted to raise a few more head of cattle starting next spring? You'll need some more room in the barn to stack hay bales. I'm going to go out there and move that old Mercury mock-up outside. It's not doing anything but wasting space."

"Son, there's no need for that," answered Papa Ourecky, slipping an LP album onto the spindle of the record player. Seconds later, the voice of Andy Williams spilled out of the speakers, crooning Christmas carols. "There's plenty of room for more hay. Besides, it's Christmas Eve. There's no need to fool around out there in that old barn."

Without speaking, almost as if he were in a daze, Ourecky donned his woolen hat and mittens, pushed open the door and went outside. He trudged toward the barn, leaving deep impressions in the otherwise virgin snow.

Rebecca tugged at Bea's hand. "Can we go, can we go?" she implored, jumping up and down excitedly.

"Please?" asked Andy.

Surrendering to their demands, Bea quickly stuffed the children into their snowsuits, hats, mittens, gloves and boots. She grabbed her own coat, slipped her feet into her boots, and escorted the children to the barn.

By the time they arrived there, Ourecky had already pushed open the big sliding doors and was wrestling the mock-up space capsule onto its side. Watching him, she remembered the Christmas Eve just four years prior, in this same barn, when he had knelt in the hay in front of the mock-up and asked for her hand in marriage.

As she and the children watched, Ourecky rolled the cone-shaped mock-up into an open patch of snow about thirty yards from the barn, and pushed it upright so that its blunted vertex pointed at the dismal gray sky. The flimsy framework was fractured by his rough handling, and large pieces of cardboard, tarred roofing shingles, and plywood had fallen off along the way. Slowly collapsing in on itself, the mock-up was damaged beyond repair.

Cursing under his breath, he slogged back into the barn and returned with a five-gallon can of gasoline, a pitchfork, and a box of "strike anywhere" kitchen matches. "Stand back," he ordered.

Bea nudged the children away and held them cautiously at her sides,

Grimacing, Ourecky doused the mock-up with gasoline. He set the gas can safely aside, struck a wooden match, and flicked it toward the mock-up. The structure caught fire without hesitation; after years of storage in the barn, its wood was dry and crisp as kindling, and needed little encouragement to ignite. In moments, it was entirely ablaze, roaring and crackling in its destruction.

Still holding the kids, Bea studied him. His face was illuminated in the brilliant orange glow of the fire, and she was alarmed to see tears streaming down his cheeks. His hands trembled as he watched the pyre consume the plywood and cardboard. Bea suspected that burning the mock-up was not something that he was just doing on a whim, to make the time pass faster, but rather an act that he had been contemplating for a while. And whatever the reason for his odd behavior, it obviously had little to do with creating room for more hay.

Bea knelt down and whispered in Rebecca's ear, "Becky, take Andy back into the house."

"But we want to see the bonfire!" declared the little girl, obviously more than curious about Ourecky's uncharacteristic behavior. "Please, Aunt Bea, please..."

"Take Andy inside *now*," said Bea sternly. She pointed at the kitchen door, where Papa Ourecky waited, a concerned look on his face. "You can watch from the kitchen. Fire is nothing to play with, especially for little ones like you two."

The children reluctantly walked back to the house, and Papa Ourecky took them inside.

Bea went to stand beside her husband as the token of his teenage years went up in flames. She was bewildered by his attitude and actions; she couldn't decide whether he was frustrated over the present situation with Carson, or whether he might be angry over his failure to achieve his childhood dream of becoming an astronaut. Deciding on the latter, she tried to console him: "Scott, I know you're disappointed, but sometimes our dreams just don't come true."

He looked at her, then at the blazing mock-up, and then replied in a quavering voice, "Sometimes they do, but just not the way we want them to."

Within minutes, the fire had reduced the mock-up to a rough circle of black soot and charred bits of wood, stark against the mantle of fresh snow.

Using a pitchfork to sift through the smoldering residue, Ourecky periodically stomped the ashes to extinguish the few embers that still glowed. Bea was pleased to see that he seemed much calmer and composed now, as if torching the mock-up had been some sort of emotional sacrifice, like burnt offerings to appease some demon that tormented him. Maybe—*hopefully*—the catharsis was complete, and the evil spirit was permanently exorcised.

Apparently satisfied that the destruction was adequate, he returned the pitchfork, matches and gasoline can to the barn, closed the doors and returned to her.

"Feel better now?" she asked, hugging his waist.

"Much," he replied, looking toward the faint sun hanging over the bleak horizon. "Hey, it should be about time for dinner. Hungry?"

"Famished," she answered. "Lead me to the carp, Mr. Ourecky."

Grinning, he leaned toward her, kissed her, and said, "*Vesele Vanoce*, Bea. I love you."

"*Vesele Vanoce*, Scott. I love you, too."

8:20 p.m.

The big farmhouse was quiet after the massive meal. After exchanging some gifts, Ourecky's siblings and other relatives had gone home almost an hour ago. Snoring quietly, dressed in his favorite Liberty Brand overalls and Pendleton wool shirt, Papa Ourecky dozed in his big recliner next to the fireplace. Lying in wait for Saint Nicholas, Andy and Rebecca were sound asleep, snuggled in thick blankets, resting on straw pallets underneath the Christmas tree.

As Bea lolled on the couch, engrossed in an old family photo album, Ourecky sipped eggnog as he reflected on the evening's news. There was scarcely anything substantial from Vietnam, other than reports about the Christmas ceasefire. A C-130 transport plane had been shot down over Laos, where the ceasefire was not in effect. Closer to home, a massive earthquake had struck Nicaragua. Managua, the capital of the

Central American nation, was largely destroyed; the estimated death toll was five thousand, with 20,000 people homeless and/or injured.

Daintily arranging *dulkove kolacky* pinwheel cookies on a lacquered platter, Mama Ourecky was in the kitchen when the phone jangled on the wall. She answered it, listened intently for a moment, and calmly announced, "Scott, it's your friend Virgil again."

Sitting up, Bea groaned as she set aside the album. "*Virgil*," she hissed. "Just when we're finally happy again, *Virgil* has to call."

Wearing a perplexed expression, Ourecky went into the kitchen, took the receiver from his mother, and said, "Yes, sir." He listened silently for a few minutes, and concluded the largely one-way conversation with, "Yes, sir. Uh…Given the circumstances, it might not be appropriate, but Merry Christmas, sir."

"I'll fetch your suitcase from upstairs, Scott," snapped Bea. "Luckily, everything is still boxed up back at our place, so I won't have much to…"

Hanging up the phone, he gestured toward the back door. "We need to talk," he said in a hushed voice. "Outside."

She took her pea coat from a peg by the door as she followed him out onto the porch. Shivering, she tugged on the heavy wool coat. As he closed the door, she asked, "How long will you be gone this time?"

"That's not why Virgil called. I'm not going anywhere," replied Ourecky quietly. "He had some bad news. *Very* bad news."

"*What?*" she demanded.

"The details are still sketchy, but Drew was shot down over Vietnam today…"

"Oh my God," she gasped. "I just *knew* that this was eventually going to happen."

"It happened several hours ago, before the Christmas ceasefire started," he explained. "He and his back-seater were able to eject. They've made radio contact with Drew. He's banged up some, but he's moving and hiding from the enemy. They've attempted to rescue him, but the helicopters ran into heavy fire and had to turn back. They've already picked up his back-seater. He's in surgery back on their carrier."

"So what will happen now?" she asked calmly. "Is Drew going to land up in a POW camp?"

"Probably not. They've sent in a special team for him. He'll be in good hands. They're the same guys who…" Abruptly quiet, Ourecky shivered as he brushed fresh snowflakes from his hair.

"The same guys who *what*, dear?" asked Bea. "And did you imply that Drew was flying from an *aircraft carrier*? Isn't that what *Navy* pilots do?"

Ourecky looked at the half-full cup in his hand and asked, "What have you been putting in this eggnog?"

"Truth serum, obviously. I guess I need to increase the dose slightly, though."

23

SWAMP

N estor Glades and his small team arrived at the link-up site shortly before dawn. After establishing a tight security perimeter, he scripted an initial entry report as *Trung si* Dinh pieced together his GRC-109 clandestine radio. Using its coder-burst device, Dinh transmitted the brief message to Da Nang to notify them that the team was in position to contact the pilot.

A few minutes later, they received a short coded dispatch from "Gull Wing," a powerful FM transmitter located on a Navy picket ship several miles off-shore, acknowledging their arrival announcement and apprising them of the current situation.

Operating in the midst of the NVA, communications were tremendously awkward. For most of his previous MACV-SOG missions, in areas where the US maintained undisputed dominance in the air, Glades enjoyed the luxury of a "Covey Rider" aircraft flying over his team on a routine basis. But unlike the skies of South Vietnam, Laos and Cambodia, which the US essentially owned, chunks of North Vietnamese airspace could be temporarily leased, but there was typically a high price to be paid, usually in American hardware, blood or both. Without the option of habitual overflights, operating in a hostile environment where the enemy aggressively exploited sophisticated technology to ferret out covert agents and teams, Glades and his men had to rely on an extensive array of equipment and specialized techniques to communicate.

The communications network was like a triangular system of one-way pipes. The team transmitted to their base station in Da Nang, Da Nang transmitted to Gull Wing, and Gull Wing closed the unidirectional loop. As the most vulnerable piece of the complicated puzzle, the team could only transmit short bursts at infrequent intervals, and even those required an extensive amount of preparation. The GRC-109 operated in the high frequency range; combining a highly directional antenna, low power transmission, and the coder-burst device that essentially "squirted" their abbreviated messages, the covert radio was virtually invisible to even the most sophisticated radio detection equipment.

By relaying traffic through Gull Wing, the Da Nang mission controllers could provide the team with updated instructions and a virtual running commentary of operations and intelligence concerning the enemy situation. Since their ship-based radio station was safely beyond the NVA's grasp, Gull Wing could transmit continuously, if need be, on frequencies that the team could monitor on a standard tactical radio.

While the communications were awkward, they were also absolutely essential. To ensure that the scheme would function under all contingencies, the team had to haul a *lot* of radios. Besides Dinh's GRC-109, they carried two redundant radio systems to receive Gull Wing's transmissions, as well as a Soviet-built field radio to eavesdrop on NVA's tactical chatter.

Glades and the other two Americans also carried small PRC-90 survival radios, exactly like the pilot's, which were only suitable for short-range ground-to-air use. If there wasn't an aircraft overhead to talk to, they were virtually useless, and it was unlikely that there were going to be many US aircraft venturing close enough for a conversation.

In all, on top of their weapons, ammo, and other essential gear, each man on the team carried roughly twenty pounds of communications equipment—radios, batteries, extra crystals, tubes and other spare parts—most of which would not be used unless a primary system failed.

Glades listened as Dinh quietly dismantled his radio. The team's radio operator was tremendously vital; their survival literally hinged on his competence. On the recon teams he had run in the past, Glades had always entrusted this crucial job only to an American. At the outset, he hadn't been confident in relying on the Vietnamese SMS sergeant, especially since he had been a former enemy, but he had grown to trust Dinh without question. The little Vietnamese sergeant obviously took great pride in his role and was always fastidious in his work.

After Dinh stowed his gear, there was little to do but wait. Glades let half the team sleep as the others kept vigil. Although he rarely allowed his thoughts to drift away from the mission, Glades remembered that it was still Christmas Eve back in Florida, and it would still be hours before Deirdre sent the children to their beds. He could just see them, rising before dawn, squealing with anticipation as they ran to the living room to open the presents that she had stashed under the tree in the wee hours. Glades sighed; he had missed more Christmases than he could remember, and he hoped that he could someday make it up to her.

As the sun rose, he heard a significant air operation underway several miles away to the southwest. Glades smiled, knowing that the action should significantly improve their chances for success and survival. Listening to the roar of jets, he was aware that the Air Force and Navy were pretending to resume their SAR effort.

It was an elaborate deception plan designed to draw attention away from the link-up site. Hopefully, the feint would convince the North Vietnamese that the pilot had evaded all the way to the coast last night. Remaining offshore and out of reach, the SAR forces would go through the motions—including mimicking the appropriate radio traffic that they knew the NVA would monitor—of rescuing the pilot. If all worked well, once they had rendezvoused with the pilot, Glades and his men would be able to stealthily slink to the coast with minimal interference.

7:15 a.m.

Snugly wrapped in his poncho to stay warm, Glades was sleeping fitfully when Henson nudged his elbow. Mouthing the words "Merry Christmas," he pointed to the west and then to his ear.

Glades pushed away the condensation-soaked poncho, rolled on his side, cupped his fingers around his ear, and listened carefully. Sure enough, he detected the distinctive sounds of someone moving through the woods. He glanced at the plastic face of his tactical wristwatch, noted the time, and then tapped Finn's foot.

With hand gestures, he dispatched Finn and *Trung si* Hieu to scout in the direction of the movement. If the pilot had been moving last night, he certainly would be near exhaustion after stumbling through the difficult terrain in the dark. With luck, the two men would find him and guide him to the team, and then they could make their way south to culminate this mission.

As he waited for Finn to return, Glades grew more apprehensive as the sounds grew progressively louder. He couldn't imagine a single man generating so much noise, regardless of how tired he might be. Thirty minutes after he had sent them out, Finn and Hieu returned with an answer to the mystery.

Finn knelt down and removed his pith helmet. In a soft voice, he calmly explained that a large NVA force was approaching. As he described it, the force consisted of at least a hundred or more men, arrayed on line, moving on an axis roughly perpendicular to the river.

"Do they have the pilot?" whispered Glades.

Finn shook his head. "I didn't see him. Damn it, Nestor, they're headed *straight* for us. Whatever we do, we had better do it *chop, chop.*"

Glades nodded his head and motioned for the two men to return to their spots in the perimeter. He gestured for the team to be ready to move immediately. Rising to a kneeling position behind a tree, he peered through his East German Zeiss binoculars at the row of NVA soldiers slowly moving through the woods about three hundred yards distant.

He was mystified by the development; it was as if the NVA knew precisely where they were. The NVA unit's progress was apparently governed by a pair of men operating what appeared to be an RDF— Radio Direction Finding—device. The two men would move forward a few paces, sweep the air with a cage-like contraption of parallel antennas, and then repeat the process.

The remainder of the NVA paced their movement on the RDF crew, so the mass of men appeared to like a giant sluggish caterpillar crawling sideways. Glades suspected that the RDF operators were sniffing the two SAR guard frequencies—243.0 and 282.8 megahertz—pre-set on the PRC-90 survival radio issued to the pilot.

Glades considered the situation for a few seconds. Although the NVA unit's relatively slow pace would buy the rescue team precious time, he saw that their options were painfully limited. The link-up site had been selected so that it could be readily located by the evading pilot but still offer him a considerable amount of concealment as he moved parallel to the river. As such, it was not a location tenable for mounting a sustained defense, particularly against the overwhelming force now moving in their direction.

Glades knew that if they ran to the north, they would have to cross the river, and it was likely that they would be chopped to pieces as they swam for safety. At a minimum, they would probably lose a substantial portion of their critical equipment. It was also tempting to flee parallel to the river, but if they did, the NVA would likely surge forward and their escape would be like running through a protracted ambush.

He saw only one practical way to escape, and that was to assault directly into the midst of the NVA force. Looking toward the other men, he issued instructions using slow motion hand signals. The three Vietnamese SMS soldiers were calm, almost unnaturally so, but Finn and Henson were now noticeably frightened. The Americans' eyes were open wide and their lips were tightly drawn, as if their grim demise was all but assured. Glades shrugged his shoulders and the five men nodded in assent, indicating that they understood his hastily conceived plan.

Slithering low, they silently dissolved their small perimeter and arranged themselves in a line parallel to the advancing rank of NVA soldiers. Spaced about ten feet apart, each man carefully positioned his heavy rucksack slightly in front of him, so that it would afford at least some modest protection against gunfire, and poked his left arm through a shoulder strap so that he could swing the pack onto his shoulder in one smooth motion. With their weapons ready, they waited in ambush. Glades wiped sweat from his brow before resting his M79 grenade launcher, primed to fire, beside him.

Carefully watching the line of approaching troops through binoculars, Glades noticed something unusual. With the exception of the two men operating the RDF equipment, the NVA soldiers appeared to be kitted out with brand new uniforms and equipment. Their pre-mission intelligence reports stated that a newly formed NVA infantry regiment was training nearby, preparing for an upcoming offensive in the South, so he suspected that they had been pressed into service to search for the evading pilot.

Soon, it was time to kill. Glades selected his target. He thought of the simple guidance he gave while training his Vietnamese recon troops. He had conditioned them to seize absolute advantage of the surprise granted by firing the first shot. He taught them that once the first shot was fired, there would be just the slightest instant of confusion and hesitation, and then people would instinctively react and take cover.

So that first shot had to count; it had to kill, and selectively so, not just dropping the common soldier. In training, Glades had jokingly told his Vietnamese counterparts, "If there's ever a question of who to shoot first, don't shoot the guy with the radio, shoot the guy standing next to the guy with the radio." And although the suggestion was humorous, to a certain degree, it also rang with truth. He scanned the line of troops and spotted a NVA soldier carrying a radio.

Looking to the radio operator's right, Glades identified a man who was clearly an officer. Although the green NVA uniforms lent few clues to distinguish superior from subordinates, Glades saw a map case lightly swinging at the man's waist. The low morning sun glinted from a wristwatch on his left forearm. Those indicators—together with the mannerisms that naturally came with leading troops—marked the NVA officer for the first bullet.

Glades signaled the others in the team. In a sign language that would be incomprehensible to anyone outside the intimate circle of the team, he conveyed his plan. In the subtle movement of his hands, he designated which target was his; from that, they knew to distribute their fire and select their own targets to the left and right. He specifically designated Finn and Hieu to engage the RDF operators; hopefully, killing them would snatch that card from the enemy's hand.

As he waited the last few seconds, patiently allowing the line of NVA troops to draw closer and closer, he carefully aligned his sights on the man he would shoot. His attention was riveted on the map case at the officer's side. He wanted it badly, not as a keepsake but because it likely contained a map or other intelligence that might lend clues to the disposition of other NVA forces in the area. The terrain was furrowed with small draws and ridges running perpendicular to the river. Glades decided that he would wait until the officer walked about forty more meters, into a slight draw, which should afford some slight cover as he grabbed the map case.

What happened next happened in a second. And if a second could be parceled into many pieces, the stories told and ended in that second would be manifold. Knowing that the timing was as close to perfect as it could be, Glades aimed steadily at the center mass of the officer's chest, so that his bullet would pass directly through the man's heart, drew a breath, exhaled half of it, and exerted gradual pressure on the AK's trigger.

It was just as if he was a boy again, hunting a rabbit for his mother's stew; the gun went off almost by itself. The copper-jacketed projectile crossed the short gap in a trajectory that was almost flat. Sure that he had touched off a clean kill, Glades was dismayed as the NVA officer pivoted slightly and stooped slightly to the right in the last split-second before impact.

Instead of smacking him square in the chest, as Glades had intended, the bullet caught the officer in his left shoulder and spun him off his feet. He pin-wheeled backwards, tumbling to the ground. In the same instant, the rest of the team fired; their weapons sounded in one report.

Other NVA soldiers fell, and the rest were obviously stunned with momentary confusion. Finn yelled for everyone to put their heads down, and then detonated Claymore directional mines they emplaced last night to protect the perimeter. A gale of high velocity metal swept through the ranks of the NVA, blasting several more off their feet.

The carnage continued. The team continued firing as Glades rose to his knees and pointed the M79 at the disintegrating green line. He was fond of the versatile grenade launcher; he didn't use its sights—in fact, he had long since removed them—but aimed it instinctively, in a manner not unlike throwing a baseball or casting a bass lure.

He fired, and the 40mm grenade detonated with a sharp *CRUMP* about twenty meters to the left of the wounded NVA officer. Even as the pale gold grenade was still in the air, he smoothly reloaded and sent a second grenade to a point about twenty meters to the right of his first target. He fired again, then dropped the M79 so that it dangled on a strand of parachute cord beside him, and picked up the AK-47. Then he was on his feet, swinging his rucksack onto his back and sprinting forward. Glades had created chaos and now ran headlong into it. The other men were up and following him instantly, maintaining their tight little line.

Glades had been in his share of ferocious firefights, but very few fought this closely and fiercely. The team instinctively slipped into the gap they had blasted in the NVA formation, and quickly fell into position as if they were securing a breach in a perimeter. Half the team oriented themselves to the left, directing their fire down the line of NVA troops to the east; the other half faced the opposite direction, firing at the green-clad soldiers to the west.

Despite the overwhelming odds against them, the team fired methodically. Instead of relying on the standard American tactic of full auto "spray and slay" to compel the enemy to panic and seek cover, the men fired precisely and discriminately, every single bullet finding a target and in most cases, instantly killing the recipient.

Brushed by scores of passing bullets, the vegetation was animate; leaves fluttered as branches and twigs twitched. The air was also alive, crackling and booming, laden with almost as much metal as oxygen, filled with the shrieks and anguished cries of dying men.

Glades's plan was simple. He wanted the two isolated wings of unseasoned NVA troops to turn towards the breach, where the rescue team was currently hunkered down, systematically decimating their ranks. Deprived of their leader, the unseasoned NVA soldiers would begin firing on themselves in the ensuing panic and confusion.

He fired four quick rounds in succession, downing three NVA soldiers attempting to rush them, hurled a pair of grenades, and then knelt beside the NVA officer he had wounded only moments before. He drew a curved magazine from the cloth pouch across his chest, and reloaded the AK. He was angry with himself for not killing the officer outright; the man's plight reminded him of the possum he had wounded in his first successful hunt as a child. He decided that since the officer was still alive, he might be of intelligence value.

He gestured to Henson, waving him towards the downed NVA officer. "We're taking him as a prisoner. Get him ready to move," he calmly shouted over the din of gunfire. "You have about a minute at best. Stay with him and keep him *alive*. You can patch him up better down the road."

Henson deftly applied a combat dressing to the dazed man's shoulder wound. As the medic exerted pressure on the wound to slow the profuse bleeding, Glades drew his Kabar knife and sliced through the strap of a map case slung over the officer's shoulder. He glanced into the case, quickly inventorying its contents, before stuffing it into one of his thigh cargo pockets. As he frisked the officer, *Dai Uy* Quan rifled the pockets of the dead radio operator. After locating a notebook, Quan fired several rounds into the radio to ensure that it was out of commission.

When it looked like they wrought enough mayhem, Glades pounded Henson's shoulder. "Secure the prisoner! We move in ten seconds! Move in ten!" he shouted above the noise. He repeated the action with Finn on the opposite flank. "We move in ten! Move in ten!" He yanked a smoke grenade from an ammo pouch on his belt, tugged out the pin and dropped it at his feet. "*Move!*" he yelled as the obscuring smoke started to billow.

Carrying the wounded prisoner, they ran approximately five hundred meters to the southwest. Glades paused momentarily so the men could sow a string of M14 toe-popper mines on their back trail, ideally to discourage any NVA soldiers who might stubbornly pursue them.

Afterwards, they doglegged to the south and then to the northeast, forming a large buttonhook. He intended to gradually make his way back to the river, winding through a series of additional anti-tracking buttonhook maneuvers, hopefully to find a hiding place in the swampy ground running parallel to the waterway as it meandered toward the southeast.

They had been moving for almost an hour before Glades ordered them to halt. There were no indications that they were being chased, so he granted the men a few minutes to catch their breath and grab a quick gulp of water.

He opened the map case to examine its contents more thoroughly. He found a map and opened it, and was stunned to find the planned link-up site—the place where they had patiently waited last night—clearly marked in pencil. He was furious, but said nothing. Obviously, someone higher up the food chain was funneling information to the North Vietnamese. The prisoner was a heavy burden to lug along, but if he could furnish any clues concerning the leak, he could be worth his weight in gold. Not revealing his discovery to the others, Glades leaned toward Henson and said, "Make *damned* sure that you keep this guy alive."

12:48 p.m.

The team sought asylum in a dense swamp adjacent to the river. Besides the need to remain silent, every step required a deliberate effort, since there was at least a foot of vile foot-grabbing muck concealed under the waist-deep water. After quietly slogging for almost three hours, Glades located a slightly elevated piece of dry ground sufficiently large enough to accommodate the seven men. Settling into a tight gaggle, the team packed onto the cramped hummock as best as they could. The moss-covered ground was damp and squishy, but at least it offered a respite from constant immersion in the murky water that surrounded them.

Although his capacity for Vietnamese was limited, Glades examined the prisoner's documents. In addition to his official identification, the man carried a ration card, a handwritten roster of about twenty personnel, and a small black-and-white photograph of a young woman cuddling an infant child in her arms. The woman's face was sorrowful, as if she had not seen her husband in a long time and probably did not expect to see him again.

The captive's blood-spattered papers identified him as Bao Trung, a twenty-four-year-old lieutenant born in Haiphong. *Dai Uy* Quan, who possessed a good grasp of English as well as Vietnamese, interpreted the documents for Glades. According to the papers, Bao Trung had been in the North Vietnamese Army for six years. He had spent three years as an enlisted soldier, apparently in some obscure technical field, before he was commissioned as an officer.

Quan reached into his pocket and pulled out the radio operator's notebook. Smiling, he handed it to Glades. "Radio frequencies, call signs and code words," he observed.

"Monitor these," stated Glades, passing the waterlogged notebook back. Quan softly issued orders in Vietnamese. Nodding solemnly, *Trung si* Dinh took the notebook before opening his soaked rucksack. The SMS sergeant extracted a Soviet-made radio from his pack, switched it on, fidgeted with some dials, and then clamped a set of awkward earphones over his head.

Although clearly in severe pain, Bao Trung's face exhibited a resolute demeanor as he watched the proceedings. He obviously expected to be tortured for information.

"Whenever you're ready, I can motivate him to talk," boasted Quan. Grinning enthusiastically, he drew a thin-bladed Gerber combat dagger from a leather sheath and brandished it toward the prisoner.

Trung si Hieu squatted behind the weakened Bao Trung, brutally snatched him by the wrists, and then stretched him out over a rucksack. Trembling and squirming, the captive's eyes opened wide, silently pleading for Glades to intercede.

"Stow that pig sticker, *Dai Uy*," growled Glades. "I don't condone that crap." He dug a green-wrapped "indig" ration out of his rucksack. "Feed him. Give him some water to drink."

Glades continued. "Tell him that we'll treat him humanely, but no matter what happens, he is coming out with us. Make sure that he knows that it's better to cooperate now instead of later, especially if he keeps us from walking into an ambush. Tell him, Quan."

"I will, but I don't know why you're being so lenient," said Quan. "And I think it is a bad idea to take him with us. He'll be nothing but a hindrance."

"It's not your decision, *Dai Uy*," answered Glades.

Wearing a stubborn expression, Hieu reluctantly released Bao Trung. The sergeant quietly tore open the indig ration. He poured water into a plastic bag filled with dried rice and shrimp, stirred it with a plastic spoon, and then set it aside to reconstitute. He unscrewed the lid from his enameled NVA canteen and held it to Bao Trung's lips. Obviously unsure of what was to ensue, the thirsty NVA officer gratefully drank.

As the Hieu fed Bao Trung, Glades examined a folded piece of black silk that he found at the bottom of the map case. He opened it, and spread it out on the damp ground. Approximately a meter on each side, the square segment of fabric was intricately embroidered with odd symbols and Vietnamese characters. Thinking that it was somehow pertinent to the mission, he nudged Quan, who quietly asked Bao Trung about the mysterious cloth.

Quan spoke quietly to the NVA officer. Glades didn't understand the conversation, but he could clearly discern from Bao Trung's face that it was a prized possession.

Quan turned to Glades and explained, "Your prisoner is an amateur astronomer. That cloth is a homemade star chart that his wife made for him. She copied it out of an astronomy book and sewed it entirely by hand. It has all the constellations and planets. It took her months to make it. It was her wedding gift to him."

"A star map?" asked Finn. He leaned over to examine the needle-point celestial map, and then started to roll it up. "Man, this will make a great souvenir!"

Bao Trung was obviously distraught that the American would confiscate his precious keepsake, but obviously realizing that he wasn't in a position to argue, he made a gesture to indicate that Finn was welcome to take it.

"Cool," noted Finn, jamming the black silk in his thigh cargo pocket.

"Not so fast," interjected Glades, yanking the cloth from Finn's pocket. He neatly folded it and placed it in Bao Trung's lap. "It's his personal property, and we're not taking it from him." Glades turned toward Henson and added, "Patch him up as best you can. He has to be hurting something awful. Give him something for the pain, but lay off the morphine until I tell you. I don't want him floating when we talk to him."

Henson zipped open his M5 medical bag and pulled out a pair of scissors. He snipped through the blood-soaked pressure dressings and cut the sleeve from Bao Trung's shirt to expose his shoulder. Using a slender metal rod and tweezers, he probed the fresh wound. Listening with a stethoscope, he squeezed Bao Trung's brown deltoid muscle.

"How does it look?" asked Glades.

"Not good, baby," replied Henson, shaking his head. "I'm hearing a grating noise. Your bullet impacted bone and probably damaged some nerve tissue as well." Bao Trung winced sharply and grunted as the medic pried the wound open with a pair of stainless steel retractors. "What a grisly damned mess this is. See all that purple goop in there, Nestor? The stuff that looks kind of like chewed-up hamburger?"

"Yeah?"

"That's devitalized tissue. It's dead, shredded by the bullet's shock wave. I can clean him up a bit and excise some of this, but if he doesn't make it to a surgeon soon, he's going to lose this arm, especially if infection sets into the bone. It's just about as likely that he'll die from shock, even though I have his blood loss under control for the moment." Henson packed the wound with bulk sterile cotton before applying a fresh dressing.

"So your diagnosis is?" asked Glades.

"Nestor, I'll tend to him as best as I can, but it's likely he'll die no matter what I do. It's almost dead certain he'll die if we try to haul his ass to our pick-up site. That's not to mention the wear and tear that our guys will sustain if you insist on carrying him out."

"So this gook's going to die?" whispered Finn sarcastically. "Good riddance, so long as we yank some good intel out of him before he kicks the bucket."

"*At ease*, Finn," scolded Glades, mashing a mosquito on his neck.

As Henson checked the prisoner for other injuries, he nudged Glades and pointed out previous wounds in the officer's abdomen and calf. "This guy's been around," he noted. "Those wounds are at least six months old."

After Henson did what he could do, Hieu lifted up Bao Trung and flopped his upper body against a rucksack. Scowling and softly gnashing his teeth, Hieu spooned moist clumps of rice and rehydrated shrimp into Bao Trung's mouth. The North Vietnamese officer ravenously wolfed down every bite as if it was his last meal. Perhaps he suspected that it was.

When the plastic bag was empty, Glades said, "Chow time is over. *Dai Uy*, he's all yours."

While Quan quietly interrogated the prisoner, Glades pondered yet another dilemma. Soon, Dinh would shinny up a tree to string an antenna and then tune up his radio, in preparation to send a status update to Da Nang. Glades would script the message, encrypt it using WW II vintage "one-time" cipher pads, and then give it to Dinh to transmit.

The process should all be straight forward, but Glades was leery about how much he should report. He didn't know the source of the leak, and if he reported his current location and plans, someone might immediately feed the information to the NVA. Glades suspected that the culprit was likely Major Lahn, and that he surely would be in the operations center with Coleman, helping to update the tactical situation map.

Obviously, the NVA wanted the pilot in the worst way. Glades decided that he would submit his report after Quan finished his chat with Bao Trung, but he would not provide his correct location unless the team fell into significantly dire straits. It was just too risky to do otherwise, since he wasn't sure who he could trust.

Glades was not one to abandon a mission, regardless of how the odds were stacked. Despite their compromise, he was optimistic that they could still salvage the rescue if the pilot had not yet been captured. It would be dark again in a few hours, so they could operate in their favored element. They could return to the area where the pilot was hiding, perhaps aided by more detailed intelligence, and grab him. There were still loose ends to tie. He had to decide what to do with the prisoner, since they obviously couldn't carry him along if he continued the mission. Of course, with what Henson described, that unpleasant situation could very well resolve itself before the sun went down.

There was yet another loose end that bothered Glades. In his years of playing at this high stakes cat-and-mouse game, he had not developed a single sixth sense, but had cultivated several. He *knew* when he was being pursued, even in situations where there wasn't an enemy hot at his heels. But now, as he reclined on this spongy ground, he just didn't feel that nagging sensation that they were being chased. Certainly, he was relieved, but in a strange way, it troubled him that the NVA must know that the team was here, in their land, and yet there was no sign that they were aggressively on the hunt.

Quan tapped Glades on the shoulder, and indicated that Bao Trung was forthcoming with some useful information. The lieutenant claimed that he was assigned to a recently formed infantry regiment stationed nearby. Yesterday, they had been yanked from the rifle ranges, loaded onto trucks, and sent to search for two American pilots who had parachuted from a jet.

Bao Trung stated that he didn't know if the evading pilots had been captured, but when and if they were, they would surely be taken to a nearby camp, at least on a temporary basis. He stated that the camp held no other American US prisoners, but was a reeducation facility primarily for South Vietnamese pilots and commandos. Communicating through Quan, he related some other details, most of which were essentially inconsequential.

"Good job, *Dai Uy,*" commented Glades. "But tell him that I have a specific question for him." He spread out the captured map on the soggy moss and pointed at the link-up site marked in pencil. "Ask him where he acquired this information."

"Damn it," muttered Finn softly. "They knew *exactly* where we were. We are *so* screwed."

Bao Trung swallowed deeply and then whispered something in Quan's ear.

"He swears he doesn't know," stated Quan. "He claims that his regimental commander gave him those coordinates, and told him that the American pilot would be located there."

"So he has absolutely *no* idea where this information came from?" asked Glades.

A brief whispered exchange passed between Quan and Bao Trung. The prisoner's face was frantic; he shook his head repeatedly.

"He still claims that he doesn't know," said Quan. "I think he's hiding something. I can pull it out of him, if you'll allow me."

"No," reiterated Glades. "He may be hiding something, but I think he's probably telling the truth about the spot on the map."

Writhing, Bao Trung groaned, clutched his shoulder and whispered something to Quan.

"He claims that he's suffering a great deal of pain," explained Quan.

"I don't doubt it," said Glades. "Matt, tag him with some morphine, but don't overdo it."

As Henson prepared a syrette of morphine, Dinh removed his earphones and held them so that *Dai Uy* Quan could listen. Consulting the list of NVA code words, Quan pronounced, "Bad news. Enemy trackers have located your pilot. They've surrounded him and are moving in other forces to block his escape if he attempts to run."

"Where?" asked Glades, checking his AK-47.

Dihn scribbled down some coordinates and handed them to Quan. The Vietnamese captain referred to the code word list and pointed at location on the NVA map. "Here," he said. "That's about five kilometers from us."

Glades studied the map, tugged his rucksack onto his broad shoulders, and slid his feet into the dark water. "*Up*," he ordered, without hesitation. "We're moving."

"What will we do with him?" asked Henson, still holding the unused foil syrette of morphine.

Finn unsnapped a leather shoulder holster and drew a High Standard .22 caliber pistol fitted with a silencer. "I'll fix this mess for you, Nestor. Someone has to do what has to be done. We'll dispose of this business here and now."

Glades swiveled to face Finn and tugged the pistol from his hand. "*No*, Ulf. He's our prisoner. Like it or not, we've taken him under our protection, so we're going to safeguard him. Period. Now, get ready to move."

Finn holstered the pistol, shrugged into his rucksack, and stepped into the water. "I think this is a bad idea, Nestor, especially if you're still planning to go after the pilot."

Flanked by the SMS sergeants, Quan frowned and said, "If you don't have the stomach for this, Glades, walk away. We'll take care of it and catch up."

Recalling what Henson had said, Glades could not help but think they were right. The North Vietnamese officer was in a bad state and probably would die before the sun fell from the sky. It was futile to think that he might live, just as it was equally pointless to believe that they might successfully rescue the pilot even as the NVA were closing on him.

Glades thought of the picture Bao Trung carried. He had conditioned Deirdre to expect an empty casket at his funeral, but he didn't think that Bao Trung's young wife deserved a similar fate. "If we leave him here, he won't be found. We'll take him to the edge of the swamp."

"But, Nestor..." whispered Finn.

"We're *not* leaving him here," hissed Glades, not wavering in the least. "Hand him to me. I'll tote him."

Henson jammed the syrette's needle into Bao Trung's thigh. After injecting the potent narcotic, he put on his rucksack, slipped into the water beside the others, and hoisted the diminutive NVA officer onto his broad shoulders. "Nestor, you can spell me when I get tired," he said softly, picking up his weapon from the hummock. "In the meantime, you can lead us the hell out of here."

12:59 p.m.

Slightly more than three miles distant, crouched in an almost impenetrable bamboo grove, Carson was not yet aware of his tenuous situation. He watched the second hand sweep the face of his Breitling chronograph. At exactly the top of the hour, he switched on his survival radio, pressed it to his ear, and listened for two minutes. He heard nothing but silence.

"Damn," he muttered to himself. He switched off the radio and stowed it in his torso harness. He had heard the nearby firefight this morning and surmised that his rescuers had been compromised. A pair of Navy attack jets had flown over exactly at noon to tell him to stay put, continue monitoring the radio on the hour, and to wait for further instructions. If nothing else, he had picked an excellent place to lie up. It had taken him most of the day to wriggle into the middle of this dense thicket. It was all but bulletproof, and he imagined that he could linger here indefinitely if need be.

Of course, there were a few drawbacks, he reflected as he scratched a maze of welts on his arms. The green bamboo was covered with a dust-like fuzz that was more aggravating than any itching powder. Besides that, it had been several hours since he had drained the last can of drinking water packed in his kit. He recalled that there was some technique to draw water from old bamboo but couldn't remember exactly what it entailed. He was sure that it involved hacking or cutting, and as parched as he was, he couldn't risk drawing his survival knife and making a racket. He looked up into the sky through the lush foliage, saw a passing cloud, and desperately wished for rain.

His temples throbbed with a headache. To pass the time, Carson lightly fanned himself as he watched a column of black ants scour the carcass of an enormous horned beetle. Sprawled on the hard black ground, he was drowsy, on the verge of sleep, when he was startled by the ominous sound of an NVA officer or a sergeant shouting instructions. He heard bamboo stalks cracking and falling as the machete-wielding NVA troops tentatively explored the thick stand.

Almost thirty minutes passed as the NVA soldiers painstakingly worked their way into the core of the bamboo. Making ready to run, Carson drew his pistol. His heart pounded as he clutched the Browning High Power in trembling sweaty hands.

Finally, he glimpsed someone through a narrow gap between green bamboo stalks. It was a teenaged soldier wearing a brand new uniform, tightly gripping a machete in one hand and a bayonet-fitted SKS carbine in the other hand. The sweat-drenched kid stopped momentarily, took off his pith helmet, and wiped his brow.

After over twenty-four hours on the run, Carson was weary and dehydrated, but not at all ready to give up. Listening to excited voices in all directions, he was certain that the stand of bamboo was encircled by enemy soldiers. He looked down at his pistol and remembered that its large magazine held thirteen 9mm rounds. Surely there were more than thirteen enemy soldiers lurking in wait, and although he might kill an unfortunate few, he speculated that it was just as likely he would accomplish nothing but to enrage the rest.

He wanted to run, but there was nowhere to go. He thought of Ourecky, and wondered if his cerebral friend could dredge up a mathematical formula or engineering solution that would deliver him from this predicament. More so than anything else, he was relieved that Ourecky had not followed him here.

And now, as he considered the odds of whether he might survive this day or even the next minute, Carson wondered what good might come from killing a few kids in a last-ditch show of resistance. Suddenly, the adolescent NVA soldier pushed aside a dead bamboo stalk; as he and Carson locked eyes, he dropped the machete and his mouth flew open in a howling shriek. Like Western gunfighters locked in a quick-draw confrontation, the kid raised his carbine and Carson raised his pistol.

1:12 p.m.

Glades signaled for the wading men to halt. Noting that the stagnant water was hitting him about mid-calf, he consulted his map; in less than half a klick—five hundred meters—they should be walking on dry land again. He turned to look at the faces of the other men. The only sounds were mosquitoes buzzing and a pair of noisy birds squawking nearby.

The stillness was shattered by several staccato bursts of gunfire to the northwest. "That's pretty damned close," whispered Henson. "Maybe two klicks away at most."

Frightened by the noise, several small deer scurried past the men toward the safety of the swamp. They bounded effortlessly, scarcely even making a splash. They paused momentarily, communicated amongst themselves in faint whistles, and then continued on.

The men cautiously knelt down to listen, but the shooting was over as rapidly as it had begun. An anxious expression passed over Dinh's face. He quietly clicked his tongue, held up his hand, and then pointed at his earphones with the other hand. He apparently had overheard something on the NVA's radio net. He listened intently for a minute and then shook his head. He studied the list of captured NVA code words and then spoke quietly to Quan.

"The enemy just killed your pilot," relayed Quan, speaking quietly. "I'm very sorry."

"Is he *sure?*" demanded Glades. His stomach sank.

"He is," replied Quan solemnly.

Glades shook his head. Struggling against the weight of his rucksack and the unsure footing, he climbed to his feet. He signaled for the men to follow him, and led them out of the water. About thirty minutes later, he located a place for them to temporarily halt on the fringes of the swamp. If the pilot was dead, their prospects weren't likely to improve any time soon. The NVA surely knew that the rescue team was still operating in their backyard, so they would likely shift their resources to kill or capture Glades and his men.

Henson slipped Bao Trung's inert body from his shoulders and lowered him carefully to the dry ground. Relieved to be free of his burden, he took in a deep breath and exhaled quietly. Glades sat on the ground next to him, looked at the captive, and asked, "How is he?"

"Still breathing," replied Henson. "I'm really surprised. He must be like one of those Timex watches you see on TV: takes a licking and keeps on ticking."

Dinh continued to listen to the NVA's tactical frequencies while Finn monitored another radio. He leaned towards Glades, held out a handset and said, "Gull Wing just came up. They said they have critical traffic in two minutes. For your ears only."

Glades nodded, took the handset, and listened. After listening to the brief message, he passed the handset back to Finn and waved for the others to draw in close. After they clustered tightly around him, he softly announced, "Gull Wing confirmed that the pilot is dead."

He paused sufficiently to allow Quan to translate for Dinh and Hieu, and then added, "We've been ordered to execute the extraction plan. We'll hunker here until nightfall, and then we'll head south to the coast to the link-up site. Once we're there, we'll call for pick-up by boat."

He continued. "The bad guys know we're here, so we need to be ready for a fight. We have a couple of hours until nightfall. I want everyone to take turns lightening their load. From here on, we'll only carry the bare essentials to fight our way out of here. We'll cache the rest back in the swamp. Finn, you and Henson will go first while the rest of us pull security."

As Quan spoke to the SMS sergeants in Vietnamese, Glades opened his rucksack and extracted the waterproof bag that contained the uniform, flag, bartering kit and other gear they would have given the pilot. Crestfallen, he tossed the bag to Finn. "Here. We obviously don't need that stuff anymore. Sink it."

Finn hefted the bag and replied, "Nestor, there's *gold* in here, remember? The bartering kit? You're not going to leave that loot behind, are you?"

Checking over his AK-47, Glades leaned toward Finn and replied, "If you expect to survive today, you need to focus on hauling your stupid ass to the pick-up site. As for the gold, I don't see us dickering our way out of any firefights. If you're going to weigh yourself down, carry stuff that's going to kill people and shed the rest."

"How about him?" asked Henson, looking towards Bao Trung. The Vietnamese lieutenant's glassy eyes were now open, but he was groggy from the loss of blood and clearly numb from the morphine. "Are you still planning to haul him out?"

Shaking his head, Glades pointed at the map. "No. There's a foot trail about four hundred meters south of here. At nightfall, we'll drop him there so he'll be found by his own people."

"Nestor, you know damned well he's going to talk to his people when he's found," replied Henson. "Then we're screwed."

"No. I figure that he won't be found until daybreak, assuming that he survives through the night. By that time, we'll either be eating breakfast chow on a Navy ship or we'll be dead."

As he sorted through his ponderous medical bag, culling those items that he considered less essential, Henson quietly replied, "Nestor, I trust your judgment, but it's still a huge gamble. We could make him comfortable and leave him here. There's nothing for you to feel guilty about. It's not like you would be…"

Glades closed his eyes, wishing that there was a simple way out of this mess. "Look, I can't explain why, but I can't leave him here," he answered. "Hell, Matt, I can't even explain it to myself. It's damned sure not the sensible thing to do. It's one of those gut things. You're just going to have to trust me on this."

"I trust you." Henson held out a soda-sized gray metal can. "I'm packing a few extra units of serum albumin," he explained. "While we're waiting for the sun to go down, I can load him up. The increased volume should keep him from lapsing into shock, at least for a while."

"Do that," replied Glades. Each of the six men carried a can of the blood expander taped to their web gear, so there should still be plenty available if someone was wounded.

As Henson opened the can with a metal key, not unlike opening a sardine can, he added, "I'm going to lay off giving him any more morphine until nightfall. Hopefully, we can keep him quiet without it, and then I can give him a large enough dose to knock him out for a few hours."

6:35 p.m.

Glades reviewed the plan in his mind. He, Henson, and Quan would carry Bao Trung to the trail. They would leave him there, then return here to pick up their gear before heading south to the pick-up site. He adjusted his damp chest webbing, turned towards Henson and asked, "Ready?"

"I suppose," answered Henson, stooping down to collect their prisoner.

"*Wait*," muttered Bao Trung.

Startled, the two men looked toward him in disbelief.

"What?" hissed Glades, kneeling down.

"Wait," said Bao Trung weakly. "I have something to tell you."

Frantic, Finn blurted, "Oh, *great*. He speaks English and he's overheard every damned thing we've said. That's just groovy. Now we *have* to kill him. We don't have an option."

Unsheathing his Gerber knife, Quan fervently concurred. "He's right. We have no choice. We cannot carry him and we cannot leave him here alive."

"You speak English?" asked Glades, furious at himself for putting the men in such danger.

"I do," answered Bao Trung. He grasped his shoulder and grimaced. "Before I went to officer school, I was in Signals Intelligence. I was a voice interceptor."

"But you never told us you could speak English," growled Finn.

"You never asked."

"This is really not a good time to be funny," interjected Glades. "Since you're likely to be *dead* soon. Now, what is it that you have to tell me?"

"A warning," replied the NVA officer. "Do *not* go to the south."

"Why?" asked Glades.

"If you go to the south, they will be waiting on you. There's at least a battalion of special troops waiting in an ambush. Their orders are to take you alive. These are not men to be trifled with. They track and kill agents and commando teams who sneak in from the South."

"And they're going to all this trouble for a single pilot?" asked Glades.

"They were *never* interested in the pilot," explained Bao Trung. "They could care less about him. We have prisons full of American pilots, so much so that they're getting to be a nuisance. As I said, they were never interested in him. They want *you*. They knew in advance that you were coming and they knew that once the pilot was dead, then you would leave. To the *south*."

"But why me?" asked Glades.

"You killed a regimental commander and his staff a few years ago. Remember?"

"I've killed a mess of folks," answered Glades. "Excuse me for not remembering specifics, but it's hard to keep track. Was there something special about this one?"

"He was General Giap's nephew. He was also very well respected in his own right."

"And why are you telling me this?" asked Glades.

"As we left to go on this operation, they warned us how fearsome you were, and that you would give no quarter." Bao Trung coughed weakly and spat out blood. "But I have seen that you are not just a fearsome warrior, but you are also an honorable man. You have shown me great mercy, so I cannot allow you to walk into a trap."

"How do I know you're not lying?"

"Simple," answered Bao Trung. "There are many points on a compass. I am not telling you where to go. I am only telling you where *not* to go."

He did have a point, thought Glades.

"What do we do, Nestor?" asked Henson.

Gazing up into the night sky through the trees, Glades was silent for a moment. Then he answered, "We drive on with the plan. We don't have time to waste. Finn, we'll be back here in about thirty minutes."

"You're *still* going to leave him alive?" asked Finn.

"Yep."

"Listen," said Bao Trung, gesturing for Glades to come closer. As Glades bent over him, the NVA officer whispered in his ear.

"Okay. If you insist," replied Glades. He stood up, and then clobbered Bao Trung with the wooden butt of his AK-47, and then walloped him again for good measure.

"I think you busted his jaw," noted Henson, kneeling down to feel the side of Bao Trung's face. "I guess we don't have to worry about him talking to anyone anytime soon."

"Which is *exactly* what he wanted," noted Glades. "Squirt enough morphine into him to keep him unconscious for a while. Then we'll haul him to the trail, drop him, and be on our way. Finn, while we're gone, I want you to script a message for Da Nang. We'll transmit it on the move tonight. I want you to confirm that we are headed south to the pick-up site. Also, I want you to tell them that we captured a prisoner, interrogated him briefly, but he escaped."

7:16 p.m.

Half an hour later, after depositing Bao Trung at trailside, they returned to the hide site. "Nestor," whispered Finn. "I thought about it while you were gone. I think the gook was right. We shouldn't go south. It's too dangerous. We should go west, back towards where we dropped, and call for extraction by helicopter."

Hoisting his rucksack onto his shoulders, Glades answered, "No. That's way too risky. All it would accomplish would be to get more guys killed."

"So you're just going to blindly head south?" demanded Finn.

"I didn't say that, Finn. We'll tell Da Nang that we're headed south. That doesn't mean we're actually doing it."

"So what will we do?" asked Finn.

"*Trung si* Hieu knows where a boat is cached," interjected Quan. "It's a rubber raft, big enough to carry seven men. It's hidden about ten kilometers from here, to the east, on the coast near a village called Tam Po."

"How does he know that?" asked Glades. They were as deep into North Vietnam as it was possible to go without venturing into China. While the boat sounded promising at this point, he wasn't overly predisposed to venture off on a wild goose chase. On the other hand, they weren't exactly overwhelmed with other options.

Quan explained, "He was on a mission six months ago and buried it there himself so that an agent could escape if necessary. He's sure that it's still there, because the agent was killed. And if it isn't, he is sure that we could steal a boat from one of the villages on the coast."

Glades considered it. If they had a boat, they could make it out to sea and then radio Gull Wing directly to be picked up by helicopter. They wouldn't have to inform Da Nang, and if Bao Trung had told the truth—and Glades was confident that he had—they would be long gone before the NVA special troops had an opportunity to re-set their forces. He looked at the faint luminous markings on his compass. "We're going east," he declared quietly.

24

TO BE MADE OVER

In a world fraught with uncertainty, certainty still remains. The earth will heave and change, floodwaters will rise and scour the land, dictators will climb to power and tumble into obscurity, but the soul-crushing gears of bureaucracies will grind and grind and incessantly grind.

Because of the considerable intelligence collected in the aftermath of his shoot-down and the subsequent rescue attempt, Carson was presumed dead. While a formal declaration could not yet be issued, the US government assumed that the intercepted radio transmissions were entirely valid, and that Carson had been killed when NVA forces closed in to capture him.

Administratively, as the years passed and until his case was finalized, he would continue to be promoted with his peers and his monthly pay would accrue in his bank account. Of course, since he had no living relatives, there was no one clamoring for access to the funds or his insurance money, nor were there any irate family members to dispute the military's findings.

Although they were strongly rebuked for allowing Carson to fly in Vietnam, Wolcott and Tarbox didn't suffer any severe consequences for their complicity. While they both thought Carson's death was tragic, they genuinely felt that they had only acted in earnest to fulfill his oft-stated desire to fly in combat. To a large degree, although they would never admit it, they were also tremendously relieved that Carson had been killed and not captured.

With the exception of the Nestor Glades, the Special Forces sergeant who led the rescue mission, everyone was entirely comfortable with the notion that Carson was deceased. Although he had no credible evidence on which to anchor his claims, Glades tenaciously insisted that Carson was likely captured alive, and that the NVA's report of his death was merely a ruse to lure the rescue team into an ambush. But no one lent any credence to *that* outlandish theory; even Glades admitted that it was based more on a "gut feeling" than anything truly substantial.

Although everyone wanted to wipe the slate clean and move on, Carson's presumed death resulted in a pesky accounting discrepancy at an annoyingly inconvenient time. In theory, he was still a Naval aviator, but except for a trumped-up dossier buried in the bottom drawer of a file cabinet in the personnel office aboard an aircraft carrier, there was *no* official evidence of Lieutenant Commander Andrew C. Scott, United States Navy. But despite his participation in the secretive Project, the US military possessed an abundance of official evidence pertinent to the life and career of Major Andrew M. Carson, United States Air Force.

Why was this inconsistency such a pressing issue? The long-fought war between the United States of America and the Democratic Republic of Vietnam was all but over. At the Paris negotiating table, the US emissaries insisted on a thorough resolution of all matters concerning prisoners of war and missing personnel.

As a condition of the treaty, the North Vietnamese were expected to account for and return all US personnel held as prisoners, as well as the remains of those who had died in captivity, combat or other circumstances. For their part, the United States would furnish a comprehensive list of the missing and otherwise unaccounted for, and the North Vietnamese—to the best of their ability—would reconcile the list against their records.

US intelligence agencies had long suspected that the North Vietnamese maintained a warehouse-sized morgue filled with the remains and personal effects of deceased US personnel. As the war creaked to a halt, specialized teams of medical investigators and forensic sleuths prepared to scrutinize a virtual onslaught of human remains, anticipated to range from skeletal fragments recovered from aircraft crash sites all the way to intact bodies disinterred from hastily dug graves. With the fates of almost two thousand men still in question, and only a small portion of those known or believed to be POWs, hundreds of families anxiously awaited some closure concerning their loved ones lost in the cruel fog of war.

As the reconciliation list was still being drawn up, a handful of high-level planners were painfully aware that the physical remains of Lieutenant Commander Andrew C. Scott—if the North Vietnamese were eventually forthcoming with his body—would not match the medical or dental records of anyone on the list. As these men agonized over this problem, they also knew that it was entirely possible that the North Vietnamese would surrender a set of remains that *would* correspond to the medical and dental records of Major Andrew M. Carson.

After grappling with the issue, they decided that it was best that Carson be correctly identified, especially since it was a rudimentary clerical matter, readily solved. Besides, since Carson was almost certainly dead, what harm could possibly come of it? So, with a simple stroke of a pen, Lieutenant Commander Andrew Scott ceased to exist—officially or otherwise—and the name of Major Andrew Carson was quietly appended to the reconciliation list.

Reeducation Camp # 4
Lang Hien, Quang Ninh Province, Democratic Republic of Vietnam
8:25 a.m., Tuesday, January 23, 1973

Kneeling in a shallow trough filled with pea gravel, with his hands and ankles tightly bound with coarse sisal rope, Carson was in abject misery.

Of course, he had no inkling that his official identity had been abruptly altered yet again. There was no way for him to know that he was no longer Lieutenant Commander Andrew Scott, having been officially reverted to his previous identity. After all, he wasn't even cognizant of his recent demise at the hands of the NVA. If he had known any of these things, his life would probably be much simpler, both now and in the coming weeks.

By his estimate, he had been kneeling in the gravel for over eighteen hours. The tiny pebbles dug deeply into his knees; he felt sure that they would leave permanent indentations. If he spoke or attempted to move to relieve the pain, a guard seated behind him swatted his head with a large stick.

His muscles burned and hunger pangs gnawed at his empty stomach. Throbbing veins pounded in his aching temples. He was so parched that he imagined that his blood was as thick as gravy. Gorging themselves on trickles of blood, several black flies buzzed around the open wounds where the rope fetters gouged into his flesh. His hands and feet were numb from the lack of circulation. He hadn't even been allowed to get up to relieve himself, so his pajama-like trousers were caked with reeking wet filth.

Reflecting on the circumstances of his capture, framed in the harrowing reality of his present suffering, he desperately wished that he could have died that day. In the heat of the moment, he just couldn't compel himself to shoot the kid soldier who confronted him in the heart of the bamboo thicket. After witnessing the death of the VPAF pilot he had defeated in combat, he had lost all interest in killing yet another human being. Nor did he have the fortitude or presence of mind to turn the Browning on himself.

Instead, he had bashed the kid's head with the pistol's butt before awkwardly fleeing through the dense obstacle of intertwined cane. As he emerged from the maze, he was pursued by over a dozen NVA soldiers shooting wildly in his wake. He was eventually tackled by several NVA soldiers. They all floundered in gooey black muck for several minutes, but eventually they wrestled him to the ground and subdued him. Instead of the dramatic shoot-out climax of *Butch Cassidy and the Sundance Kid* he had envisioned only moments beforehand, his last moment of freedom was more like a slapstick scramble reminiscent of the Keystone Kops.

In slow motion, he subtly shifted his weight to alleviate the pressure on his right knee and was immediately rewarded with a painful rap to the back of his skull. Although he wasn't sure if he could bear any more pain, he was acutely aware that this was only the warm-up session for yet another excruciating marathon of interrogations. Hearing the door open, he watched four men enter the room and realized that the long-dreaded quiz session was about to commence.

Carson immediately recognized the interrogator, a thin Vietnamese officer who appeared to be in his early thirties. Two other men were apparently the inquisitor's trusted minions; they devised and meted out the physical unpleasantries while he focused on the questioning. Carson suspected that only the interrogator spoke English. The fourth man, whom he didn't recognize, took a seat in a corner and opened a writing tablet, apparently preparing to take notes.

The scribe looked like he had been in a terrible accident. His face was badly swollen and his left arm, encased in a white plaster cast, was suspended in a gray muslin sling. His mouth was tightly clinched, almost like he was terribly angry, but he said nothing; Carson suspected that his jaw was temporarily wired shut.

At this point, having undergone several similar rites, he knew that the pain of the coming hours would far surpass his present agony. Watching as they arrayed the instruments of their profession, he was morbidly curious concerning what today's torture would entail. In the last iteration, they had flayed his legs and back with pliable strips of wet bamboo. The bamboo flails had left stripe-like welts that took several days to heal. Before that, they had almost drowned him by forcefully and repeatedly dunking his head in a metal pail filled with filthy water.

The nefarious tormenters went straightaway to their labors. They wrapped his wrists with yet another length of sisal rope, although much thicker than the lashings that currently encompassed his wrists, and then threaded the rope through an iron eye-bolt anchored in a wooden beam in the ceiling. Then the duo hoisted Carson until his body was wrenched completely free of the gravel-filled trough. His wrists had been tightly bound in such a manner that he could not bend his elbows, so his full weight was suspended by his arms, awkwardly outstretched behind him. In moments, hanging at the end of the taut rope, he was delirious with pain.

Satisfied with their handiwork, the pair cinched off the rope and stepped to the side. They stood there, absolutely impassive, watching him for several minutes as he struggled in anguish. The Vietnamese were nothing if not patiently methodical. Carson felt like a live bug pinned to the wax-filled bottom of a dissection pan, squirming, waiting to be sliced open.

Slowly twisting in the air, he was conscious that it was only a matter of time before his body weight would dislocate his shoulders. In a few minutes, he heard a sickening pop as his left shoulder snapped out of socket. Seconds later, he winced sharply as his right shoulder separated in the same manner. Violently trembling, he resisted an almost over-powering urge to scream or cry out.

Writhing in pain, Carson knew that the worst was coming. He decided that it was time to board his imaginary submarine to ride out the remainder of the session. As he dangled there, he escaped deep into the recesses of his mind, patiently closing every watertight door and hatch, working his way through the different compartments inwards toward the control room where he could watch the proceedings through his periscope.

Regardless of the mechanism of suffering, the interrogator's agenda rarely changed. His preliminary questions were exactly the same, as if he was always starting anew. The scowling man stood next to Carson, adjusted his wire-rimmed spectacles, and curtly asked, "What is your name and rank?"

"Lieutenant Commander Andrew C. Scott, US Navy," grunted Carson defiantly.

"And what kind of aircraft do you fly, Lieutenant Commander Scott?"

At that instant, aboard Carson's make-believe vessel, the last hatch clanged shut; he madly spun the wheel to dog it down, and then he was finally isolated from his pain. He heard the inquisitor's question like a hushed voice from a distant room, but did not respond. He wasn't sure how long he could linger here; his endurance had been eroded considerably with every new torture session. Still, he knew that rescuers would eventually swoop down on the camp to extricate him from these horrors. In the meantime, he had to strive to remain calm under intense duress, endure whatever he was confronted with, and stick to the story he had been given.

10:15 a.m., Friday, March 9, 1973

Several weeks had elapsed since Carson's untimely arrival in the North Vietnamese camp. His initial treatment had been brutal, but for some inexplicable reason, his captors had lightened up significantly in the past month. He was no longer being tortured; perhaps he had convinced them that he was immune to physical coercion. His diet had improved dramatically; once given only a small bowl of rice a day, he was now fed a generous serving of rice in the morning and an even more substantial meal in the late afternoon.

Although he had settled into a predictable routine and his conditions were immensely better, it was still captivity. He spent the largest part of each day cooped up in his small cell. Constructed entirely of concrete, it was square, roughly eight feet on a side, with two small windows set just below the ceiling to let in light. Earlier in his stay, he had been clapped in manacles and leg irons every night, but lately he was allowed to sleep without the chafing restraints.

He had a thin mat of woven straw to shield him from the rough floor, and a moth-eaten wool blanket to keep him warm. The flimsy blanket stank of worn-in sweat and grime, and was obviously sized for the Vietnamese; when the nights were cool and damp, it was a struggle for him to cover his entire body, almost like trying to sleep under a handkerchief. One metal bucket served as a chamber pot and another held his daily allotment of water for bathing and drinking. All told, that was the extent of his accommodations.

One of the most aggravating aspects of his ordeal was that he had not yet seen or spoken to any other American prisoner. Recalling his SERE training at Aux One-Oh and intelligence briefings that he had received aboard the carrier while en route to Yankee Station, he knew that the North Vietnamese maintained an elaborate network of POW facilities, so he assumed that he was being temporarily held in solitary confinement to break his will, and that other Americans had to be close by, perhaps in another part of the compound.

Maybe since his captors had realized the futility of torturing him, and now that he was being fed and treated so much better, then surely he would be allowed to mingle with other US servicemen. While he assured himself that he could endure pain and deprivation indefinitely, the seclusion was wearing on Carson, and he ached to be among others like himself.

With nothing to read or otherwise occupy his idle time, Carson had plenty of time to think. He thought of his childhood and wished that he had been granted the opportunity to know his parents better. He thought of his days at West Point, learning to fly in the Air Force, the grave disappointment he had felt when he was selected to fly interceptors after flight school, and the marvelous exhilaration he had experienced when he was chosen to become a test pilot.

He reflected on all the money he had squandered on sports cars and expensive watches, and he thought about the scores of women he had known but barely knew. He wondered what his life might be like if he had made an effort to cement a meaningful relationship with someone, and whether he might be married now with a wife and children who anxiously awaited his return.

More than anything, he reminisced about the very best days of his life, when he and Scott Ourecky had flown into space together. He thought of him often and hoped that he would see his friend again soon. He missed Ourecky more than he could have ever previously imagined, and hoped that his friend had finally found his way to MIT.

Considering the circumstances, he thought that he was doing a good job of keeping up his spirits, but occasionally he would become disheartened and lapse into a melancholy funk that stretched on for days. What motivated him to remain alive was the certain knowledge that he would eventually be rescued. *It's only a matter of time,* he assured himself over and over, especially when he heard the roar of jets nearby. All he need do was to be patient and endure.

He knew that other prisoners were housed in nearby buildings. He regularly heard their muffled voices at night, but their Vietnamese was an incoherent babble to his ears. He never saw them, but knew their routine by heart. Early in the morning, even before the sun rose, they assembled in a central courtyard between the buildings. Judging by their voices, it sounded as if they answered a roll call.

Every morning, after he had finished his breakfast of rice, he was escorted to the same courtyard for his daily exercise. He spent most of the time walking laps around the perimeter of the yard. As he exercised, he remained attuned to his surroundings, looking for any possible weak spots that he might exploit in an escape attempt.

Apparently the other prisoners worked while he exercised. Because of the camp's layout and the other buildings, he couldn't see them, but he heard the sounds of digging and scraping sounds, like earth being broken with pick axes and shovels, and assumed that they were tending to some sort of garden.

After their morning labors, after he had been locked back in his cell, the unseen Vietnamese prisoners were brought back to the central courtyard for classes or lectures of some kind. The material was delivered in a painfully boring monotone, amplified over a blaring speaker. Typically, the lectures dragged on for hours, and then the prisoners were apparently sent on labor details.

Carson stiffly stood up and kneaded the lumbar region of his aching back. Although the torture sessions had exacted a debilitating toll on his body, he suspected that the trauma of his ejection had also caught up to him; he now endured a chronic backache where his spinal disks and vertebrae had been explosively compressed.

Stretching as he listened to the incomprehensible drone of the Vietnamese orator, Carson studied the angle of the shadows formed on the cell's gray floor. His heart beat slightly faster in anticipation of the high point of his otherwise oppressively monotonous routine. Every day at this time, he received a visitor, an English-speaking NVA officer named Bao Trung. The officer had been present for his interrogations, but had not taken part in torturing or questioning.

When he visited, Bao Trung did not press him for information; most of their time was spent talking about places in the States, history, and other innocuous topics. Bao Trung occasionally asked him about his upbringing and family, but Carson diligently avoided revealing anything of a personal nature. While Carson relished the opportunity to converse with another human being, he didn't entirely trust Bao Trung, and assumed that the officer was being sent in to subtly extract information while portraying himself as a sympathetic listener.

Early on, Bao Trung had attempted to teach him rudimentary Vietnamese, but abandoned the effort when Carson showed no enthusiasm or aptitude for learning a new language.

Lately, the Vietnamese officer had begun teaching him to play chess, which Carson enjoyed immensely. So, they had fallen into a routine where they played and chatted, and Carson had become comfortable with it. Interestingly enough, just a few weeks ago, they discovered that they shared an interest in astronomy. Often, Bao Trung brought a detailed cloth celestial map that his wife had embroidered for him, and Carson taught him the English names for the constellations and stars, and described the Greek mythology behind the names. In turn, Bao Trung told him the Chinese names for the familiar shapes in the heavens, and the various legends associated with them. Carson regretted that he wasn't able to view the night sky, except for the fragment through his tiny cell window. He frequently begged the Vietnamese officer to escort him outside at night, but citing security reasons, Bao Trung told him that he could not.

As he wondered what would be in store today, Carson decided that he wanted the answer to a question that had been troubling him for weeks, and he elected to forgo the daily pleasantries and casual conversation until Bao Trung was forthcoming with information.

Carson heard the clicking sounds of the lock being unlatched. The heavy wooden door opened, and Bao Trung stepped in from the bright sunlight. Carrying a pasteboard box in his good hand, he announced, "I brought my chessboard, Drew. Perhaps we can work on your game again. You've been making excellent progress."

"If you don't mind, I would rather just talk today," said Carson. "I have a question for you."

"Talk? Certainly. Is there something pressing on your mind?"

"Yes. Why am I kept from other Americans?" asked Carson.

"Honestly, I don't know, Drew," replied Bao Trung, sitting down on the floor and crossing his legs. "Perhaps you know the answer yourself."

9:25 a.m., Monday, March 12, 1973

Major—*Thiếu Tá*—Han Ngoc Thanh was the administrator for Reeducation Camp # 4. The camp's inhabitants were mostly recalcitrant South Vietnamese officers, almost exclusively former military pilots, clandestine agents and commandos. They were sent here to be made over. The camp's primary focus was to fracture their spirits, obliterate their capitalist tendencies and the other loathsome habits they had accumulated in associating with the French and the Americans, and to educate them in the pristine virtues of socialism. All this was done in the hope that they might eventually be rendered into productive members of a progressive society.

To be made over, mused Thanh, reflecting on the irony. He would probably have better luck leading rabid dogs to water. A hand-painted sign at the camp's entrance proclaimed it to be a glorious site of socialist redemption, but Thanh knew that he was actually overseeing a one-way railroad into oblivion.

Every few weeks, yet another consignment of healthy men trudged through the gate. For the first two weeks, they were crammed into dark cells, receiving no food or medical treatment. After the initial softening-up phase, the bewildered survivors were let out to fall into the camp's numbing routine.

In relatively short order, they were systematically starved and worked to death, eventually finding their places in the camp's burgeoning cemetery. In the interim, as they waited to expire, they attended indoctrination lectures and labored under the scorching sun, chiseling shallow graves out of hardened clay.

As any loyal socialist soldier would, Thanh had taken his duties seriously when he first arrived here. Triumphantly standing before the vanquished, even though he had never personally set foot on a battlefield, he obediently recited the carefully scripted lectures, sincerely believing every word that he expounded. He even convinced himself that he glimpsed hope in at least a few of their faces, a chance that at least one or two would come to their senses and abandon their old ways.

Now, he was just weary of this sordid tedium, watching as men faded away and eventually disappeared. As he delivered the lectures, all he could see was the exhausted resignation in the gaunt faces of those wretches who had been here for several weeks, those already on the verge of death, and restrained anger on the faces of the newly arrived who realized what their abbreviated futures held.

While he first disliked being so removed from Hanoi or Haiphong, he since had come to appreciate the virtues of administering a camp that was far from the beaten path. Camp # 4 rarely hosted official visitors. Consequently, while Major Thanh and his men enforced discipline and maintained their stern façade, they had settled into sort of a symbiotic equilibrium with their charges. Unlike the other reeducation camp administrators, Thanh did not go out of his way to torture or otherwise punish his prisoners for minor infractions. Their unspoken agreement was that he would do nothing to make the prisoners' stay any more painful than what it had to be, so long as they reciprocated by dying politely and on schedule.

This morning, reviewing a packet of documents just in from Hanoi, he was in a terrible quandary. He glanced at a small chalkboard hanging by the door; it reflected the official headcount from morning muster.

The numbers were constantly in flux, but consistently tended to trend downwards as each day passed; when the sun rose this morning, Camp # 4 housed one hundred and fifty-six men. According to the senior guard's estimate, that tally would likely dwindle to one hundred and fifty by nightfall, since six sick prisoners—four with malaria and two with dengue fever—were expected to die today. Thanh knew that the count would slip another six or seven notches by dawn, as others succumbed to starvation or sheer exhaustion.

Of the one hundred and fifty-six men currently in his charge, four were not South Vietnamese. Three were Chinese soldiers who had accidently—or perhaps intentionally—strayed across the nearby border. The fourth was a US Navy pilot named Lieutenant Commander Drew Scott.

How the American came to be here was no profound mystery. Although its primary mission was ostensibly rehabilitation, Camp # 4 also served as a way station and consolidation point for Americans captured in the region. If their conditions mandated, they received cursory medical treatment in the camp's modest infirmary. They remained here, strictly segregated from the South Vietnamese prisoners, until suitable transportation was available to move them to Hoa Lo or one of the other POW facilities that housed Americans.

Since he had briefly trained at Hoa Lo before being assigned here, Thanh knew that some US prisoners were occasionally earmarked for "special treatment." In such instances, the chosen unfortunates were kept strictly isolated from any other Americans. Thanh was aware that there were three principal categories of US servicemen to be set aside.

The first category comprised prisoners of potential political value, such as men related to political figures or high-ranking officers, who might be valuable as bargaining chips in the future. The second category were US prisoners who had become such disruptive troublemakers that they could not be mixed in with the general POW population, and had to be relocated to special camps elsewhere.

Obviously, the third category were those prisoners suspected of possessing "special knowledge" that was of particularly high intelligence value. This category included pilots who flew new or unique aircraft types, had undergone special training in classified systems or nuclear weapons, but also included personnel who had been assigned to the US Strategic Air Command.

Almost without exception, according to what Thanh had been told, "special knowledge" prisoners were so designated by Soviet GRU *spetsgruppa* advisors in Hanoi. The *spetsgruppa* was an intelligence organization tasked with exploiting high value American personnel and equipment. He had once heard that so much new American technology and expertise fluttered down from the skies over North Vietnam, the *spetsgruppa* joked that it was like receiving manna from heaven.

Most "special knowledge" prisoners were questioned by specially trained Vietnamese interrogators, operating under the supervision of the *spetsgruppa*. But as rumor had it, a select few "special knowledge" prisoners were immediately delivered into the waiting hands of *spetsgruppa*, and spirited away to God knows where after that.

But Lieutenant Commander Drew Scott was in a category unto himself. From the information Thanh had gathered, Scott was set aside not because he was thought to have special knowledge or political value, but rather because of some peculiarities surrounding his capture. Specifically, the Americans had mounted an unusually intensive effort to rescue him. In particular, the high command in Hanoi was perplexed why the Americans would go so far as to send a special team—led by one of their most experienced and capable commandos—deep into North Vietnam. In retrospect, it appeared that the team probably had not been dispatched for Scott, but rather for the man who shared his aircraft.

With an edict that Scott was to be kept strictly secluded, Hanoi periodically dispatched a special interrogation team to work him over. After five grueling sessions, the trio found Scott to be exceedingly disciplined and virtually impervious to even the most barbaric tortures, but they were fairly confident that he was nothing but a common Navy pilot.

After all, the GRU *spetsgruppa* had expressed no particular interest in him, and they seemed to have an acutely developed knack for identifying prisoners worthy of special treatment. Consequently, as of last week, because there was no evidence that he was anyone extraordinary,

Scott's status was on the verge of being downgraded so that he could be treated as a regular US prisoner.

But now there was a problem. Although Scott had revealed no information of value, the interrogators had recently come to realize that he was not correctly identifying himself. This may have never come to light, except for a recent development that emerged during the Paris peace talks. The arrogant Americans demanded a full accounting for prisoners of war, as well as the missing and otherwise unaccounted for, in Vietnam, Laos, and Cambodia. To this end, they had provided a list of 1925 names for resolution. The problem was that the name of Lieutenant Commander Andrew Scott did not appear on the reconciliation list.

Thanh read a terse missive from Hanoi. Until his situation was clarified, Lieutenant Commander Scott would remain a guest of Camp # 4. Thanh was dismayed by the distraction and desired nothing else but to wash his hands of the American. Worse yet, since Hanoi was swamped with the details of the ongoing prisoner reparations, they tasked Thanh to resolve the situation with Scott. Glancing into the flimsy plastic pouch that contained the documents, he saw that Hanoi had at least had the foresight to send him some resources to assist in the process.

Gazing at the ceiling fan slowly spinning above his desk, he groaned in frustration. Apart from being confounded by the American's stubbornness, there were other aggravating issues as well. To look after Scott, he was compelled to routinely allocate at least three guards from his understaffed security contingent.

That left him barely enough to adequately guard the South Vietnamese prisoners; he was sure that if they ever recognized the lapse, they would quickly seize control of the camp. To make matters even worse, he expected another influx of at least one hundred prisoners at the end of the week, and the current crop had proven themselves considerably more hardy than most, so he already had a backlog to contend with.

Thanh was weary with the grim drudgery of administering the camp, and he didn't relish any additional duties. If anything, he wanted to be far away from here, to be assigned to an infantry or tank regiment preparing for the impending last invasion of the South. He scratched his nose and took off his glasses; the ponderously thick lenses were a constant reminder why he had been rejected for the glories of combat duty.

He finished the letter from Hanoi and formulated his plan. As serendipitous fortune would have it, he had been dealt a unique card. At the same time Scott was delivered, a badly wounded infantry lieutenant was evacuated to the camp's infirmary. The lieutenant—Bao Trung—had been captured by the American rescue team, but valiantly escaped after being tortured. Since he knew English and was too badly injured to return to combat duties, Thanh had arranged for him to remain at Camp # 4 to assist with the handling of the American.

Thanh had been extremely impressed with Bao Trung's tenacity and dedication. Even though the American commandoes had brutally smashed his jaw, Bao Trung volunteered to assist with interrogating Scott even before he was able to speak. During the questioning sessions, he sat in as a recorder, which enabled the interrogator to more effectively use his time and energies.

As Bao Trung recuperated, he spent more time with the American, becoming his de facto caretaker. He talked to him daily, making patient inroads to hopefully learn more about Scott's background and training. Additionally, since he had sustained such serious injuries and would not return to his infantry duties, Thanh requested that he be permanently assigned to Camp # 4 to oversee the guard force. In fact, Bao Trung had been authorized a furlough to go to his home, in far western Lai Chau province, to collect his family and bring them here to his new posting.

"Corporal!" he yelled to the soldier standing guard outside his office door. "Bring me Bao Trung! *Now!*"

Ten minutes later, Bao Trung appeared at the door. Although his jaw had healed, his face was still badly swollen and discolored. His right arm was still in a sling and would likely remain so for several more weeks.

Thanh gestured for the lieutenant to take a seat and then quickly described his dilemma with the American. Waving the letter, he concluded, "And so long as we don't kill him, we are authorized to do whatever is necessary to determine his true identity."

"You mean that we're authorized to torture him?" asked Bao Trung quietly. "I thought that only the interrogation team was permitted to employ physical measures on him. I'm not questioning your word, Comrade Major, but Hanoi has actually cleared *us* to beat him?"

Thanh nodded. "Whatever it takes," he clarified. "Provided that he can be healed up sufficiently afterwards so that he can be sent to his country."

"Well, sincerely, I don't think you can pound the truth out of him, Comrade Major," observed Bao Trung candidly. "I've witnessed him being interrogated, you know. He has an immense threshold for physical pain."

"I know that you spend a considerable amount of time with him. You've become friendly with this American?"

"No, but I admire him for his toughness, Comrade Major," answered Bao Trung. "But since he is secluded from his own kind, he's terribly lonely. It's obvious that he enjoys talking with me in his own tongue, but I must caution you that he's still very cagey. In all the times we've conversed, he has divulged nothing of significance about his background, military or otherwise. I don't think it would suit our interests to coerce him physically."

"Here," said Thanh, proffering the thick packet that had been dispatched from Hanoi. "Use these items as you see fit."

Bao Trung opened the packet to find two mimeographed copies of the reconciliation list—one printed in English and the other in Vietnamese—along with a glossy American news magazine. Frowning, he flipped through the magazine. Although it covered some news items, its slick pages seemed largely devoted to disgusting advertisements for automobiles, cigarettes, liquor and other products of America's excessive culture. "Why on earth would Hanoi send this capitalist propaganda?" he asked, closing the journal and slapping it on Thanh's desktop.

"I don't read English, but I think it's to help convince him that we're sincere about the prisoner repatriation." Thanh pointed at the cover, which showed former POWs arriving home. "I rely on your judgment in this matter. If you think it's not wise to show it to him, then don't."

"And I can show him this list?" asked Bao Trung.

"You can let him see the English version. The other is strictly for our use." Thanh explained that the other list was annotated with a distinctive code next to each name if the man had already been identified as a POW to the American authorities. In other instances, a different code reflected the man's ultimate disposition. A sizeable portion of the names bore no codes at all, indicating that their status was truly unknown.

"I don't understand," said Bao Trung. "Would it not be enough for him to just identify himself correctly? If we find a corresponding name on the list, wouldn't that suffice?"

Thanh slapped at a mosquito and answered, "No. First, our superiors in Hanoi strongly suspect that Lieutenant Commander Scott may have been sent here to deceive us."

"How so?"

"The overall timing is very suspect," explained Thanh. "It has to be more than a coincidence that we captured an American pilot who refuses to correctly identify himself just weeks before the American negotiators provided us with this list of names. Surely, their intent might be to trick us into releasing someone whose name is *not* on the list. Then they could claim that we are not cooperating or that we're holding more prisoners than we're admitting to."

Thanh continued. "And even if you convince the American to give you a name, it must still be vetted against *this* list to ensure that he is not trying to pass himself as someone already identified to the Americans as a POW."

"Or someone *else*," noted Bao Trung, scrutinizing the other codes on the list. "So what happens if the American correctly identifies himself and he's legitimate?"

"Then he will be sent directly to Hanoi for repatriation with the other prisoners. That simple. The arrangements have already been made. All you need do is motivate him to give his correct name. Can you do that, Hero Bao Trung?"

"I will do my utmost," stated Bao Trung. "But, Comrade Major, you should be aware that my furlough has been granted by Central Headquarters. I depart for home on Monday."

Thanh nodded. "How long has it been since you've seen your wife?"

"Not since I was wounded."

"But that was only in December."

"Not since I was wounded the *first* time," clarified Bao Trung, rubbing his stomach. "That was down South, over two and a half years ago. And I've never seen my son."

"Then you will go on your furlough," asserted Thanh. "I cannot ask you to do otherwise."

"Thank you, Comrade Major," replied Bao Trung. "I greatly appreciate your kindness."

"Do you think you can persuade him to reveal his true identity?" asked Thanh. "In the time available before you leave on furlough?"

"I think so, comrade," replied Bao Trung, nodding. He tucked the lists and magazine into the packet. "And you are authorizing me to do anything that I see fit?"

"*Anything*," affirmed Thanh, nodding.

"And what happens if I fail in this task?"

"Then he will remain here, Bao Trung," answered Thanh, gazing out a window towards the dusty cemetery. "With the others."

10:00 p.m., Monday, March 12, 1973

In what had become a nightly ritual, after lying down at dusk and sleeping only a few hours, Carson rose, stretched, and went to the solitary window of his cell. With the exception of a few guards who patrolled the grounds with flashlights, periodically pausing to peer into shadows, the camp compound was pitch black at night. He would stand there for hours, looking up through the tiny portal, gazing at the stars as he contemplated his circumstances.

Although he could only glimpse a sliver of the night sky, the tableau progressively changed as the stars traced their arcs through the heavens. Even though the transition occurred at a snail's pace, it was motion and change nonetheless, so much different than his present environment, painted in immutable and immoveable shades of gray.

In the course of his brief captivity, Carson had discovered much about himself. He learned that he could bear loneliness. He could tolerate confinement. But after spending most of his adult life travelling at speeds not even imaginable to most mortal men, he hated the sensation of being motionless. So this became his special time, when his world did move, if but gradually.

As the hours passed, old friends came to visit. For most of his life, they had largely been strangers until Ourecky introduced him and taught him their names and their stories. Now, there was King Cepheus, husband to Cassiopeia, who herself would arrive shortly afterwards. Cepheus was also the father of beautiful Andromeda, who unfortunately would remain outside his limited field of view. The familiar stars beckoned to him, reminding him of the days when he and Ourecky rode into orbit together.

As he watched, a meteor streaked across his narrow patch of sky, etching a bright trail across the heavens. The fleeting celestial event triggered a distant memory. He was reminded of his childhood, in the years after his father died in a crash and his mother killed herself—and another motorist—while driving drunk. He was sent to live with his uncle, a pilot like his father, who was aggressively engaged in climbing the Air Force career ladder. His uncle was a confirmed bachelor who didn't have the slightest clue about how to raise a young boy, nor did he appreciate the potential interference with his social life. His solution was to pack young Carson off to a string of military schools.

With the exception of their names and locations, the boarding schools were essentially the same: headed by pompous retired officers, they universally featured apathetic teachers, sadistic drill instructors, bland food, an abundance of teenaged bullies, and a corresponding profusion of the vulnerable boys who were their prey.

As his uncle ascended the ranks and transferred from one base to another, Carson likewise migrated from one dismal academy to the next, and became incredibly adept at swiftly fitting into their regimented lifestyles and interchangeable cultures.

He ridiculed the kids who couldn't similarly adjust and adapt. He recalled how ruthless he had been, and remembered a pathetic eleven-year-old boy at a military school in rural Alabama—Lyman Ward—who had received his greatest scorn. A new arrival, the dreadfully homesick kid lay atop his GI wool blanket on his Army surplus bunk, crying as he stared out the window. Feigning sincerity, Carson had asked him what he was looking for. The gullible kid admitted that he longed to see a shooting star that he could wish upon; he ached for his parents' broken marriage to be healed so that he could go home again. After that night, for the next several months, Carson and his mates had rendered the hapless kid's life into a living hell, until he was finally taken in by his merciful grandparents.

Now that he had a greater insight into that homesick kid's anguish, Carson looked at the falling star's fading trace, closed his eyes, and wished that he could also go home again.

5:35 p.m., Thursday, March 15, 1973

Smelling the cooking odors wafting from the kitchen, Carson's stomach rumbled in anticipation of his evening meal. It was a long time since breakfast, and he was ravenous. He cocked his head to listen for the footfalls of the approaching guard who would deliver his repast.

Unexpectedly, the door creaked open and Bao Trung stepped inside the cell. "Rise up," he ordered. "Come with me, Drew. I'm taking you to eat with the others."

Shielding his eyes from the unfamiliar glare of the late afternoon sun, Carson clambered to his feet and replied, "The others? Americans? I'm finally going to see *Americans*?"

"No. There are *no* other Americans here. You are the only one."

Walking gingerly on bare feet, Carson ambled behind Bao Trung into the bare earth courtyard adjacent to the communal kitchen. The courtyard was crowded with Vietnamese prisoners, obediently waiting in line to be served their evening meal.

Standing behind a rough wooden table, two prisoners doled out the meager rations, ladling each man an exacting share of rice, some weak broth, and a damp lump that resembled boiled cabbage. At the end of the line, two other men—apparently the ranking officers amongst the South Vietnamese—inspected each bowl to ensure that the portions were strictly uniform. Carson was amazed at the disciplined approach, and speculated why it might be necessary.

As they approached the entrance to the kitchen, Bao Trung shouted something in rapid-fire Vietnamese. A prisoner scurried inside and returned with a covered metal pan that he carefully handed to Carson.

As a malnourished prisoner shuffled by, Carson stole a glance into his bowl. The bottom was scarcely covered by rice. "There's not much in their bowls," he noted, looking toward Bao Trung. "I guess that they're fed more at breakfast?"

Bao Trung sniffed, shook his head, and answered, "This is their only meal for the day, Drew. As meager as it is, that is all that they eat every day, unless they are fortunate enough to catch a rat, a bird or some insects."

Carson realized that the gaunt Vietnamese prisoners subsisted on rations that scarcely kept them at the fringe of starvation. Suddenly consumed with guilt, he lifted the lid of his pan and looked inside; it overflowed with a heaping bed of brown rice garnished with glistening chunks of pork fat and fresh vegetables.

"Here," said Bao Trung, pointing at a spot in the shade of the kitchen. "Have a seat."

"Can't I just go back to my cell to eat?" asked Carson self-consciously.

"Not today."

Carson sat on the hard ground. As he uncovered his dinner, he noticed that the emaciated Vietnamese men were glaring at him. Drooling at the tantalizing odors emanating from his pan, they looked more like animated skeletons than human beings. Most wore only simple loin-cloths; their exposed skin was afflicted with expanses of festering sores and scaly patches of ringworm.

Despite their shared misery, there was no fighting or bickering over the meager rations; instead, the men consumed only what they had been given, and the more able-bodied fed the weak. Not too far from Carson, two men sat facing one another, picking lice from each others' scalps.

"I brought you out here for a reason, Drew," explained Bao Trung. "Most of these men will *never* go home. That's one reason why they are not fed very much. Starving men are more—*what's the word?*—compliant."

Carson gestured at his pan and asked, "Can I give this to them? *Please?* I don't feel very much like eating anymore."

"*No.* That meal is yours, not theirs. Besides, Drew, they would literally die if they were permitted to eat food that rich. And if you feel so guilty that they have so little and you have so much, do you intend to donate *all* of your meals to them? Now, eat, and then we'll go back to your cell and talk."

Averting his eyes from the suffering, burning with shame and no longer hungry, Carson devoured his meal quickly. Bao Trung sent a prisoner to return the empty pan to the kitchen, and then escorted Carson out of the courtyard.

The two men entered the cell. Carson took a seat on the straw mat that served as his bed. Still and humid, the air smelled of clay dust. Birds chirped in the distance.

"Drew, may I ask you something?" asked Bao Trung, placing a gray plastic envelope on the cement floor.

Carson nodded.

"I've asked you this before, but do you have a wife? Do you have a family?"

Carson said nothing.

Bao Trung held out a small black-and-white photograph. "This is my wife and son."

"Your wife is beautiful," observed Carson, looking at the woman's hauntingly sad face. "Thank you. My son was still an infant in that picture, but he's three years old now. He was born after I departed to fight in the South. I have not been home since." Bao Trung tucked the picture away. "I don't know if you have a family, Drew, and I'm not going to continue pestering you about it. I will tell you this, though: if you ever want to see your family or go back home, you need to listen to me very carefully."

Bao Trung continued. "Drew, the war between our nations, is *over*. Right now, as we speak, my government is repatriating American prisoners of war to your government. This has been going on for several weeks. In fact, we are very close to releasing the last of the prisoners."

"Then I'm going home!" exclaimed Carson.

Bao Trung shook his head. "If only it were that simple."

"It is. I'm an American POW. If what you're saying is actually true, then your government must send me home."

"That would be true if your government was *asking* for your return," explained Bao Trung. "But they're not."

"They're not? How could that be?"

Bao Trung squatted down next to the mat. "Here," he said, handing several mimeographed pages to Carson. "That is the reconciliation list that your government gave to us. They want an accounting of every single name on that list. So far, a few hundred of those men have been identified as POWs, and they *are* being repatriated. Drew, there are 1925 names on this list, but the name of Lieutenant Commander Andrew Scott is *not* on it."

Slowly flipping through the pages, Carson glimpsed his name—Major Andrew M. Carson, USAF—and swallowed deeply. He regained his composure and continued to the end of the document. As the Vietnamese officer had asserted, his Navy pseudonym—Lieutenant Commander Andrew C. Scott—was *not* on the list.

"You saw something?" asked Bao Trung, carefully watching his reaction. "Your *real* name, perhaps?"

Carson handed the list back. "How do I know that this is not some sort of scheme?" he asked. "I'm obviously being isolated from other Americans. How do I know that your government is actually releasing POWs? How do I know that the war is over?"

"Here," replied Bao Trung, handing Carson the glossy American news magazine. "The cover story is about something called 'Operation Homecoming' back in your country."

"How do I know that this is not a trick?" asked Carson. The magazine looked genuine, but it could also be a carefully manufactured fake. He opened it and turned to the article about the POW reparations.

One of the first pictures he saw almost caused him to gasp. It showed one of his squadron mates from flight school, gaunt-faced and several years older, being greeted by his wife and children in California. Carson knew that the man had been shot down three years ago, and he had personally called his wife to console her, so he knew that the magazine must be genuine.

He studied the other pictures, hoping to glimpse other familiar faces. But he was confused; what would happen if he deviated from the story he had been given? What if they already knew who he was and had even the faintest notion of what he had been doing for the past five years? Surely, the US government knew that he was a prisoner of the North Vietnamese and would not leave any stone unturned until he was repatriated.

"Put down that magazine and look me in the eyes, Drew," ordered Bao Trung, turning to face Carson. "Let me be clear, so there is *no* confusion in your mind. Although we've shared some pleasant conversations and amusing moments together, I *am* your enemy. I *have* killed Americans, *many* Americans, and if we were in different circumstances, I *would* kill you without the slightest hesitation. But with that said, I'll tell you something else, and you need to listen."

Lowering his voice considerably, Bao Trung continued. "Without going into details, I am alive only because one of your countrymen showed me great mercy. Now, I feel obligated to pass that kindness on to you. Drew, I am *not* trying to trick you. I sincerely want you to go home, but I can assure you that you will *not* be released if you do not correctly identify yourself. Do you understand me?"

Carson nodded weakly.

"Drew, I am leaving tomorrow morning. I will be gone for two weeks."

"Where are you going?" asked Carson.

"I have been granted a furlough. I'm going home to see my wife and child. But that does not matter. Soon after I return, you will meet with the same interrogation team that questioned you when you first arrived here."

"The men who tortured me?" asked Carson.

"Yes," replied Bao Trung. "But they are not coming back to torture you. They are returning strictly to determine your true identity. I'm warning you, Drew, you need to tell them your *correct* name. If you do, there's a good chance that you will be released back to your country. If you don't correctly identify yourself, then you will remain here. *Forever*. Do you understand?"

"Yes," croaked Carson.

"Then I will leave you here to reflect on your circumstances. You may keep the magazine, if that helps you to remember who you are." Bao Trung slipped the reconciliation list back into the folder, stood up, and abruptly departed the cell. The heavy wooden door closed behind him, and Carson heard the lock's hasp click into place.

Wiping away tears, he opened the magazine and studied the picture. It was definitely his friend from flight school. Although the picture and magazine had to be real, he was still wary of being lured into a trap. He was incredibly frustrated and desperately wished that there was a clear-cut answer to his dilemma. Should he accept the potentially dire risk of revealing his true identity to the North Vietnamese interrogators, or should he rely on the US government to deliver him from this debacle?

As he pondered the question, a mosquito flitted in through one of the cell's small windows and landed on Carson's bare arm. When he noticed it, he crushed it under his thumb, but it was too late; in the process of seeking a meal of his plasma, the miniscule insect injected something into Carson's bloodstream, a microscopic parasite that would alter his fate forever.

25

SPETSGRUPPA

Reeducation Camp # 4
Lang Hien, Quang Ninh Province, Democratic Republic of Vietnam
7:15 a.m., Friday, March 30, 1973

Carson was left to ferment in solitary confinement while Bao Trung went home on furlough. He was not allowed to venture outside, even for his customary exercise periods. Every morning, a guard brought his ration of rice and replenished his water bucket. Every evening, another guard delivered his dinner meal and swapped out his waste bucket.

Unbeknownst to Carson, as he stewed quietly in seclusion, his body was under attack by a microscopic army of relentless protozoan parasites. A tiny contingent of plasmodium had infiltrated under the cover of the mosquito's bite, and then subtly established a tactical foothold in his liver. From there, the patient invaders breached his red blood cells in successive waves, pausing for forty-eight hour intervals to feast on his hemoglobin and reproduce, exponentially increasing their numbers as they cascaded through his bloodstream.

As he contemplated whether to reveal his true identity, Carson fell ill. Initially, he suspected that it was merely the onslaught of a bad cold or the flu. He was first afflicted with a crushing headache, fever, and nausea. In time, he endured cycles of intense chills on every third day. Slightly more than a week after he was bitten by the infected mosquito, he was suffering with full-blown malaria.

Malaria? The mere word struck fear in most civilized ears, but for a sizeable portion of the world's inhabitants, malaria was a daily fact of life. Epidemiologists estimated that at any moment, over 300 million humans were infected with malaria, predominately with the most common strain, *Plasmodium vivax*. In fact, many residents of tropical regions were chronically afflicted with the disease, periodically suffering bouts of alternating chills and fevers throughout their entire lives.

But of the four types of plasmodium that caused malaria in humans, Carson was stricken with the most virulent—*Plasmodium falciparum*— which typically killed over half of the people it infected. The strain had a unique trait not shared by the other three variants: when it attacked red blood cells, it also altered the properties of their surface proteins, which caused them to stick to the interior walls of blood vessels like discarded chewing gum on a hot city sidewalk. This tendency, known to scientists as cytoadherence, caused localized obstructions to the blood flow that could be especially detrimental to oxygen-hungry organs like the kidneys, liver and brain.

Swaddled in his flimsy and undersized blanket, Carson knew he was sick, but no one else did. The two guards who daily attended to his cell were ordinary soldiers. They had no medical training, so they didn't notice that something was dreadfully wrong with the American prisoner. After all, he didn't complain or cause them any undue trouble; he seemed content to sleep all the time. And in the past several weeks, they had witnessed him occasionally sink into similar unresponsive funks, so this state was nothing new. Moreover, the two attendants quickly discovered that the American left most of his prodigiously large meals untouched, although the pails were scraped spotlessly clean by the time they were returned to the kitchen.

As woefully sick as Carson was, when the chills and fevers periodically subsided, he was clearheaded and coherent. During these intervals, he made up his mind that he would correctly identify himself when questioned by the interrogation team. As luck would have it, he was relatively lucid on the morning when Bao Trung returned from his furlough.

Slouched in the corner of his cell, forcing himself to eat some plain rice, Carson heard a truck pull into the compound and knew that his day of reckoning had arrived. Minutes later, the cell door swung open and Bao Trung stood in the threshold. He looked appalled at Carson's dreadful appearance. He sniffed the dank air and announced, 'You're sick, Drew. How long?"

"Over a week," muttered Carson, setting aside his bowl.

"Chills?" asked Bao Trung, stepping into the cell and kneeling down to touch his forehead.

Carson nodded.

Bao Trung sniffed the air again. "It's malaria." He gritted his teeth, clearly becoming angry. "I can't believe those idiots didn't catch this. How do you feel now?"

"Better, I suppose," answered Carson. "I'm probably getting over this."

"You will *never* get over this," chided Bao Trung.

10:15 a.m., Saturday, March 31, 1973

As he waited for the interrogation team to arrive from Hanoi, Bao Trung spent almost every waking moment with Carson, trying to nurse him back into some semblance of health. When the American was relatively lucid, they talked like they used to before Bao Trung went on furlough, but it was obvious that the malaria could readily seize control of his thoughts. To Bao Trung, the conversations were like talking with a drunk or perhaps a precocious child who was prone to teasing. The

American could be speaking with absolute clarity in one moment, then suddenly drift into delusional nonsense, and it was up to Bao Trung to sift through what was valid and what was not.

The humid air was filled with the scent of clay dust, the sounds of tools scraping dry earth, and the shouts of guards. Bao Trung entered the small cell and sat cross-legged on the floor opposite the American pilot. "Here's some rice, Drew," he said, holding out a small bowl. "I brought it from home. My wife fixed it especially for you. There are big pieces of chicken in there, along with onions and other vegetables."

"I'm not hungry," replied Carson. "And isn't your wife in..."

"I am permanently stationed here now, so I received permission to bring my family to live with me."

"I am happy for you," said Carson.

"You must eat," urged Bao Trung, holding a damp clump of rice to Carson's lips. "You're very weak. You need to regain your strength for your journey home."

Carson grudgingly took the rice and chewed it. He grimaced as he swallowed. "That's enough," he said. "Leave the rest here, and I'll try to eat more later. I promise."

"I'm not going anywhere. I'm staying here to keep you company." Bao Trung studied him; his speech was very coherent, and he didn't seem to be suffering from chills or fever. Maybe the malaria had finally lapsed. That was good; his final interrogation session would be easier to bear if he was able to regain some strength. Not only that, but the long journey to Hanoi would be more than arduous if he was still sick.

"Well, if you insist on staying, then maybe we can talk about the stars again, especially since you can't see fit to actually let me look at them."

"I wish that I could, Drew," answered Bao Trung. He unfolded the silk star map and spread it out between them on the concrete floor.

"It wouldn't be that hard to accomplish. All you have to do is leave that door unlocked one night and..."

"The time is coming soon when you'll be able to look at the sky again," claimed Bao Trung. "I promise. All you have to do is reveal your true name, and you'll be free to gaze at the stars anytime you desire."

"We'll see."

"How about this constellation?" asked Bao Trung, attempting to change the subject. He gestured at a tight clump of white points sewn into the black cloth. "The little one near Orion. The Chinese call this group the blossom stars and the flower stars, and it also is called the hairy head of the white tiger of the West. What is the English name for it?"

"That one?" asked Carson. "That's the Pleiades. It's also named the Seven Sisters. It's not a constellation, it's a star cluster. Scott taught me that."

"Scott?" asked Bao Trung. "Isn't *your* name Scott, Drew?"

"My friend's *first* name is Scott," replied Carson warily.

"What a coincidence. And does Scott also enjoy astronomy?"

Carson closed his eyes, smiled, and nodded. "He's the one who taught me to enjoy it. I'm very grateful to him for that. I'm very grateful to him for a lot of things."

"He sounds like a good friend."

"He is. He's my *best* friend. He taught me a lot, even when I wasn't so eager to learn. Here's some trivia: Do you know how many stars are in the Pleiades cluster?"

"Seven, I suppose," replied Bao Trung. "After all, it is called the Seven Sisters, right?"

"Good guess, but incorrect. Most people only see the *six* brighter stars. The seventh star is Merope, which some astronomers call the Lost Pleiade, since the average person can't see it," replied Carson authoritatively.

"So Merope is the seventh sister?"

"Yes. According to Greek mythology, the seven sisters are the daughters of Pleione and Atlas. Merope is the only sister who didn't marry a god, so she had to hide her face in shame. She was married to a mortal, Sisyphus."

"Sisyphus? I know that name," said Bao Trung. "He was the king who was forced to push a boulder up a mountain, only to watch it roll down, over and over."

"That's right," said Carson. He shivered, as if a chill was coming over him, and pulled his blanket over his shoulders. "Anyway, even though most people see only six stars in the Pleiades, there are many more. There are actually over a thousand stars in this cluster, but under the best of conditions, most people—even trained astronomers—can see no more than twenty with their naked eye."

"Interesting," commented Bao Trung. "And so how many have you seen?"

Carson's hands trembled as he closed his eyes momentarily. Almost a minute passed before Bao Trung quietly said, "Drew?"

"Sorry," mumbled Carson. "A big chill just passed over me. What did you ask?"

"We were talking about the Pleiades."

"Oh, yeah. The Seven Sisters."

"Yes," replied Bao Trung, trying to nudge him back on track. "You said that the average person could only see six stars. I asked you how many you have seen."

Carson looked at the gray ceiling, as if looking for a clue in its dullness, and quietly replied, "I'm not sure. I counted one time before it drifted out of view, and I got up to a hundred."

"A *hundred*?" replied Bao Trung. "Well, Drew, I have to admit that I am envious of you. I would give anything to look at the sky through one of those big telescopes that you have in the United States."

"When I saw those hundred stars, I wasn't looking through a telescope," bragged Carson, almost nonchalantly. He shivered, coughed, and pulled his blanket tighter.

"Oh, really? Then, pray tell, how did you see them? Perhaps you eat a lot of carrots? I've heard rumors that carrots are good for your eyesight."

"No carrots," replied Carson. "And no telescope, either. You have to get very high up to see that many stars. Way up above the atmosphere."

"So you saw them from an *airplane*?" asked Bao Trung. He couldn't conceive of any planes that flew at that high an altitude, except one, and even the *possibility* that the American may have flown *that* craft raised considerable cause for alarm. "Drew, are you implying that you flew the X-15 rocket plane?"

"No. When I counted those stars, I was flying much higher than the X-15 can go."

Bao Trung gulped. *Who could possibly know what the Americans had built?* After all, he had long heard rumors of a black spy plane that looked like a manta ray, that flew so fast and so high that it could not be shot down, and then recently had seen pictures of it actually flying. If the Americans had built something that amazing, and now it was public knowledge, who could guess what sorts of things that they kept hidden from view? Moreover, there was always the possibility that Drew—in his less than lucid state—was actually telling the truth. Now, it made absolute sense why the Americans had undertaken such an intensive effort to rescue him. Only minutes ago, Bao Trung had been anxious to deliver the American into the hands of the interrogators, so that he could provide the key—his true name—that would allow him to go home, but now he was concerned about what other secrets the fever-addled prisoner might inadvertently reveal.

"So, you don't believe me?" asked Carson quietly, like a child innocently trying to explain the mystery of a missing cookie to his parents. His voice was raspy and faint, like he was teetering on the brink of unconsciousness. "You don't believe that I could have flown that high?"

"Drew, don't even joke about things like that!" hissed Bao Trung, admonishing his American charge. "If the interrogators catch wind of something like that, you may never go home, even if you do identify yourself correctly."

8:43 a.m., Monday, April 2, 1973

"The interrogation team is here," announced Bao Trung. Are you well enough to talk to them?"

"Am I well enough?" muttered Carson. "Does that *really* make any difference?"

"I suppose not." Bao Trung bent over, slipped his good arm around Carson's thin waist, and helped him to his feet. "Are you *ready* to talk to them?"

Carson nodded.

"How do you feel?"

"Weak. And I have the chills again."

"Sorry. Do you remember what we discussed before I left?" asked Bao Trung. "All you have to do is correctly identify yourself, and this will all be over. You should say no more, and it should pass very quickly."

Squinting against the glaring sunlight, Carson nodded. *All over*, he thought. *All over. One question, one answer, and it would all be over.*

Helping the frail American to walk, Bao Trung escorted Carson to the inquisition cell. The three men from Hanoi were already present, but appeared to be in especially good spirits today. Breathing a quiet sigh of relief, Carson immediately noticed that there were no ropes, flails, canes or other implements of torture. Instead of forcing him to kneel in gravel, the interrogator gestured towards a chair. "You look thirsty," observed the interrogator. "Water?"

Carson shook his head. He *was* terribly thirsty, but he was also anxious to put this ordeal behind him before the next round of chills racked his skeleton.

"As you wish," said the interrogator. "Are you ready?"

Carson nodded. "I am."

"What is your name and rank?" asked the interrogator.

"*I am* Andrew M. Carson, Major, United States Air Force," divulged Carson truthfully and without hesitation.

Seated at a small table in a corner, Bao Trung consulted the mimeographed list and smiled. He announced something in Vietnamese, incomprehensible to Carson. The three members of the interrogation team grinned and congratulated themselves, as if it was a truly momentous occasion, and then one of the interrogator's assistants left hurriedly.

"I am proud of you, Major Carson," said Bao Trung, standing up to join the interrogator. "It's all over. You're going home."

"What happens now?" asked Carson. He felt tremendously relieved, like the weight of the world had been hoisted from his weary shoulders, but wanted nothing else but to return to his cell to sleep.

"The arrangements are being made as we speak. Because you are ill, I have asked for an ambulance to be dispatched to pick you up. It will likely take a day or two, but we will care for you in the meantime."

The interrogator closed his notebook, wound the stem on his Soviet-made watch, turned as if to leave, and then paused as if he had forgotten something. He leaned towards Bao Trung and whispered something in his ear. The lieutenant shook his head vigorously. The two men quietly argued for a moment. Carson had no idea what the polite disagreement concerned, but Bao Trung apparently conceded. He nodded and gestured towards Carson.

The aloof interrogator stepped forward, removed his wire-framed spectacles, smiled, and asked, "And what kind of aircraft do you fly, Major Carson?"

Since he had correctly identified himself and thought he was home free, the interrogator's follow-on question took Carson entirely by surprise. Since he had mentally programmed himself to truthfully answer *one* question, to provide the *one* solitary answer that would grant him his passage to freedom, he panicked. In all of his adult life, even in years of flying under some of the worst conditions imaginable, Carson had never once panicked. Certainly, he knew what fear was, but not panic, at least not until now, not until this very moment.

Dazed, he scrambled for the steel refuge of his mental submarine, hoping that he could lock himself within, but his imaginary vessel had foundered on treacherous and unfamiliar shoals. Now, instead of viewing the proceedings at a distance through his pretend periscope, it seemed as if he was trapped on the *outside,* as if he were physically separated from his mind and body, hovering somewhere off to the side, unable to control his thoughts and actions.

Grasping his notebook, the patient interrogator repeated the question: "What kind of aircraft do you fly, Major Carson?"

"I've flown just about everything in the US inventory," Carson heard himself boast. Since he had programmed himself to answer truthfully, once the floodgate was opened, the minute trickle became a surging torrent.

"Really?" asked the interrogator. "That's interesting. If you've flown almost everything in the US inventory, then you must have flown some very fast aircraft. What's the fastest speed you've flown? Mach One? Mach Two perhaps?"

"Faster."

The interrogator's assistant consulted a reference book, and then excitedly uttered something in Vietnamese.

"You have flown *faster* than Mach Two? That's *very* fast. Have you ever gotten close to Mach Three?"

"Faster," replied Carson, slightly slurring his words. "Much faster."

"*Faster* than Mach Three?" asked the interrogator incredulously, scratching his head.

The interrogator's assistant flipped through several pages of his reference book, leaned forward, and whispered into his superior's ear.

The interrogator smiled broadly, as if the answer to a mystery had been abruptly revealed, and then asked, "Tell me, Major Carson, do you have flight experience in the SR-71 Blackbird?"

Carson shook his head vigorously. He felt very sick, but still felt compelled to answer the inquisitor's questions.

"You're not rated in the SR-71? Well, if not the SR-71, then what?" asked the interrogator. "I can't imagine that many aircraft fly as fast as the SR-71, except perhaps the X-15, and it's rocket-powered. So, let's approach this from another perspective: How *high* have you flown, Major Carson?"

"*Very* high. Much higher than fifty miles," disclosed Carson spontaneously. His admission yielded a bitter memory, and he added, "You know, the Air Force awards the astronaut badge for flying over fifty miles."

"So, Major Carson, you're an *astronaut?*" asked the interrogator sarcastically. Snickering, he punched Carson lightly in the arm. He was obviously aware that the American was woefully ill; he spoke like a condescending parent playing along with a child's silly joke. He turned and chatted with the other man in Vietnamese; the two shared a laugh, but Bao Trung's impassive expression swiftly became a scowl of disapproval.

"I *am* an astronaut, but I've never received the astronaut badge," declared Carson angrily. "Even though I am qualified for it." His resolve was completely diminished by illness. He wished that he could compel himself to quit rambling, but he couldn't.

"Really? You haven't received your astronaut badge? I imagine that must be quite a disappointment. How *unfortunate*," said the skeptical interrogator, gripping his hand over his mouth to stifle a laugh. Grinning, struggling to maintain his composure, he scribbled some notes. "You know, Major Carson, I'll concede that we live in a backwards country here, but occasionally we receive news from the outside world. I don't ever recall seeing your name mentioned as an astronaut, so maybe the US Air Force has some logical reason for not awarding it to you."

Unable to contain gales of laughter, the interrogator was literally in tears as he scrawled down a few more comments. He shared something with his assistant, and both laughed uproariously. The interrogator spoke to Bao Trung, waved at Carson, and departed the cell.

"I hope you enjoyed your little antic," scolded Bao Trung, closing the door. "That was *stupid*, Major Carson. Your childish prank obligates them to file a formal interrogation report, which means you'll be sitting here at least another day before you can be transported."

"I'm sorry," said Carson meekly, staring down at the dirty floor. He saw a fragment of gravel and cringed.

"You can be sorry with yourself," snapped Bao Trung, lifting the frail pilot out of the chair. "You'll have plenty of time for introspection back in your cell."

Headquarters of the *Glavnoye Razvedyvatel'noye Upravleniye (GRU)* Khodinka Airfield, Moscow, USSR 10:32 a.m., Tuesday, April 3, 1973

After more than two years of laboring within the Encyclopedia, Anatoly Morozov was finally on the verge of becoming a Second Class Analyst.

Although the official promotion list was not scheduled for release until Friday, his advancement was a foregone conclusion. His tireless Section was the most productive of all in the Encyclopedia, and everyone—especially the bosses—knew it.

As his four archivists excitedly whispered about the various perks that would be theirs next week, Morozov submerged himself in a historical account of the great patriotic battle of Stalingrad. As much as he relished the thought of the promotion, he steadfastly refused to dwell on it. After all, in this gloomy basement, fortunes could change in the blink of an eye.

From a nearby table, Popov—his friend and fellow analyst—guffawed. Morozov tried to ignore the distraction, but Popov persisted in his uproarious laughter. Finally, Morozov set aside the heavy Stalingrad tome, jumped to his feet, and angrily strolled to Popov's table. "Dmitry Anatolyevich, what is so *damned* funny?" he queried. "I'm trying to concentrate on my reading, and your…"

"I'm sorry, but I just could not restrain myself," confided Popov, wiping tears from his eyes. "Please excuse my skepticism, but our stalwart Vietnamese comrades have just informed us that they captured an *astronaut.*" Because of his previous background in Cosmic Intelligence, despite the fact he had been ousted from that august bureau for technical incompetence, Popov was routinely tasked to review any space-related dispatches that arrived in the Encyclopedia. Consequently, any scrap of intelligence that mentioned space flight, in even the vaguest sense, would eventually land on his table.

"What?" asked Morozov. "*Astronaut?* I haven't seen anything in the news or reports."

"Your old boss from Washington sent this down. Surely he should know better than to waste my time."

"My old boss?" asked Morozov. "Federov? The Crippler?"

"One and the same," replied Popov, nodding knowingly. "Federov has been promoted to general, and has just taken over the Bureau that oversees the *spetsgruppa* in Vietnam. You know: The Bureau of Special Cooperation. They receive all the reports about captured American technology and prisoners."

"Oh. I wasn't even aware that he had returned to the Aquarium." Morozov's face burned even as the lie slipped his lips. He knew full well that Federov had been posted to Moscow and was doing his utmost to avoid him.

"Anyway, this astronaut business came out of an interrogation report from the questioning of an American in their custody. I wager that Federov's staff shot it down the tube just to aggravate me. Those inconsiderate bastards! It's not as if I don't have better things to do."

"May I, Dmitry Anatolyevich?" asked Morozov, pulling out an empty chair and sitting down.

"I suppose there's no harm in you looking at it." Popov glanced around before guardedly slipping the document to Morozov. "Especially since it's obviously just harmless fluff. Clearly, one of their American prisoners is playing some childish prank on them."

Morozov flipped the card over to examine it.

"Those Vietnamese monkeys should have never forwarded this to us," jeered Popov. "I deplore these sorts of ludicrous reports. My time is much too valuable to be squandered on this moronic prattle."

Nodding in agreement, Morozov covered his mouth and snickered as he scanned the poorly written report. It was downright hilarious, right until he saw the American prisoner's name. "*Carson?*" he muttered aloud.

"What is it?" asked Popov. "Please don't tell me that there's actually something in there. I don't have time to be chasing this inane crap this morning."

Carson, thought Morozov, studying the name on the document. There wasn't much to the report—just the name, a date, and a few relevant numbers—so he quickly committed the pertinent details to memory. *What serendipity!* It was finally a chance for vindication.

"So what is it?" asked Popov. "Is it something I should be concerned about?"

Morozov noted the control number in the header of the document before casually sliding it back across the table. "Nothing for you to worry over," he answered. "Do you have plans for lunch today? Perhaps you would like to join me in a game of chess?"

"Perhaps," replied Popov, applying a rubber stamp to the document. He blew softly on the fresh magenta ink before scribbling his initials across the header.

With mental gears aggressively spinning, Morozov pivoted away and slowly ambled back to his section, pausing momentarily at the samovar to draw a glass of tea. "Snap to," he ordered quietly, sitting down at his Section's table. "There's work to be done."

"But we have not received a tasking in over an hour," observed his senior retriever, the middle-aged woman from Stalingrad. She adjusted her faded blue cotton blouse, the same one she wore almost every day. "What on earth are you talking about?"

"What on earth? I am not talking about *anything* on earth," replied Morozov, jotting down notes on a card. "Listen to me. I want you to find me these things without drawing attention to yourself. *Now*. Get cracking."

The woman read his notes, rolled her eyes, and quietly growled, "You *idiot!* Can you not think of anyone but *yourself*, Anatoly Nikolayevich? We are dependent on *you*. Here you are, just about to be promoted, and you're still willing to put all of your accomplishments at risk because of this damned obsession of yours. Your frittering will put *all* of us at risk."

"Watch your tone. Do as you are ordered."

She shook her head and softly balked. "*Nyet*. We are not obligated to follow an order that is invalid and illegal. *Nothing* good can come from this goose chase. The best we can hope for is that our lives will be spared and that we will remain trapped on this floor forever, but more likely we'll all be dragged upstairs for *your* indiscretions."

"Listen to me, woman," he said sternly, squelching her with a flat voice that was just barely above a whisper. "Perhaps you're right, and this is just a wild goose chase. That said, I promise you this: if you do not cooperate with me *now*, I will immediately file a request that you be transferred from this section. So, it's up to you. You can roll the dice and bet on my instincts, and perhaps be promoted with me, or you can rot on this floor forever as you attempt to curry favor with your *new* section head. Now, what will it be?"

3:02 p.m.

Even though the reluctant woman and his other retrievers put forth their best efforts, there was little to be found that Morozov had not already pored over in the past. He had repeatedly read Carson's personnel dossier and intimately knew the pilot's career. He could quote from memory the types of aircraft that Carson was qualified to fly, his certifications on various systems, his previous duty assignments, his stints at various professional development courses, etc., etc.

There just wasn't any new information to be digested. Although he had a virtual copy of Carson's official military personnel transcript, a folder that was effectively a duplicate of the one maintained by the United States Air Force, the records had obviously been cleansed of even the slightest bits of information that might be of value.

On paper, Carson was just another pilot; granted he was an exceptionally qualified and competent pilot, but he was just a pilot. The interrogation report from Vietnam indicated that Carson had claimed that he was an astronaut, by virtue of flying *something*—perhaps some sort of new reconnaissance craft—higher than fifty miles and faster—*much faster*, the report stated—than even the American's SR-71 Blackbird. Perhaps that be true, but although he was qualified as a test pilot, there was not a shred of evidence in his records that indicated that he had flown anything particularly unusual.

Besides Carson's personnel file, Morozov's retrievers had discovered a copy of Carson's military pay records in the Encyclopedia's stacks. The Encyclopedia was presently being modernized, but the process was slow. Although most of the Encyclopedia's archival documents were currently being reduced to microfiche film, the copies of Carson's official personnel file and pay records were still maintained in unwieldy paper format.

Grasping at straws, almost at the point of surrender, Morozov closed the personnel record and shoved it aside. Yawning, he sipped lukewarm tea as he flipped open the cover to Carson's pay records. He knew it was a futile gesture; surely, there was nothing of value that could be found in the copies of pay stubs, vouchers, income tax withholdings, and other minutia.

As he reviewed the musty papers, Morozov was terribly frustrated. Unlike the scarce extract of his personnel record, Carson's financial record contained too *much* information. As the insidious Crippler had himself insinuated years ago in Washington, an effective clandestine agent must develop an intuitive sense of when he is being led astray; in the same vein, an agent must cultivate a knack concerning what information is relevant and what is not, and focus his energies only on that which is vital.

The nincompoop who had accumulated *this* information obviously lacked that degree of insight, because he or she had clearly devoted a tremendous amount of energy to amassing a useless stack of drivel and trivia. Moreover, since the Encyclopedia's available space was at such a premium, Morozov was surprised that the record wasn't more effectively screened to cull the documents that were of little use.

Although the Americans had been fastidious about erasing any vestiges concerning operational matters from Carson's records, their accountants had been simultaneously fussy about how money was handled and accounted for. Buried in the back of Carson's otherwise innocuous pay records was a fairly thick stack of travel vouchers.

Morozov was familiar enough with the American's pay system to know that whenever an Air Force officer travelled anywhere on official orders, he was mandated to submit a travel voucher—a summary accounting of all his travel-related expenses—upon his return to his regular duty station. Because there was so much fraud and abuse associated with travel expenses, the vouchers were regularly subjected to intense scrutiny to flush out fraudulent claims.

As a lark, right at the cusp of calling it quits on this fiasco, Morozov decided to skim through the accumulation of travel vouchers to determine if there was anything that might shed additional light on Carson's activities. Strangely, in the swollen pile of administrative trivia, he discovered something especially odd. He located a travel voucher that showed Carson departing his normal duty station—Wright-Patterson

Air Force Base—on June 9, 1969—for a temporary duty assignment at Hickam Field, Hawaii. Curiously, although the document theoretically placed Carson at Hickam Field, it did not reflect his return travel to Ohio, even though it was filed the following week on June 16.

Intrigued, Morozov sifted through the stack of vouchers and swiftly discerned an unusual pattern. There were thirteen travel vouchers showing Carson on temporary duty at Hickam Field. They almost invariably cited that Carson travelled from Ohio to San Diego, and then proceeded from San Diego to Hawaii on government-owned aircraft. The corresponding travel orders uniformly stated that he was authorized or may be required to alter his itinerary as necessary to accommodate his mission.

Oddly, with only five exceptions, the travel vouchers did not reflect a homeward leg from Hawaii back to Wright-Patterson. It was as if Carson ventured to Hawaii and disappeared, only to magically reappear in Ohio to file the requisite paperwork to receive his travel pay. Morozov laughed quietly, thinking that it was like that popular American television show where future space travelers were "beamed" hither and yon.

Suddenly, as if a brilliant floodlight had been switched on, there was clarity. Morozov gasped as everything meshed. Amazed that the Americans could be so sloppy, he was astonished at how simple it was to assemble the pieces. He was angry with himself for not spotting the trend earlier. Painstakingly flipping through the vouchers, he wrote down a series of date sequences for the temporary duty periods in Hawaii. Giddy with anticipation, his heart thumped in his chest. He then curtly directed one of his archivists to make copies of the vouchers, and then hustled to Popov's table.

"Please go away," begged Popov. "I've watched you toiling for hours now, and I know that you're up to no good."

"Listen," said Morozov quietly. "Months ago, you hinted that seven of our satellites had mysteriously vanished. I have to assume that you must also know *when* they vanished."

"*Please* leave me," pleaded Popov. His hands trembled and his broad face glistened with a greasy sheen of perspiration. "*Puzhalsta!* Nothing good can come of this."

Ignoring him, Morozov consulted his notes and matter-of-factly recited his list of date sequences. With every interval that he cited, Popov's eyes opened progressively wider as his brow furrowed deeply. Swallowing deeply, he nodded at almost every sequence.

"So, did I find all seven?" asked Morozov, flipping the card over.

"Not all of those dates are correct," confided Popov. "There are two extras that I cannot account for, but yes, Anatoly Nikolayevich, you found *all* seven."

"*Spasiba*, Dmitry Anatolyevich," replied Morozov, thanking Popov. "You've been immensely helpful." He returned to his table, and softly ordered: "Fetch me a typewriter! Now! I have an important report to write!"

5:36 p.m.

Pounding on the keys in a clattering frenzy, Morozov finished the last page and tugged the flimsy paper from the *Groma* portable typewriter. He quickly collated the report, placed the autographed image of the Type Four reconnaissance satellite's data plate inside the front cover, and signed his name to the document header. As the members of his section looked on in horror, he purposely strode to the elevator, clutching the report to his chest, and jabbed the button for Federov's floor.

Morozov emerged from the elevator and explored the hallways until he found the Bureau of Special Cooperation. An attractive Kazakh woman in a sergeant's uniform occupied the desk in the anteroom. "I am Major Anatoly Nikolayevich Morozov," he announced loudly. "Third Class Analyst, Department of Archives and Operational Research. I am here to see General Federov. *Now*."

Revealing a dazzling collection of stainless steel caps, the pretty sergeant smiled and politely replied, "The general is in his office, but he is not to be disturbed."

"I have a report for him," declared Morozov.

"Leave it with me, Comrade Major. I'll hold it for the general."

"It's Eyes Only," he answered. With no patience or time for sergeants, pretty or otherwise, he turned toward the heavy oak door, twisted the knob and threw it open.

The sergeant gasped. "You *cannot* do that!"

Ignoring her, Morozov swaggered into the office like a colonel leading his victorious battalion on parade. "*Here*," he said, slapping the papers on the desk. As he waited, he looked at the memorabilia and keepsakes scattered throughout the office. There were several objects he had not seen in Federov's office back in Washington, including a collection of judo trophies and a control stick recovered from the crash site of an American U-2 in Uzbekistan, but the same battered *spetsnaz* hatchet was displayed prominently over the fearsome Crippler's credenza.

Despite the commotion, the Crippler didn't even look up from his paperwork to acknowledge his presence. Morozov's face burned as anger surged up within him. In a gesture worthy of Nikita Khrushchev, he reached down, slipped off one of his shoes, and then pounded its worn heel on Federov's desk blotter. "*You…will…pay…attention…to…me… Comrade…General.*"

Furious, Federov finally gazed up. He looked as if he was ready to detonate.

"Here are your damned flying saucers from Ohio, Comrade General!" declared Morozov, hammering the report with his battered shoe. The heel snapped off and skittered away on the wooden floor. "Now it should be abundantly clear what the Americans were really doing in that damned hangar of theirs."

Not waiting for Federov's comment, Morozov turned and left. Awkwardly walking in the uneven shoes, he brushed right past the Kazakh sergeant and went directly to the elevator.

The creaking elevator began its agonizingly slow descent. By now, the Crippler certainly had digested at least the executive summary of his hastily prepared report. For a brief moment, Morozov contemplated pressing the button for the first basement sub-floor. After all, there was no time like the present to become accustomed with his new environs.

Moments later, as the elevator's doors slid open, the Encyclopedia's Director was waiting for him. "*Anatoly Nikolayevich*," he said, shaking his head as he clicked his tongue. "That was a very brave gesture. And very stupid."

"I assume that General Federov's office has called."

"Oh, yes," answered the Director, blowing on the surface of freshly poured tea. "He insisted on speaking to me directly. Anatoly Nikolayevich, I cannot adequately convey to you just how much I dislike communicating directly with the Crippler. Anyway, he has already consumed most of your report. To say that he was angry would be an understatement."

"So I've obliterated my chances at becoming a Second Class Analyst?" muttered Morozov.

"*Da*," observed the Director. He took a sip of tea and added, "That's true, but since you are being advanced directly to First Class Analyst, that should no longer be of concern to you."

Morozov's jaw dropped. "*First* Class Analyst?" he stammered. "Can this possibly be true? And what of my section? What will be become of them?"

"Silly man," said the Director as he chuckled and clapped Morozov on the shoulder. "They are being promoted with you. You will retain them in your new assignment, unless you desire new personnel to work under you. And there's another matter, something far more pressing."

"*Da?*"

"You are to proceed to your apartment immediately and pack for a journey. General Federov has directed that you will accompany him to Vietnam in the morning. He intends to personally handle this situation."

"Really? Surely you jest. I thought you said that Federov was angry with me."

The Director chuckled. "Anatoly Nikolayevich, the Crippler is indeed furious and many heads will surely roll, but you can rest confident that yours will remain where it is."

Reeducation Camp # 4
7:15 a.m., Wednesday, April 4, 1973

Major Thanh had been away, attending a regional meeting in Haiphong, since Thursday. As he arrived back in his office at Camp # 4, he was met by a furious Bao Trung.

"We learned the American's true name yesterday," said Bao Trung. "He is an Air Force officer, Major Andrew Carson."

"I am very aware of that development," answered Thanh, setting down his Soviet-made plastic attaché and riffling through a stack of papers. He looked out the window to see the South Vietnamese prisoners laboring in the cemetery. "So, is there anything else?"

"Yes. Are you aware that Carson is fortunate to be *alive*?" asked Bao Trung. "He has come down with severe malaria, and your idiots neglected to recognize that he was deathly ill."

"*My* idiots? Don't be so hasty to judge them," replied Thanh, wagging an official set of transfer orders that he was given in Haiphong. "Because as of this morning, you are now officially in command of the guard detachment, so they are now *your* idiots. And one more thing, Comrade Lieutenant Bao Trung, I hold you in the highest respect because you are a decorated combat veteran and I am not, but I strongly caution you to watch your tone. You are no longer a guest in this camp, you are my subordinate and will act accordingly."

"But Comrade Major, those guards should have realized that there was something amiss."

Thanh nodded. "Possibly, but in their defense, your American made no effort to let anyone know he was sick. If we had known he was ill, he would have been put in the infirmary."

As if that would have helped, thought Bao Trung. Lacking any decent medicines and devoid of competent medical personnel, the infirmary could do little to help Carson. All they could have done was pump him full of quinine. Besides that, the infirmary was like a festering Petri dish filled with a broad spectrum of infectious diseases; Carson would have likely emerged from there much sicker than when he was admitted.

"He's *sick*. He needs to see a *doctor*. A *real* doctor," snapped Bao Trung. "Not a half-trained imbecile like the quack who oversees the infirmary.

"I agree, Bao Trung. But not to worry," said Thanh, reading from a freshly printed cable transcript. "The Russians are bringing their own physician."

"The *Russians?*" gasped Bao Trung.

"Yes. A *spetsgruppa* team is coming. They are apparently being rushed directly from the GRU headquarters in Moscow, under the direct command of a Soviet general no less. As I said, they're bringing their own physician. She's supposed to be a renowned specialist in tropical diseases. I suppose that they are very motivated to keep your American alive. Anyway, our headquarters has directed me to surrender Major Carson into the Soviets' custody upon their arrival. So, Lieutenant, he will no longer be our concern."

"So they'll bring him back to Moscow?" Two listless prisoners lack-adaisically swept the floor in Thanh's outer office. Concerned that they were eavesdropping, Bao Trung glared at them, pointed at the door and tersely directed them to immediately leave.

"Moscow? What do *you* think?" asked Thanh, shrugging his shoulders. "We're holding an American who has refused to correctly identify himself, and when he finally does identify himself, he also lets slip that he's some sort of space traveler. The Russians learn of this, and they're flying straight from Moscow. I don't think they would go to all that fuss for a casual chat."

Bao Trung swallowed and said, "Obviously."

"Now, Lieutenant, in the interim, what should we do with your Major Carson?"

"What do you mean? Surely you're not implying that we should question him some more."

"No. The Russians should not arrive in Hanoi until late tonight, and probably won't arrive here before the early morning. Certainly, I intend to spruce up the camp a bit, but what should we do with Carson? Should we stick him in the infirmary?"

Bao Trung closed his eyes and considered the situation. Wringing his hands, he opened his eyes and replied, "No, Comrade Major. I just came from visiting him. He is very sick, but he is comfortable in his cell. I think we should leave him there until our Soviet brothers arrive."

11:23 p.m.

Tightly wrapped in blankets, convulsing from chills, Carson slept on a pallet made of straw.

A faint noise disrupted his stupor, and he awoke to glimpse a shadowy figure entering his cell. "Huh?" he asked, slowly realizing that his nocturnal visitor was Bao Trung. Faint moonlight shimmered through the cell's solitary barred window; it was still very dark outside.

"*Quiet!*" whispered Bao Trung, squatting down as he prodded Carson with his finger.

"Is it morning?" asked Carson. He was terribly lightheaded and confused. "Is it time for me to go home?"

"It is nearly morning," whispered Bao Trung. He placed a small candle in the corner and lit it. "We must be quiet, Drew. The other prisoners are sleeping. We must not awaken them."

Drenched with sweat but still shivering violently, Carson sat up, drawing a flimsy cover around him. His bones ached, as if he had been run over by a steamroller. His skin was riddled with weeping sores and abrasions, especially on his joints and buttocks, from constantly lying on coarse concrete. Even though he had been provided a new mat and extra blankets, they were of little comfort. "What do you want?" he demanded in a raspy voice, stifling a yawn and rubbing his eyes. "Why are you here?"

"It's time," said Bao Trung quietly. "The Russians will arrive soon."

"The *Russians?*"

Bao Trung nodded solemnly in the flickering candlelight. "They're coming for you," he said sternly. "You wouldn't stop blabbering, so now there are consequences."

"But I was just joking," muttered Carson. "You know that, don't you?"

"Be *quiet!* Take off that prisoner uniform," hissed Bao Trung, pulling a bundle of folded clothes from a small sack. "*Quickly.* Put these on."

Still confused, Carson slowly stripped out of his uniform. Made of cotton, bearing thick vertical stripes of maroon and gray, the shoddy pajama-like garments were a snug fit when he first arrived at Camp # 4, but now they sagged loosely on his emaciated frame. His hands trembled so much that he couldn't loosen the trousers' drawstring; Bao Trung had to assist him.

As Carson dressed, Bao Trung reached into a breast pocket and extracted a small shard of dried white clay. Using the kaolin as chalk, he scrawled a name on the concrete wall.

"What are you writing?" asked Carson, reading the block-lettered name as he struggled to button his unfamiliar shirt.

"*None* of your concern," answered Bao Trung sternly. "Face the door while I finish here. Quit dawdling and get dressed."

26

PALE BLUE

Cambridge, Massachusetts
1:25 p.m., Tuesday, May 28, 1974

Andy fussed and squirmed in his booster seat as Bea futilely tried to feed him macaroni and cheese left over from last night's dinner. He evidently had eaten his fill, so she wiped his messy face with a damp cloth and lowered him to the floor to play with his favorite Tonka toy firetruck.

As she scrubbed dishes in the sink, listening to Elton John singing "Candle in the Wind" on the radio, she surveyed their disheveled kitchen. The rickety table was stacked high with textbooks. On the wall, a cork bulletin board was covered with graphs, diagrams, and class schedules. It seemed like their snug little apartment was in a perpetual state of upheaval.

They had been in Cambridge for over a year. It had taken a while for her to adjust to the college town atmosphere and the foreign culture of academia, but despite the chaos, she had grown to enjoy this life. After the past five years of constant uncertainty, with Scott often disappearing with little or no warning, this was a happy time in their marriage, possibly the happiest time that they had ever enjoyed. It brought back memories of her early childhood, when her father was still alive and going to college himself, before he was killed in Korea.

As a stand-by flight attendant with Delta, Bea still flew at least three times a week, almost exclusively on short regional hops from Logan Airport. She occasionally made it back to Dayton to check in on Jill's mother and Rebecca. Bea had been bitten by the college bug herself; pursuing a business degree, she also started classes at Boston College last winter.

Since Andy had started kindergarten, their entire little family was in school now. As much as she enjoyed it, her life was a complicated existence; between her commitments as a student, part-time flight attendant, wife and mother, her schedule was tight. Thank goodness there was always an ample supply of coeds willing to babysit Andy.

Even though Scott would remain in the Air Force, their future was at least somewhat more predictable. After defending his thesis and graduating next spring, he was slated to return to Wright-Patterson for a three-year research assignment. Planning ahead, she was already making arrangements to transfer to a college in Dayton to finish her degree. She also hoped that they could find a house near Jill's mother.

Scott had found his place at MIT; he was in his natural element in this high-powered academic world. Far removed from the Air Force, he blended into the college environment; sporting hair down to his shoulders and a scruffy beard, he rarely wore anything but jeans, dark T-shirts, and a well-worn pair of Vibram-soled Vasque hiking boots. All of his uniforms, accoutrements and military memorabilia were buried deep in the back of their bedroom closet.

Although he didn't conceal his Air Force affiliation, he didn't exactly go out of his way to aggressively advertise it, either. If they didn't know any better, most observers would likely guess that he was returning to MIT for an advanced degree after spending the past few years employed in the aerospace industry. With the space program on the downswing, there were more than a few MIT grad students who fit that bill.

Their little apartment was a revolving door; Scott's undergrad students came and went, frequently stopping by for a cup of coffee or to mooch a meal. He held court in the living room almost every evening, as they gathered to excitedly debate the advanced mathematics and other concepts associated with travelling to other planets and galaxies.

Although she didn't comprehend the esoteric ideas they discussed, and probably never would, she appreciated their youthful enthusiasm and enjoyed listening to them. When they ventured out too far into the theoretical, Scott always seemed ready to point out practical concerns to bring them back into focus.

Listening intently, the students hung on his every word as he described potential issues with propulsion, fuel consumption, consumables, radiation exposure, effects of cosmic rays, skeletal degeneration, power supplies, and instrumentation. He spoke with such august authority, it was as if he had flown into outer space himself. Scouring a frying pan from breakfast, she smiled to herself, thinking of his space capsule mock-up amid the musty hay bales in his parents' barn back in Nebraska. Then she frowned, recalling the Christmas Eve when he inexplicably burned the mock-up, on the same day that Drew Carson was shot down over North Vietnam.

As Elton John played his last poignant notes, Bea thought of Carson. For whatever reason, whenever she tried to remember his face, she could never see him entirely, but only seemed to glimpse his pale blue eyes. In her thoughts, his eyes always seemed filled with sadness and longing, like he was gazing at her across a broad void, pleading for a chance to return. She knew that Scott missed him—terribly so—and now she missed him also. She sighed, wishing that he hadn't gone to Vietnam. Although his remains had never been recovered, the military assumed he was dead; if Scott knew otherwise, he wasn't letting on.

As she rinsed Andy's bowl, Bea heard a faint tap at the door. Expecting to shoo away an obnoxious salesman, she strolled into the living room and cracked open the front door to glimpse the grinning face of Death.

It was a horrific nightmare in broad daylight. She slammed the door with enough force to rattle the entire apartment. Screaming at the top of her lungs, she whirled around and scrambled into the kitchen. Crying, she sagged to the floor next to Andy. Clinging tightly to her, even though he could not possibly understand why she was so upset, the child bawled also.

The tapping persisted, progressively growing louder, accompanied by a voice: "*Bea?*"

"*Please*, please," she blurted. "Please just go away."

The door creaked open to reveal the gaunt specter. Outfitted in his customary cowboy garb, Virgil Wolcott tipped his white Stetson, grinned broadly, and drawled, "Mornin', Bea."

"*Please*, please," she pleaded, standing up. She blindly reached into the soapy water in the sink, grabbed the first object that came to hand, and brandished an unwashed soup ladle toward him. "Scott's not here. He's teaching a lab this morning. Please, Virgil, just go away. *Please* leave us alone."

"May I come in?" he asked, removing his hat.

"*Please* go away," she begged. "Hasn't he given enough? We're happy now. Just leave us *alone*. Please just let us be, Virgil. Go find someone else."

And then Wolcott uttered the words that chilled her to the bone: "I'm not here for Scott, darlin'. I'm here for *you*, Bea."

"*Me?!*"

He nodded, stepping into the living room. Bea could glimpse someone else in the hallway, just outside the threshold, standing motionless in the shadows.

Closing her eyes and breathing rapidly, she couldn't speak for over a minute. Her fear was gradually dispelled by sheer curiosity. As she calmed down, she opened her eyes and asked, "Me? What on earth do you want from *me*, Virgil Wolcott?"

"If truth be told, Bea, I'm here on behalf of Mark Tew."

"But Mark is dead," she replied, bending down to comfort Andy. Just as the toddler calmed down, Wolcott smiled at him, and he started whimpering again.

"True," he replied, walking slowly toward the kitchen. "Mark *is* dead, but with his dyin' breath, he passed on his last wish to me, and I promised that I would do my utmost to make it so. I'm beholden to him, Bea. I hope you understand."

He handed her a small white envelope. She opened it, and found an embossed card. She read it; it was a formal invitation to the opening of a new exhibit at the US Air Force Museum at Wright-Patterson. "This is nice, but I'm a little confused, Virgil. What does this have to do with me?"

"Well, Bea, it's a *long* story," he replied. "Would you mind if I asked my friend to come in? He's here to help with the explanation."

Wiping her eyes, Bea nodded.

"Ted, come on in," said Wolcott. The man in the hallway came in. Dressed in a Harris tweed jacket, a blue button-down Oxford shirt and corduroy trousers, the handsome man could easily pass as an English Lit professor at any of the campuses that dotted the Boston area. The three sat down around the book-cluttered table.

"This is Colonel Ted Seibert," explained Wolcott. "He's going to help me explain the situation." It took the two men almost an hour to present their case, and even then they said scarcely little.

"And so, that's it," Wolcott concluded. "I hate to be vague, but I can only say so much until you're formally cleared. Can we count on you?"

"I'll think about it," she replied. "But if I decide to come to this opening of yours, I'll make my own travel arrangements. And I also have a request for you."

"Name it, darlin'."

"If I'm going to be there, I would really appreciate it if you could send out a couple of additional invitations. I know a couple of other people who deserve to be there as well."

"We'll consider it," said Wolcott. "But they would have to be cleared, just like you. Now, Bea, I'm sorry to pester you, but I really need an answer today. Can we count on you?"

"I suppose."

Wolcott grinned. "Well, then, Ted has some paperwork for you to fill out. I know it's a nuisance, Bea, but we can't tell you any more until after your temporary security clearance has been approved."

"How long will this take?" she asked. "I'll need to start dinner soon."

"The paperwork will take us a couple of hours, at least," explained Seibert, proffering a multi-page form and a government issue Skilcraft black pen.

Looking at the form, she smiled. "So does this mean that you're going to *finally* tell me what Scott's been doing for the past five years?"

"Well, not exactly," answered Seibert, slowly shaking his head.

"So are you still at Wright-Patt, Virgil?" asked Bea, as she started to fill out the form. "I was under the impression that your office had been closed down. That's what Scott said, anyway."

"That's true," replied Wolcott. "But there are still chores to be completed, junk to be packed away, and records to be filed. The Devil's work is never done."

Bea shuddered. "I suppose you're right."

Cambridge, Massachusetts
5:15 p.m., Friday, May 31, 1974

Nudging the door closed with her hip, Bea sorted through the day's mail. Chewing on a No. 2 pencil, cross-referencing two thick textbooks and three notebooks, Ourecky was deeply immersed in his studies. She stood there for a moment and watched him. He finally looked up and smiled. She smiled back. In the living room, the evening news was on the television. Virtually all of the stories concerned the Watergate scandal and the Middle East.

"Did you stop at the post office?" he asked, tapping digits into his new Hewlett Packard handheld calculator. "My paycheck should have landed in the box today."

"I got it," she replied, nodding. "I put it straight into the bank. The deposit slip is on your desk. Did Andy ever settle down for his nap?"

"About an hour ago. Sound asleep."

"I really don't like him napping this late in the afternoon," she observed. "He'll be awake late again. Are your undergrads coming over tonight?"

"They are. They're bringing tomato pies from that Greek pizza place."

"Pizza? Yummy. I guess there's no need for me to cook, then. This came in today's mail," she said, handing him a large manila envelope. It was franked with an official stamp. "It was registered mail, so I had to sign for it. It looks important."

He tore it open and examined its contents. "This is odd," he commented, scratching his bearded chin. "This is from the Secretary of the Air Force. It's an invitation to attend the opening of a new space exhibit at the Air Force Museum at Wright-Patt."

"*Really?* That *does* seem odd. Are you going?"

"I don't have much choice. I know that you would like to forget it, but even though I'm a doctoral candidate here, I'm still an officer in the Air Force. There's an instructional letter and a set of orders in here. And a round-trip plane ticket."

"Interesting. Well, on the plus side, you do love museums, dear. When is it?"

"June 7. Next Friday. Aren't you flying to Dayton next week to see Jill's mother?"

"You know, dear, I think you're *right,*" she replied. She pivoted around to refer to a calendar on the wall by the refrigerator. "Yeah, Scott. Wow, what a coincidence. Do you think maybe I might be able to see that museum with you?"

"You probably won't be able to go to this opening," he replied, studying the letter. "It's apparently invitation only. If we stay for the weekend, maybe we can go the next day, after things settle down a bit."

"Sounds good to me. So you'll need to go in uniform?"

He nodded, and pointed at his beard. "I guess these whiskers will have to come off. And I'll need to pay a visit to the barber."

She grinned and bent over to hug him. "Well, I *really* wouldn't miss the beard if you decided not to grow it back. I might even be motivated to kiss you more often."

"Then consider it gone."

United States Air Force Museum
Wright-Patterson Air Force Base, Ohio
9:35 a.m., Friday, June 7, 1974

If he wasn't aware that he was just an average Air Force officer, Ourecky would have believed that he was some sort of VIP or show business celebrity. An official sedan, complete with an escort officer, had picked him up at the Dayton Airport to whisk him to the museum.

As he watched the familiar landmarks of Dayton pass by outside, he reflected on the events of the past few months. Shortly after Carson had been reported killed by NVA soldiers in North Vietnam, Ourecky was finally released from the Project. With Carson gone, he was finally able to convince Wolcott and Tarbox that he was no longer interested in flying in space, nor did he harbor any overwhelming desire to be publicly acknowledged as an astronaut.

It was just as well that he bowed out of the Navy's MOL flight, because the program was cancelled before the mission ever flew. Shortly after Ourecky bowed out, the National Command Authority had rescinded their tentative blessing to allow the MOL to fly in public view. Just a few months later, any and all hopes of another clandestine mission were dashed as well. The last handful of military astronauts had already disbanded, and all of the MOL hardware had been summarily stashed in mothballs at some facility in California. Ourecky had an excellent source of information concerning the program's demise; he ate lunch regularly with Gunter Heydrich, who now worked at MIT's Draper Laboratories, the prestigious facility that designed the Gemini and Apollo spacecraft navigation computers.

From what Gunter had told him, the entire joint Air Force-Navy manned space program was also on the cusp of folding. Although Wolcott remained to finalize the close out, Tarbox had already moved on to another highly classified program within the Navy. Gunter confided that the joint program was deemed superfluous in the aftermath of a series of high-level secret meetings between US and Soviet officials, in preparation for an Apollo-Soyuz space mission slated to fly next year. In the course of the meetings, the militarization of outer space was discussed extensively. In response to very pointed questions from Soviet officials, who were perplexed about the untimely demise of several of their satellites, the US negotiators admitted to successfully fielding a system that could readily destroy *any* suspect object in orbit. And while they didn't reveal *how* they did it, they did admit to knocking down a significant number of Soviet space vehicles and implied that the system was still fully operational. They also made it emphatically clear that a threat of orbiting nuclear weapons would not be tolerated, and the consequences—if it ever happened again—would be significantly greater than a warhead being harmlessly sent to the depths of the Indian Ocean. Both parties agreed to limit—although not entirely curtail—the militarization of space.

The sedan dropped him off at the curb directly at the front entrance, next to a white sandwich board sign that claimed the museum was temporarily closed for a special event. He noticed that the parking area entrances were sealed off with wooden barricades, and that all the vehicles entering the lot were stringently checked by SPs. He also observed three men in dark suits standing by the museum doors, and assumed that they were some sort of undercover security agents.

As he slid out of the sedan and closed the door, Ourecky recognized a familiar face. He strolled up to a broad-shouldered black man wearing Air Force blues and a distinctive red beret atop a closely cropped Afro. "Henson, right?" he asked.

"Yeah. I mean, yes, *sir*," replied Henson, pivoting around and snapping up a sharp salute. "It's been quite a while, hasn't it? Haiti was over four years ago."

Returning the salute, Ourecky clasped his stomach and grimaced. "Haiti? Afterwards, they told me you were there, but all of that was like a dream to me. A *bad* dream. Forgive me if I don't remember too many pertinent details."

Henson laughed. "I've tried to forget most of it myself."

"So long before that, when we met on the plane, didn't you tell me that you were interviewing for some sort of mining company? Was that just some sort of cover story?"

"Apex Exploration? More or less," replied Henson. "But I did marry the boss's daughter." He pointed at a strikingly attractive woman waiting at the museum's entrance. The mocha-skinned beauty, who appeared to be in her late twenties, was standing next to an elderly white man leaning on a cane. "That's my old boss, Abner Grau, and my wife, Adja."

"She's pretty," commented Ourecky, shielding his eyes from the bright morning sun. "He's *her* father?"

Grinning, Henson nodded. "It's a very long story. I also made enough money at Apex to send Adja to medical school. She's in her second year right now."

Ourecky studied Henson's uniform. "So I take it that you left Apex Exploration and went back into the Air Force?"

Henson nodded. "I did. Not too long after Haiti, I went back on active duty and volunteered to transfer into the pararescue field. From there, I went straight to Thailand, and then..."

"Then what?" asked Ourecky.

"Uh, nothing, sir."

"So, PJ's, huh?" Ourecky pointed at a silver badge on Henson's uniform and read its inscription: "*That Others May Live.* Do you believe in that?"

"Very much so. I can't quite explain it, but I heard the calling after I worked with Steve Baker in Haiti. He's the guy who kept you alive between there and Cuba. Anyway, the PJ training pipeline took me about a year. After I rotated back from Thailand, I went to Lackland, near San Antonio, training PJ recruits. That's where I am now, but I get out in two years, roughly at the same time Adja finishes medical school. I plan to finish my degree and go to medical school myself."

"Good for you," said Ourecky. "What sort of medicine you want to practice?"

"I've thought about going into the emergency field. Adja and I have even seriously talked about going overseas, kind of like missionary doctors, maybe to Africa or even back to Haiti. There's a French outfit called *Médecins Sans Frontières*—*Doctors Without Borders*—working in Biafra and Nigeria. We might throw in with them, or maybe even set up our own operation. My father-in-law has a lot of experience all over the world, especially in Africa, and he said he would help us get started."

"Interesting."

"So whatever happened to that Delta stewardess?" asked Henson. "The really cute blonde who had such a shine for you?"

"Bea? I married her."

"Well, Major Ourecky, good for you. Try not to lose her."

Smiling, Ourecky replied, "Good advice. I'll make a point to work on that."

Henson started to say something, and then fell silent. A strange expression passed over his face, like he was recalling a particularly bad experience.

"Something wrong?" asked Ourecky.

"Maybe, sir," said Henson. "Are you going to be around later?"

"I assume that I am. There's supposed to be some sort of reception after the ceremony. It's on my invitation, so I guess that I have to attend it."

Henson nodded. "If you don't mind, sir, if you're going to be there, I would sure like to talk to you. And there's someone else you need to talk to."

10:00 a.m.

Once inside the huge museum, his escort officer guided Ourecky into a dimly lit auditorium and showed him to his designated seat in the front row. The seat, like several others in the first two rows, was marked by a diagonal strip of red fabric. As he waited, he looked around; the space looked like it could accommodate roughly two hundred spectators. It was rapidly filling up, mostly with people who had previously been assigned to Blue Gemini, along with their families.

As he spotted Gunter Heydrich entering the auditorium, it finally dawned on him that the ceremony probably had nothing to do with a new exhibit and obviously had something to do with the Project.

As more attendees filed in, he was joined on the front row by Parch Jackson, Mike Sigler and Ed Russo. Wearing dark glasses, apparently almost blind, Russo looked terrible; Ourecky had heard through the grapevine that he was currently suffering serious health problems.

After the auditorium filled to capacity, Virgil Wolcott—uncharacteristically wearing a dark gray suit—climbed the four steps to the low stage, stood behind the podium and announced in his Oklahoma drawl, "Ladies and gentlemen, if I can have your attention. I would like to welcome you to this awards ceremony to honor former members of the Aerospace Support Project here at Wright-Patterson. Please rise as we welcome the Secretary and Chief of Staff of the United States Air Force."

In awe, Ourecky jumped to his feet and joined the audience in applauding the Secretary and Chief. Oddly, since the past five years had gone by in such a frenetic pace, this was only the first awards ceremony that he had ever attended in his military career.

Compared to most of the other uniformed men in the auditorium, even though he was a major, he looked like a neophyte to the Air Force. He wore only two ribbons—the red and yellow "Everyman" National Defense Service Medal and an Air Force Commendation Medal for his research work at Eglin—above his left breast pocket. Ironically, the Commendation Medal had never been formally presented to him; it had been mailed from Eglin over a year after he had transferred to Wright-Patt.

As Ourecky sat patiently for over an hour, the Secretary and Chief personally presented formal awards to the men and women who had labored on the Project as military personnel, government civilians or contract workers. The awards were significant; they were easily some of the highest accolades that could be presented in a peacetime environment, but the accompanying citations—which were read in full for each presentation—were noticeably devoid of any specific facts that would explain why the recipients merited such prestigious awards.

Ourecky had heard similar citations in the past, particularly in relation to classified ordnance projects at Eglin. His old boss, Colonel Paster, referred to their cryptic language as "precise vagueness." The citation was couched in terms that were suitably precise to indicate that the recipient had accomplished something truly momentous, but also sufficiently vague so as not to reveal what the accomplishment was, and in some cases, where or when it happened. Ourecky was sure that more than one spouse would leave the ceremony scratching their head, aware that their significant other had done something tremendously important but not having a clue what it might be.

While most recognized noteworthy achievements, three of the awards were bestowed for bravery. Matt Henson and Steve Baker received the Airman's Medal, the highest award that could be given for heroism in peacetime. In the same presentation, a scary-looking Army master sergeant—Nestor Glades—received the Soldier's Medal, the Army's equivalent award.

The Army green dress uniform worn by Glades bore so many decorations that Ourecky wasn't sure that there was room for any more. It wasn't until the vaguely worded citations were read that Ourecky realized that the three men had been instrumental in rescuing him from Haiti.

Finally, the DFC—Distinguished Flying Cross—was awarded to Howard, Riddle, Jackson, Sigler, Carson, Ourecky and Russo. Only four of the seven men came to the stage; the others were awarded posthumously. The DFC's for Ourecky and Carson were the last awards presented; their citations—virtually identical—read in part: "For heroism and extraordinary accomplishment in flight, for eight special reconnaissance missions executed in an extremely hostile operational environment during the period June 1969 through October 1972. These eight missions were of the greatest strategic significance, resulting in the timely collection of extremely valuable intelligence unobtainable by any other means..."

Smiling to himself as he returned to his seat, Ourecky wondered how he would eventually explain the award to Bea, since she would undoubtedly be curious how a non-pilot could receive such a prestigious award for flying. He reminded himself that since she was unlikely to see him in uniform for at least another year, it wasn't something he needed to be immediately concerned with. Looking to his right, he saw that Jackson and Sigler—with their two failed missions in orbit—appeared almost embarrassed to be here.

Colonel Seibert, the Project's former intelligence officer, strolled to the podium and tersely spoke. "Ladies and gentlemen, this concludes the *public* portion of the awards ceremony. At this point, unless you are seated in one of the chairs bearing a red stripe, we invite you to a reception in the Assembly Hall. If you are seated in a red stripe chair, then we request that you remain in your seat so we can adhere to our timetable for the remainder of the presentation."

A shuffle took place as most of the family members were escorted out. Once they were gone, another group of attendees entered through a side door. In short order, the seats of the auditorium were filled; the spaces vacated by the Project workers' families were occupied by high-ranking officers, mostly generals.

Ourecky was astounded at the amount of rank in the room; he hadn't seen so many stars in one place since he had walked in space almost two years ago.

Security agents in plain clothes moved through the auditorium, verifying invitations and checking identification cards to ensure that the remaining men were authorized to be seated in the red stripe chairs. Ourecky uncomfortably felt as if he didn't belong here, that he wasn't worthy to sit in one of these chairs. As he began to stand up, he felt something pushing against his chest, like a massive hand gently nudging him back into the chair. Sitting back down, he could swear he heard "Big Head" Howard's gruff voice whispering in his ear: "*You've earned the right. You've flown over the line.*"

Once everyone was seated and their credentials verified, Colonel Seibert spoke again. "The remainder of this ceremony is classified. The proceedings are not to be discussed outside this forum. While we apologize to the recipients, the awards presented today must be physically returned after this ceremony is concluded, *before* you exit this auditorium. They will be stored here at Wright-Patterson in classified archives. They will not be disclosed in your official records nor may they be worn in public without express authorization of the Secretary of the Air Force."

With that said, the Project's former pilots and Ed Russo were called forward to receive the Air Force Astronaut Badge from the Secretary of the Air Force. Even though they could never wear the wings in public, they were now part of an extremely exclusive fraternity, consisting of the Air Force pilots who had flown on NASA space missions as well as five Air Force men who had piloted the X-15.

Ourecky was the last man called to the stage. The ceremony was being filmed, and the camera crew's glaring klieg lights made it difficult for him to see. Facing the audience, he took his place beside the Secretary and the Chief, while Virgil Wolcott held a velvet-lined case with two sets of the astronaut wings. He held his breath as the Chief solemnly read the official orders authorizing the wings to be posthumously awarded to Drew Carson, and then the Chief read his citation. "Having met the requisite qualifications by flying in excess of fifty miles above the earth's surface, Major Scott E. Ourecky is hereby awarded the Air Force Astronaut badge."

Wolcott grinned, leaned toward the Chief and whispered something in his ear. The Chief smiled, nodded and announced, "I've just been reminded that we have a special guest here to make this presentation." He nodded toward a captain standing beside a curtain. The captain, standing a few feet to Ourecky's right, disappeared behind the curtain. In seconds, he reemerged, escorting Bea from behind the curtain.

"Bea?" said Wolcott softly, offering her a set of silver wings. "Bea, darlin', would you do the honors, please?" He leaned toward Ourecky and said, "This is Mark's doin'. This was his last request. We had to process her for an interim security clearance to pull this off."

Grinning, Bea fumbled with the wings, pinning them slightly crooked above Ourecky's left breast pocket. As the audience applauded, she hugged him and softly whispered, "Surprised to see me, dear? You look a bit stunned."

Flabbergasted, barely able to stand, Ourecky nodded. "Uh, that's putting it *mildly*."

"Look out there, Scott," she said softly, pointing back toward the seats. "When Virgil asked me to be here, I told him I wouldn't come if your parents couldn't come also. I thought they deserved to see this, too."

Ourecky shielded his eyes as he looked out over the audience to see his parents seated in the third row. Beaming, his father cradled Andy, barely conscious, in his lap.

As tears came to his eyes, he asked her, "So how much do you know?"

"Exactly enough not to ask any questions," she replied. "*Fifty* miles, huh? I suppose Drew's dinky little T-38 sure could fly much higher than most."

An athletic-looking man in a dark suit came out from the wings and quietly spoke to the Secretary. The Secretary looked at Ourecky and said "Major, you and your wife need to take your seats."

"Two minutes," announced the man in the dark suit.

As they stepped off the stage, Wolcott leaned toward Ourecky, palmed him a tissue, and whispered gruffly, "Your duty day is far from over, pard. You are a commissioned officer in the United States Air Force. Knock off that danged blubberin' and *act* like one."

Ourecky blew his nose and wiped his eyes as he made his way to his seat. Bea sat down beside him. He had just barely regained his composure when the man in the dark suit declared, "All rise for the President of the United States."

Without the playing of "Hail to the Chief" or any other ceremonial fanfare, the President walked out from behind the curtain. Eyes down, haggard, he walked slowly, as if he were in great pain. Like most of the men in the room, he looked almost exhausted from the demands of the past half-decade. Rumors abounded that he was on the verge of surrendering the reins of the Presidency to Vice President Gerald Ford.

He stepped to the podium and spoke. "Please take your seats. I apologize that I can only be here briefly. I am sure that all of you are cognizant of current events, so you are probably well aware that my Presidency is under siege. So, officially, although I'm supposed to be at Camp David today, gnashing my teeth and wringing my hands as I contemplate certain indiscretions, I have a promise to keep to Lieutenant General Mark Tew, who led many of you gentlemen until his untimely death."

He held up a handwritten letter. "Just a few hours before he passed, Mark wrote to request my assistance in expediting appropriate awards to recognize and honor the gallantry of two of his officers.

"As I read Mark's letter, it struck me that we possess few ways to measure or adequately recognize courage in the absence of armed conflict. As I pondered this thought, I was reminded of a story from World War II, one that many of you will recognize."

Clearing his throat, the President reached into his pocket and drew out an index card. Referring to it, he stated, "Early in 1943, four Army chaplains—- George L. Fox, a Methodist reverend; Alexander D. Goode, a Jewish rabbi; Clark V. Poling, a minister of the Reformed Church of America, and John P. Washington, a Roman Catholic priest—-were aboard the troopship *Dorchester*, in a convoy bound for Greenland. On the night of February 3, near Newfoundland, the *Dorchester* was torpedoed by a German U-Boat.

"Rather than seek safety for themselves, the four chaplains helped panicking soldiers into lifeboats and tried to instill calm in an extremely desperate situation. Although it meant their certain deaths in the frigid waters of the North Atlantic, they willingly surrendered their life vests to men who had lost theirs.

"As men of peace travelling on their way to war, three of the chaplains had never heard a shot fired in anger before their vessel was attacked. The fourth, Chaplain Fox, was very familiar with the nature of war. Fox had served in the Ambulance Corps during World War I, where he earned the Purple Heart, as well as a Silver Star and the French Croix de Guerre for bravery.

"As the *Dorchester* went down, the four chaplains remained on deck, praying for and comforting the wounded. At the very end, as the ship sank below the waves, they joined arms and sang hymns together."

The President placed the index back in his pocket and continued. "What those four chaplains did, in unison, was profoundly selfless. In their immortal deeds, we should recognize that when faced with dire circumstances that require decisive action, what a man does is not just an abrupt response formed in the moment, but a reflection of character forged over a lifetime. For their shared sacrifice, the four chaplains were posthumously awarded the Distinguished Service Cross, the second highest award that this nation can bestow."

The President sipped from a glass of water as he nodded toward the Secretary. The Secretary quietly moved to Ourecky's side and escorted him and Bea to the stairs. As the three came to the stage, the President wiped perspiration from his chin with a handkerchief.

"I assume that most of you are aware of the details of this event," continued the President. "For those of you who are not, without delving into the sensitive particulars, when called upon to undertake a high risk mission to rescue a fellow officer, Majors Carson and Ourecky willingly placed themselves in the most dire circumstances that any of us could possibly imagine."

The President continued. "They did so without hesitation. Moreover, they had over a week to contemplate their actions and the potential consequences before they actually executed their mission, so this was definitely not a situation where they acted in an impulsive manner.

"I would be remiss if I were to imply that it was easy for me to decide upon an appropriate award to recognize their actions. Faced with this challenge, I called on the assistance of my military advisors, and we struggled to determine what award was commensurate with the gallant sacrifice of these two men.

"As a former Navy officer, I am compelled to uphold tradition and heritage, but as an attorney, I look to the past to seek precedent. To this end, I asked my military advisors to assist me in determining what award was most appropriate. In the process, I learned some interesting things about our military's awards. As an example, let's consider some misconceptions about this nation's highest award, the Medal of Honor.

"Since the Medal is routinely referred to as the "Congressional Medal of Honor," most people assume that it is awarded by the Congress. It is *not*; although the Congress may *recommend* that the Medal of Honor be awarded to a recipient, it is awarded by the President on behalf of Congress.

"Most people would assume that the Medal of Honor is exclusively awarded for exceptionally courageous acts during armed conflict against enemies of the United States. This is also not true. In fact, only five years ago, the Medal was presented to Captain William McGonagle for his heroic actions during the attack on the USS *Liberty* on June 8, 1967. Most of you present are probably aware that the *Liberty* was mistakenly attacked by Israeli forces.

"Furthermore, the Medal of Honor has not always been awarded for actions during armed conflict. Prior to World War II, the Medal was awarded to 193 men for their heroic actions in peacetime. Virtually all of the recipients were Navy men. Amongst those who received the Medal in peacetime, you might recognize the names of Charles Lindbergh, who was a Captain in the US Army Air Corps Reserve when he flew solo across the Atlantic, as well as Richard Byrd and Floyd Bennett, who were honored for making the first heavier-than-air flight over the North Pole.

"So, as my advisors and I concluded our research, I felt confident that we had established ample precedent for an appropriate award. So without further ado, I call upon the Secretary of the Air Force to read the citations."

The Secretary cleared his throat and said, "Please rise as these men are honored. Since the President's time is limited, I will read these two citations as one. *Major Andrew M. Carson and Major Scott E. Ourecky are hereby awarded the Medal of Honor for conspicuous gallantry and intrepidity in action at the risk of their lives and beyond the call of duty. Upon learning that a fellow airman was seriously ill and trapped at an undisclosed location, they volunteered, without hesitation, to execute an exceptionally complicated and potentially perilous rescue mission in the most hostile of environmental conditions. Not only were they successful in their rescue mission, but they were also able to concurrently complete the timely recovery of immense volumes of strategic intelligence which otherwise would have been lost. Their extraordinary heroism, selfless devotion to duty, and supreme dedication to their comrade are commensurate with the finest traditions of the military service and remain a tribute to themselves and the United States Air Force."*

As the President draped the pale blue ribbon around his neck, Ourecky was surprised at how heavy the Medal was. He glanced toward Bea; she seemed to be in shock.

"Since he is currently missing and presumed dead in North Vietnam, Major Carson's award is posthumous," stated the Secretary.

"Major Ourecky, since Major Carson has no living relatives, would you accept this award on his behalf?" asked the President. The Secretary handed him a blue case approximately the size of a hardback book, which he offered to Ourecky.

"I will, sir," answered Ourecky, accepting the case and framed citation. He looked down at the closed case; on this day, more than anything else, he wished that Carson could be standing here with him.

As the Secretary presented a bouquet of roses to Bea, the President said, "Mrs. Ourecky, Bea, we honor your sacrifices as well. I'm very aware that the past few years haven't been particularly easy for you."

Smiling politely, clasping the flowers tightly as if she was afraid to drop them, she answered, "It has been somewhat of a challenge, sir."

The President asked, "Is there anything you would like to say, Major Ourecky?"

Before Ourecky spoke, the Secretary leaned toward him and quietly advised, "Son, this is a mixed audience, so tread *very* lightly. Take caution not to reveal anything beyond what was stated in that citation. Don't be too specific. Got me?"

Ourecky nodded. For several seconds, he was silent, composing his thoughts, and then he spoke. "Although I feel that I don't deserve this, I am greatly honored." Looking directly at Ed Russo, he added, "I hope that I can speak for Drew Carson when I say that we did what we had to do, we did what we thought was right, and that if the circumstances warranted, we would do it again."

Minutes later, the ceremony was concluded. After posing for a few photographs, the President was whisked away by the Secret Service. The Secretary and the Chief personally congratulated Ourecky before he was confronted with an onslaught of generals and other senior officers anxious to shake his hand.

As the crowd eventually filtered out, Ourecky realized that his parents had already been escorted to the reception in another part of the museum. He started to rush out of the auditorium as a captain approached with a polished wooden box. "Your awards, sir," said the captain. As Ourecky removed the Medal, Bea helped him take off the wings. He handed them to the captain and started to leave, but Wolcott stopped them.

"Sorry we had to surprise you, son," said Wolcott. "And I'm even sorrier that your left-seater couldn't be here. Look, pardner, I ain't sure that you fully appreciate the ramifications of what just happened, but the folks sitting in that audience are goin' to be the very ones callin' the shots on the rest of your career in the Air Force. You may never wear those wings or that Medal in public, but *those* men know that you earned both, and they ain't going to forget. So, in short, Major Ourecky, so long as you elect to remain in the Air Force, you can pretty much count on a *stellar* career. Am I making myself clear, pard?"

Ourecky nodded.

"You can do as you wish, but I think you owe it to Carson to make the best of it," added Wolcott. He turned to Bea and said, "And I have something to say to you, little lady. I know that we ain't always been on the best of terms, but hear me when I tell you that *this* man has always kept his promises to you. And he kept his promises even when it would have been much easier to break them. I'm going back to Oklahoma tomorrow, and if I never see you again after this day, I wish the best for you both always."

Bea embraced Wolcott tightly and thanked him. Afterwards, Wolcott grinned, loosened his tie, and walked out of the auditorium.

"*Wow*," observed Ourecky. "I would have *never* imagined that happening. Not in a million years."

"That you would someday get the Medal of Honor?" she asked.

"Well, that *too*, but I never dreamed I would ever see you hug Virgil Wolcott." He hugged her and said, "Hey, I'm sorry I could never tell you anything. I hope you understand now."

Bea kissed him and said, "I really have no idea what you did to earn that medal, but clearly it was very dangerous but very important. I hope that you never have to do anything like that again, but most of all, I'm happy that you kept your promise and came back to me. So, Scott Ourecky, are you home now?"

"I *am*. I am home now."

12:45 p.m.

Bearing a paper plate stacked with sandwiches, chips and hors d'oeuvres, Ourecky escorted Bea to a secluded corner of the conference room where the reception was being held. After almost an hour of shaking hands with generals and other dignitaries, he was famished. He was halfway through inhaling his second tuna fish sandwich when Matt Henson strolled up.

"Mrs. Ourecky?" asked Henson.

"That's me, but call me Bea. Please," replied Bea, using the corner of a white cloth napkin to dab a fragment of pimento cheese from her lower lip.

"I don't know if you remember, but we've already met. It was a few years ago, on a Delta flight from Atlanta to Dayton."

"Really? Honestly, I don't recall," replied Bea. "Your memory must be a lot better than mine, but to be frank, I've met so many people on planes that there's no possible way that I could remember every one of them."

"Well, I apologize for not being more memorable," said Henson. "Bea, would you mind if I borrowed your husband for a moment? We need to catch up on some unfinished business. I promise that I'll bring him back unharmed."

"Unharmed? If that's the case, how could I possibly refuse?"

Curious as to what Henson might reveal, Ourecky grabbed a couple of sandwiches and followed the pararescueman to another corner of the room, where an Army master sergeant—the one who had received the Soldiers Medal during the ceremony—waited.

"Major Ourecky, this is Nestor Glades," said Henson. "You might not remember him, but he was with me that night in Haiti." Ourecky shook Glades's hand, and said, "You know, it's funny, but I do vaguely remember you, but it sure seems like you were a black man, like Henson here."

"I was that night," replied Glades.

"Yep," interjected Henson. "We were definitely soul brothers on *that* day."

"Major Ourecky, there's something I need to tell you," confided Glades.

Ourecky chewed up a mouthful of sandwich, swallowed it, and asked, "About Haiti?"

"No. It's about your friend Carson," stated Glades. "I guess that you're aware that there was a rescue team sent into North Vietnam after he was shot down."

"I am," answered Ourecky. "General Wolcott told me that, when he called to tell me that Carson had been shot down. He said that a team from the 116th had been dropped in to link up with him. I didn't get any more details, other than they weren't successful. Carson had apparently been killed before they could get to him."

Glades nodded and quietly said, "Henson and I were on that team."

"You?" asked Ourecky. "*And* you?"

Both men nodded. "I don't think that your friend was killed that day," asserted Glades.

"Go on."

Glades explained, "Before Carson went to Vietnam, General Fels asked me to put together a contingency team to execute a search and rescue mission. That's how Henson and I, and the others, ended up at Da Nang. As you said, when Carson was shot down, we were inserted to pull him out, but then we were told that he was killed. We came out, and shortly after that, I went back to Eglin."

"So that's it?" asked Ourecky. "That's your whole story?"

"Not exactly," replied Glades. "Before I left Eglin, Fels had made arrangements to where I could get any post or assignment after I got back from Vietnam."

"And?"

"I chose Hawaii."

"That sounds nice."

"It was. It is. I'm still there," said Glades. "After I had hung my hat in Florida for almost six years, swapping back and forth between the Florida Ranger Camp and MACV-SOG, I figured that Deirdre..."

"Deirdre?"

"My wife. I figured that Deirdre and my kids were overdue for a change in scenery, maybe someplace tropical and exotic, so I picked Hawaii. It's turned out well: Deirdre loves it, and so do the kids. They're even learning to surf."

"Well, that's just great. I'm happy for you."

"That's only *part* of the story," declared Glades. "Hawaii is also the location of an intelligence analysis office of the JCRC, the Joint Casualty Resolution Center. The JCRC is the military agency that is responsible for resolving POW and MIA issues. That's the assignment I asked for, and General Fels personally saw to it that I got it. As you might figure, our main focus is on Southeast Asia: specifically the status of unresolved POWs and MIAs in Vietnam, Laos, Cambodia. Most of the JCRC is in Thailand, so I spend a lot of time travelling between Hawaii and Southeast Asia."

"I'm still at a loss here," stated Ourecky. "If we're here to talk about Carson, then he was never a POW. He was reported killed by North Vietnamese forces in December 1972."

"I don't think so, Major," replied Glades. "I think those reports were erroneous. I think he was captured, and I also think he's not dead."

Momentarily incapable of speech, Ourecky closed his eyes. Glades's assertions reinforced his suspicions that Carson was still out there, in need of assistance. "And why exactly would you think that, Master Sergeant Glades? Why are you so convinced that he's still alive?"

"In Hawaii, I'm in a position to review any and all intelligence traffic from Southeast Asia, particularly any messages that pertain to POWs and MIAs. As you might figure, after that SAR operation to grab your friend, I paid special attention to any traffic that even remotely looked like it might apply to him."

"Did you find anything?" asked Ourecky.

"I did," replied Glades. "I found *four* things. I assume that you know that Carson deployed under an alias, Lieutenant Commander Drew Scott."

"I know that."

"I discovered one message between Hanoi and a Reeducation Camp that mentioned a US prisoner named Drew Scott, where the North Vietnamese were trying to resolve Scott's true identity and status. The name Drew Scott had not shown up in the reconciliation list provided by the US, and the NVA apparently were concerned that we were trying to hoodwink them. That message was transmitted in January 1973. I wasn't able to find a subsequent reply from Hanoi."

"You said that you had *four* things?" asked Ourecky impatiently.

"Yes, sir," answered Glades. "Second, although there's no further mention of Drew Scott, I found an excerpt from a teletype message from the Soviet Union to Hanoi that directly references an Andrew Carson. The message is from the Soviet GRU headquarters, specifically from the Bureau of Special Cooperation, which coordinated intelligence exploitation of captured equipment and personnel. The teletype stated that a special team would be travelling from Moscow to North Vietnam, where they were supposed to interrogate Carson, and if he proved to be of value, to escort him back to the Soviet Union. That message was intercepted on April 3, 1973."

"Wasn't that after all the POWs had already been released by North Vietnam?"

"Yes, sir," replied Glades. "The last POW—Navy Lieutenant Commander Alfred Agnew—was released on April 1, 1973. He was the last man released, and also the last man captured: he had been shot down on December 28, four days after Carson."

"So you think Carson was taken to the Soviet Union?" asked Ourecky angrily.

Glades shrugged. "I'm not sure."

"Not sure?" demanded Ourecky. "What do you mean?"

"Bear in mind, sir, that it gets a little strange from this point," asserted Glades. "I read two more messages after that. One was from the commander of the Reeducation Camp to his bosses in Hanoi. That message implied that the Soviets took Carson. The second was from the commander of the Soviet team to the GRU headquarters in Moscow."

"And what did it say?" asked Ourecky.

"Sir, it said...the message implied that an American commando team somehow infiltrated the camp in the middle of the night..."

"And?"

"And the message mentioned *me* by name, and implied that *I* took Carson."

Ourecky shook his head. None of what Glades was saying made sense. "Well, *did* you take Carson?" he demanded.

"No, sir," answered Glades. "After our headquarters told us that Carson had been killed, they directed us to stand down and begin movement to an extraction site. Less than twenty-four hours after we made it back to Da Nang, I was on an airplane headed back to Florida. So if someone slipped into that camp to rescue Carson, it wasn't me. I swear."

"Then what do you think happened to him? Is he alive? Is he dead? Is he in the Soviet Union?"

"Honestly, sir," replied Glades. "I really don't know where he is, but I think that he's still alive."

27

TSUNAMI

Tan Son Nhat International Airport
Ho Chi Minh City, Democratic Republic of Vietnam
Present Day

L ess than twenty-four hours after the tsunami had devastated the coastal areas of Vietnam, Doctor Matthew Henson and four members of his disaster relief group—Apex Global Response— were on the scene. Working in collaboration with the Vietnamese government, the United Nations, other PVOs—private volunteer organizations—and several non-governmental humanitarian assistance agencies, Henson and Apex were here to assist with the delivery of crucial relief supplies and medical treatment teams to the most heavily impacted areas.

Although it was a disaster of almost unprecedented magnitude, with almost a half-million coast-dwelling Vietnamese losing their homes, there had been virtually no immediate loss of life. The new Tsunami Early Warning System, recently established by the Global Physics Institute in Hanoi, had sounded alarms in sufficient time to urge people to safety, even though they were compelled to abandon their homes and belongings as they raced inland. But as Henson well knew from harsh experience, although the initial fatality count had been minimal, widespread sickness and death would soon follow. The displaced populations would pack into crowded refugee camps, adequate sanitation and clean drinking water would be almost nonexistent, and disease would spread rapidly.

Henson and his wife, Adja, had started Apex Global Response over three decades ago, not long after he graduated from medical school. In their first decade of operations, they quickly discovered that there was almost always an abundance of physicians and other volunteers anxious to rush to the locus of a disaster scene, but there was a shortage of people who were willing to quietly work behind the scenes in support and administrative roles. As Adja progressively became more involved with managing medical response operations for Apex, Henson gravitated toward the role of managing logistics and overall operations. In fact, although he was still frequently addressed as "Doctor" Henson, he had allowed his medical licenses and qualifications to lapse long ago; it had been over ten years since he had even laid hands on a patient in a medical setting.

Henson had essentially come full circle, since the lessons long ago conveyed by Abner Grau, his mentor as well as Adja's father, eventually became the basis for his current actions. In effect, he was acting in virtually the same capacity that he had performed in Gabon, Haiti, and countless other Third World countries, when he worked for Grau in coordinating logistical support for a worldwide rescue and recovery network for a secretive Air Force project.

In time, Apex had joined in relief missions around the globe, gradually evolving into an organization expressly tailored to function in dangerous places plagued by war and political unrest. Apex gradually established a reputation for slashing through the red tape, graft and greed endemic to the governments of Third World nations. They were also tremendously effective in thwarting thugs and hijackers intent on stealing relief supplies, so that the lifesaving resources could be pushed through to the people desperately in need of them. Apex's unorthodox methods were sometimes called into question, but there was never any doubt that they achieved consistent results. Henson took pride in knowing that with Apex, virtually every spoonful of rice or high-protein flour found its way into a needy child's belly instead of a thug's warehouse.

Unlike most PVOs, Apex did not actively seek funding, but was financed by private donors with exceptionally deep pockets. Most were entertainment personalities who sought publicity for their altruistic actions; moreover, they craved recognition as being part of a solution that actually worked. More than content to allow the celebrities to be the public face of Apex, Henson and Adja preferred to quietly remain in the background. Drawing only modest salaries, they maintained an unpretentious lifestyle. They owned a small house in Los Angeles, where Apex was based, but rarely spent any time there, since they effectively existed as nomads, migrating from one crisis site to another.

Besides their extraordinary logistics solutions, Henson and Adja had developed a unique perspective for recruiting medical personnel to staff Apex's missions. Over time, they had realized that the majority of the physicians and nurses drawn to disaster response volunteered because they sought a personal adventure. They ventured into Third World countries hoping for some form of enlightenment, excitement or both, and rarely lasted more than a year in the field before they grew tired of poor living conditions, danger and uncertainty. Granted, some performed admirably, but a sizeable percentage were high maintenance prima donnas who demanded comfortable accommodations and decent meals even as they worked with destitute patients in the midst of terrible turmoil and suffering. Additionally, it seemed as if every relief mission gradually devolved into a soap opera of sorts, replete with serial romances amongst staff members, as well as the jealousy and pettiness that accompanied such affairs.

To effectively accomplish its mission, Apex strived to field medical workers who wanted a job, not an adventure. Adja, with considerable backing from their donors, aggressively recruited physicians and nurses from Third World countries. Apex's typical doctor signed on for a five-year commitment, which included at least three years of formal residency training, usually at a prestigious institution, and two years of deployment on medical missions. They received a living stipend, which was usually very generous in comparison to their wages at home, but most were motivated by the opportunity to become better at their craft. After their deployment phase—twenty-four months of practicing "mud medicine" in the bush—they returned home, armed with excellent medical training that they probably could not afford or even receive in their home countries.

1:35 p.m.

The massive international relief operation was headquartered in an aircraft hangar at Tan Son Nhat International Airport. Apex had been allocated a remote corner of the vacant hangar. Henson's advance team of five disaster specialists were busily engaged in setting up an array of computers and communications terminals.

He looked around, examining his surroundings. The disused building smelled of mildew and aviation fuel. A fine layer of green mold coated the bottom of a brick wall beside him. Approximately a hundred international disaster workers occupied the hangar; that number would likely double by the end of the day. When that happened, the air inside the hangar, now stifling, would be unbearable and—more importantly—decidedly unhealthy. The reason was simple: as a multitude of well-meaning volunteers arrived from other countries, they would bring with them an array of germs. The tight quarters and humid air would provide a perfect petri dish for the exchange and exponential growth of viruses and bacteria. In a day or so, virtually everyone would be ill, at least to a certain extent, with coughs, sore throats and other flu-like symptoms.

Yawning, he sipped coffee as he checked his watch: The bulk of Apex's response effort was currently en route from the States and should be on the scene in a matter of hours. Their equipment included a collection of large pop-up tents intended to set up a "bare base" operating center in the most austere of conditions. To protect his personnel, Henson made a mental plan to establish a sub-camp for Apex's operations, and leave just a skeleton crew of his hardiest workers here in this hangar.

As he waited, he worked with Apex's air operations planner to establish priorities for their small fleet of unmanned aircraft—three Kaman "K-Max" cargo helicopters, three reconnaissance drones, and two blimps—that would arrive aboard the US Air Force C-17 bearing the second wave of Apex responders, supplies and equipment.

The unusual aircraft were key to Apex's operations. Flying without a pilot aboard, each K-Max helicopter could deliver 6000 pounds of relief supplies, slung underneath in an external bundle, with pinpoint accuracy. Normally paired with the helicopters, the reconnaissance drones were typically used as robotic scouts to identify suitable drop-off sites, as well as to verify conditions in the impacted areas. The blimps served as communications relays, and also provided constant surveillance platforms to monitor relief operations in progress.

As he studied a map of planned supply shuttle routes on a computer screen, Henson suddenly felt uneasy. The hairs on the back of his neck stood up, and he had the unsettling sensation of being watched. He slowly looked up from the display and locked eyes with an elderly Vietnamese man in an officer's uniform. The man—part of an official delegation arriving for a briefing at the operations center—looked vaguely familiar, and it was obvious from his expression that he thought the same of Henson. They stared at each other for a few seconds, as if trying to place the other in space and time, and then the Vietnamese man went on his way, following his escorts into a conference area.

3:15 p.m.

Henson was briefing a pair of newly arrived team leaders when he was abruptly interrupted by a Vietnamese officer. He wasn't yet familiar with the emblems of rank on the uniforms worn by the VPA—the Vietnamese People's Army—but guessed that the man, who appeared to be around thirty years old, was probably a senior captain or junior major.

In excellent English, the VPA officer brusquely demanded, "What is your name?"

Henson held out the laminated credentials dangling from a red cord around his neck. "Matthew Henson," he replied. "I'm here with Apex Global Relief." He contemplated what was next. In past instances, even though Apex had been formally invited by the local government officials, he was often approached by unscrupulous bureaucrats demanding an "entrance fee" or some similar tribute. In other cases, particularly when some certain publicity-conscious PVOs were part of the overall effort, Apex was shoved into a mediocre role in areas far from the actual disaster area. He was here to help people in need; he didn't come here to be shaken down by greedy officials or to waste time and resources in idiotic turf squabbles.

Consulting a small notebook, the VPA officer said nothing.

"Is there a problem?" asked Henson impatiently. "Is there something wrong with our paperwork?"

Looking up, the VPA officer shook his head. Finally, after a few seconds, he tore a page from the notebook and handed it to Henson. "Do you know these names? Do you know these men?"

Two names were written on the scrap of blue-lined paper: Nestor Glades and Scott Ourecky. Henson had known both men for years, all the way back to the rescue mission in Haiti in 1970, and had seen them both frequently in the past few years. Both men had worked for Apex in conjunction with a large relief mission in a strife-ridden African country. Glades had coordinated security for delivery of humanitarian supplies and medical missions. Ourecky, now a retired two-star Air Force general, had acted as a consultant to develop Apex's innovative air operations, including the acquisition of the K-Max helicopters.

"I do know them," answered Henson, nudging his bifocals up on the bridge of his nose.

"If it were necessary, could you get in touch with them?" asked the VPA officer.

Henson had current contact information for both men, but also knew that Glades was now extremely sick. In fact, he had been in Africa only a few weeks ago, working on an Apex project, when he first fell ill. He had returned home to Florida, and as best as Henson knew, he was still in the hospital, undergoing an extensive battery of tests.

"Perhaps, but why are you asking about them?" demanded Henson.

"I really don't know," confessed the Vietnamese officer, tucking the notebook back into his pocket. "My commander directed me to talk to you, Mister Henson. He sends his regrets; he is leaving for Haiphong within the hour to coordinate the distribution of donated cholera vaccine, but he intends to communicate with you tomorrow, and more information will be forthcoming."

28

THE LAST MISSION OF NESTOR GLADES

Off the coast of Vietnam
Present Day

Flying roughly five thousand feet above the churning South China Sea, the gray C-17 cargo plane resembled an enormous winged whale as it plowed through scudding clouds. Large gaps occasionally opened in the dense cumulus banks, revealing a once beautiful shoreline ravaged by last week's fearsome tsunami. Wrecked fishing junks could be seen miles inland; only days ago the pride and livelihood of their owners, now abandoned and fit only for kindling. The devastated landscape was a splotchy mosaic of muddy brown flooded areas and isolated pockets of green vegetation.

Deep in the bowels of the massive transport, Nestor Glades grunted as he was bent double by overlapping waves of nausea and severe pain. Groaning, he gagged on caustic bile surging up in his throat. Subsisting solely on pills and lukewarm water, he hadn't eaten since departing the States yesterday.

Closing his eyes, he tried desperately to sleep, but getting any real rest was a futile effort. He recalled a philosopher's famous declaration that what doesn't kill us only makes us stronger. While Glades had already outlived his doctor's dire predictions, as well as the hushed expectations of his family and friends, he sure didn't *feel* any stronger, but was just too damned stubborn to die yet. Now he was beginning to doubt that this excursion was really such a wise idea, and whether he might have been better off spending his last days propped up in a hospital bed, tethered to a numbing drip of intravenous narcotics.

Glades looked to his right; cuddled against his shoulder, with her head supported by a familiar pillow taken from their Florida home, Deirdre snored softly. Sound asleep, clutching a large plastic prescription bottle in her freckled hands, she had served as his nurse for the entire journey. He smiled in spite of his anguish; if he had accomplished nothing else in his time on earth, he had married well. She had scarcely changed since he met her in England over fifty years ago; her hair was still fiery red and her waif-like figure lent little evidence that she had carried and delivered five children. After nearly losing her to breast cancer over a decade ago, he cherished her more than ever. She bravely endured two mastectomies followed by seemingly endless rounds of chemotherapy, radiation therapy and experimental drug trials. Now, overjoyed that Deirdre was still at his side, he accepted her scars in the same manner that she had long accepted his.

To her right, in the adjoining troop seats, was another couple: Scott Ourecky, a retired Air Force major general, and his wife, Bea. They nestled into each other like a pair of love birds, naturally seeking each other's familiar warmth and comfort. Like Glades, Ourecky wasn't a big man, perhaps just three inches shy of six feet, but it was obvious that he and his wife—Bea—kept themselves in excellent physical condition. Both were long into grey and carried their share of wrinkles and creases, but Bea was still a statuesque and beautiful woman.

Convulsing as a knife of pain ripped through his abdomen, Glades grimaced as he waited for the throbbing to subside. His agony was not new; diagnosed with inoperable pancreatic cancer eleven weeks ago, he endured suffering that he could not possibly have imagined when he was still a healthy man.

Gradually wasting away would have been an unfitting end for Glades, a consummate warrior who had evaded death's grasp on countless missions across the face of the globe. But less than a week ago he had reconciled himself to just such a fate. He was literally settling into his deathbed when a strange invitation arrived, summoning him once more to action. The promise of the invitation—an opportunity to make amends with his past—was sufficient for the dying soldier to undertake what had become such an arduous journey.

The note, borne by a special messenger from the White House, written under the President's own hand, was a personal request for Glades and Ourecky to accompany a consignment of tsunami relief supplies to Vietnam. The note was dispatched with a set of special travel orders and official instructions that gave a time and place for a more detailed briefing. Their spouses were also invited to accompany them. If the timing and circumstances permitted, the two couples would tour the flood-ravaged areas to observe the disaster relief operations currently in progress.

Quietly grunting at the onset of yet another stabbing pain, Glades swiveled his head and looked to his left. The C-17's cavernous interior was crammed with sixty very tangible tons of American good will—humanitarian rations, bottled water, tents, tarps, plastic sheeting, blankets, medicine, baby formula—all stacked neatly on flat metal pallets, tidily arranged in three rows, six large pallets to a row. If he didn't know any better, he could easily convince himself that he was in a meticulously clean supply depot, except for the faint but relentless pulsations of the shiny aluminum floor and the muted roar of turbojet engines, subtle evidence that this particular warehouse was hurtling through the sky at nearly five hundred miles an hour.

Leaning forward in his seat, Glades looked at a metal casket lashed to the aircraft's floor with thick nylon ratchet straps, and reflected on the true purpose of their mission. He was en route to Vietnam at the express invitation of the President, but his true task had nothing to do with humanitarian relief efforts. Rather, he had left his deathbed for a chance to close the books on a mission left unfinished almost four decades ago.

In the waning days of America's active involvement in the war in Southeast Asia, Glades had participated in a clandestine mission to rescue a Navy pilot shot down over North Vietnam. He later discovered the Navy aviator—Lieutenant Commander Drew Scott—was in fact an Air Force pilot—Major Drew Carson—flying under a disguised identity. Questions concerning Carson's capture were never entirely resolved, and his ultimate fate was still unknown. Haunted by the notion that he had left someone behind, Glades had personally worked behind the scenes to investigate Carson's circumstances, but the trail had grown cold decades ago.

In the aftermath of the recent tsunami, a Vietnamese colonel came forward, stating that he could provide information to clarify Carson's status. Not only did he claim to possess personal knowledge of the missing pilot's outcome, but also asserted that he could provide indisputable proof. But the potential disclosure came with strings attached: for whatever reason, the Vietnamese colonel specifically declared that he was only willing to talk to Ourecky or Glades. That was all Glades knew, but it was just as well; after a lifetime spent working in the shadows, he was quite accustomed to obscure information and incomplete orders.

Glades looked at the aluminum casket again. Technically, in military-speak, it was not even a casket; officially, it was a "human remains transfer case." It had been hastily loaded aboard the C-17 during a brief stop in Hawaii, with the assumption that there might actually be skeletal fragments or other remains to be brought back to the United States. In truth, it was just as likely that the glistening metal box would return empty, since the Vietnamese colonel's cryptic message did not tacitly state that there were remains; in fact, it did not go so far as to say whether Carson was dead or perhaps still alive.

Glades looked toward the rear of the plane, where six men dozed in troop seats. One was a forensic specialist from the Armed Forces

Institute of Pathology, who was accompanying the mission in the expectation that there would be remains to identify. Another was an intelligence analyst with the DIA—Defense Intelligence Agency—and the remainder comprised a personnel security detail—two US Secret Service agents and two Air Force security officers—charged with protecting Ourecky and Glades.

The personnel security detail—PSD—wasn't simply a formal precaution dictated by protocol regulations; there was a legitimate need to safeguard the two men, albeit for different reasons. During the war, besides slaughtering scores of other North Vietnamese soldiers, Glades had personally killed an NVA regimental commander who was a favorite nephew of General Giap. Although the United States and the Democratic Republic of Vietnam were now at peace, some festering wounds never healed completely, and there may be a former NVA soldier lurking somewhere in the background, vindictively eager to settle an old score.

Ourecky's situation was different: Glades was aware that the retired general carried an old secret of tremendous magnitude, but there was no indication that it had ever been even slightly compromised over the years. Of course, on the other hand, there was just no way of really knowing. At this point, with the international tsunami relief effort in full swing, a multitude of nations—including a large contingent of Russians—were represented at Tan Son Nhat. Consequently, if there was even the slightest indication of a confrontation or an attempt to kill or seize either man, the entire party would retreat to the C-17, and they *were* coming out, come hell or high water. If they were blocked from taking off, the PSD would barricade the cargo jet and they would wait for the arrival of a more substantial tactical force to effect an extraction.

Minutes later, the massive jet arrived at Ho Chi Minh City. After years of ferrying cargo and troops into war zones, the pilot was apparently in the habit of never letting the transport use any more runway than absolutely necessary. As the tires barked on the pavement, she braked hard, smoothly reversed thrust on the engines, and executed an abbreviated landing that would have earned the crew bragging rights at the shortest dirt strip in any combat zone.

It took a few minutes for the C-17 to taxi into the busy cargo terminal area. The plane eased to a halt and the rear ramp slowly swished down. Within moments of their arrival, a small army of trucks and forklifts descended on the plane. The C-17's aluminum floor was crosshatched with an intricate network of rollers and guides, which allowed the massive supply pallets to be pushed and prodded without mechanical assistance. Glades was fascinated with the casual ease at which the Air Force cargo handlers were able to manipulate tons of supplies.

Stretching and yawning, Deirdre rose to her feet. She peeled off her blue nylon windbreaker and stuffed it into a daypack. She and Bea chatted as they crammed their packs with bottles of water, snacks, and tubes of sunscreen. As the women gathered their belongings, Ourecky hurriedly walked toward the back of the plane, anxious to get on with the mission. Moving as quickly as he was able, Glades shuffled stiffly afterwards.

"Ness, my darling," Deirdre called out after her spouse. Her voice was a soft Irish lilt, only gently tempered by decades of living in America. "Please pace yourself. I'm sure there's time, and you needn't make yourself even sicker in this infernal heat."

Almost by instinct, like a perpetual tourist who glimpsed the wonders of the world through the narrow confines of a viewfinder, Glades stood on the C-17's ramp and rapidly snapped pictures with his digital camera. Then he realized the futility of his actions: he probably wouldn't live long enough to view the images, and Deirdre had never demonstrated any interest in learning how to download the pictures.

Deirdre joined him at the rear of the plane. Shadowed by the C-17's aircraft commander—a petite major who kept her strawberry blonde hair pinned into a tight bun—the two couples stepped down from the ramp and strode out to a white van waiting planeside. Heat waves shimmered off the sunbaked tarmac, and the pungent smell of burning wood and debris hung in the air. Dogs barked in the distance.

"General," stated the aircraft commander. "I'm Major Janet Smith. I have some papers for you." After handing Ourecky a folder of documents, she self-consciously smoothed and adjusted her sage green flight suit.

Ourecky removed his sunglasses, squinted, donned reading glasses, and reviewed the paperwork. Glades had already seen the documents; they were orders for the aircrew that clearly stated—in the least vague of any military language that he had ever seen—that the C-17 was not to depart Vietnam until Ourecky was absolutely satisfied that they had accomplished the task that they were sent to do.

"Thank you, Major Smith," said Ourecky, climbing into the van. "Ride with us, please."

A Secret Service agent drove them toward an operations building near the main terminal. There was a moment of awkward silence before Smith spoke. "General, I couldn't help but notice your name on the manifest. You're not related to…"

Chuckling, Ourecky answered, "Yeah, good catch, Major. We're related all right: he's my son. Everyone asks that. He's made quite a name for himself."

"I guess you're really proud of him," said Smith, fanning herself. "Is he still with NASA?"

"No. He's still involved in space flight, except strictly on the military side. He transferred from NASA back into the Air Force, and works with unmanned platforms now," explained Ourecky. "It's funny, though; when he was flying the Shuttle, he was always so casual about it. I could never imagine that riding rockets would become such a routine job. But it's almost as if he was born to fly into space. It all comes so naturally to him."

Looking out the window, Smith nodded. "So I suppose that this place probably brings back a lot of memories for you."

"How so?" asked Ourecky.

"Well, I just assumed that you spent some time here during the war, sir," replied Smith.

"No, I never set foot here." Ourecky grinned, shrugged his narrow shoulders, and added, "Passed over several times, maybe, but that's another story."

In the front passenger seat, a gruff-looking Secret Service agent cupped his ear, apparently listening to a radio through an earpiece. He pointed toward a blue Toyota carry-all van parked beside the operations building. "Sir, that's the Vietnamese colonel who wants to talk to you."

Glades watched an elderly man step out from the passenger side; the man looked to be in his seventies and was dressed in the distinctive camouflage uniform of the Vietnam People's Army ground forces. Even from a distance, the man looked vaguely familiar. Seeing that the man's withered left arm hung loosely at his side, his suspicions were confirmed.

Their own van pulled to a stop and the small party disembarked. The DIA analyst consulted a PDA and spoke quickly as they approached the building. "General, I just received more detailed information. This guy's name is Colonel Bao Trung. He was initially assigned as a Lieutenant in the 22nd NVA Infantry Regiment. Seriously wounded in Quang Nam province in August of 1970. They evacuated him north. After medical treatment and rehabilitation, he was assigned to another infantry regiment in training in Quang Ninh province up north, but was wounded again in December 1972. After recuperating from that, he was assigned to a POW camp as a guard force commander and then later became the camp administrator."

Ourecky nodded. "Anything else?" he asked.

"Sir, our records indicate he's a dyed-in-the-wool, Communist hardliner. He fought against our guys in South Vietnam, and there's substantial evidence that he was personally involved in interrogating a few of our pilots at the tail end of the war. I would be cautious with this guy, General. Abundantly so."

"Duly noted."

As they approached, the Vietnamese officer faced Glades and Ourecky, drew himself to attention, saluted, and bowed slightly at the waist. Years seemed to fade away as he straightened his spine and became like a younger image of himself. "I am Colonel Bao Trung," he announced proudly.

His name had barely departed his lips when Glades staggered forward and threw his arms around Bao Trung, and the elderly Vietnamese man reciprocated. No words passed, but they gripped each other tightly, as if they were long lost brothers. Tears streamed from their eyes. Ourecky watched without emotion, obviously not knowing what bond connected the two.

Taken aback by her husband's tears, Deirdre crumpled against Ourecky and began to sob.

"Are you going to be all right, Deirdre?" asked Ourecky, putting an arm around her shoulders to comfort her. "Is there something I can do?"

Stammering through her tears, Deirdre replied, "I'm sorry. I'll be okay. It's just that…it's just that after being married fifty years, burying two children, and even with this damned death sentence hanging over him, I've never *once* seen that man shed a tear."

29

THE NAME ON THE WALL

Tan Son Nhat International Airport
Ho Chi Minh City, Democratic Republic of Vietnam
Present Day

With tears flooding their eyes, Bao Trung and Glades embraced in silence for well over a minute. Taken aback at the sight of her husband's uncharacteristic display of emotion, Deirdre Glades leaned against Ourecky and sobbed. The others stood to the side and watched without speaking, obviously not knowing what to make of the strange reunion between two former enemies.

In the background, the tsunami recovery effort was still in full swing. Like squadrons of worker bees busily tending to their hive, a constant stream of military and civilian cargo planes landed and departed, delivering a steady flow of humanitarian supplies.

The hulking C-17 was quickly being unloaded. Trucks and forklifts bustled about, ferrying consignments of relief supplies to staging areas marked by color-coded pennants. A Soviet-built "Hip" helicopter hovered low over a depot, hoisting a cargo net swelling with drab-painted

fuel drums. Wearing matching jumpsuits and accompanied by burgeoning heaps of rescue gear, squads of international disaster specialists lolled
in the scarce shade available as they awaited transportation to outlying
areas. The harsh sun glared down as if in stern judgment.

Finally, the two elderly men slowly released their clench and stepped
back, scrutinizing one another to see how the long decades had passed.
Over forty years had elapsed since Glades had left the wounded NVA
officer unconscious beside a foot trail. The former American commando
looked horrible; his countenance was drawn and gray, and he walked with
the stooped posture and demeanor of someone crippled by immense pain.

Bao Trung remembered Glades as robust and vital, fully in command
of his circumstances; the frail man he saw before him was more like a
faint vestige, like a poorly developed photograph from a long-forgotten
roll of film. "Forgive me for being so honest, but you look awful," he said,
wiping tears from his eyes. "Are you not well?"

"No, I am *not* well," answered Glades in a faltering voice. Abruptly
gripping his stomach, he grimaced as he bent sharply at the waist.
Deirdre scurried forward to assist him, but he shook his head as he held
out his hand.

"Are you going to be all right, Glades? Do you need to see a physician? I can make arrangements if…"

"Thank you, but there's nothing a doctor can do for me now. I'm
dying."

"Dying?" said Bao Trung. "We all are. Surely as rust eats iron, death
will eventually consume us all, every one."

"I suppose you're right," smirked Glades, struggling to pull himself
erect. "Excuse me for not being too philosophical about kicking the can."

Standing beside Glades, Ourecky cleared his throat. "Uh, sorry,
sir," muttered Glades. "I thought you two might have already met."
Embarrassed, he offered an introduction. "Colonel Bao Trung, this is
Major General Ourecky."

"*Scott,*" said Ourecky, stepping forward as he extended his hand. "I'm
retired now, so we can dispense with the formalities. Like my old boss
used to say: the stars fell off when I shed my blue suit."

Bao Trung reached into his pocket and withdrew a piece of faded black silk, the hand-embroidered star map that his wife had made for him decades prior, tightly folded into a small square. He held it out towards Ourecky and asked, "Do you recognize this?"

Ourecky looked as if he was being called to verify a shibboleth. He examined the cluster of carefully sewn points, and then said, "That's the Pleiades. The Seven Sisters. That's my favorite group of stars. How could you..."

A faint hint of a smile came to Bao Trung's otherwise inscrutable face, and a twinkle lit his eyes. He held out his hand and said, "Welcome, Scott Ourecky. I am deeply honored to finally meet you."

"Bao Trung, I would like to introduce you to my wife, Deirdre," said Glades.

Bao Trung lightly clasped her hand, bowed cordially, and said, "I am greatly honored to make your acquaintance, madam. Your husband is truly a great and highly respected warrior. I know that you are very proud of him."

"And this is *my* better half," said Ourecky, glancing to his side. "Bea."

The Vietnamese colonel bowed slightly and said, "I am deeply honored."

"Bea, Nestor and I will probably be busy for a few hours," noted Ourecky, turning toward his wife. "Would you and Deirdre like to take a look at the relief operation? I'm sure that we can have one of the security guys escort you around."

"That sounds fine, Scott," replied Bea. "I know that you boys have work to do."

"That sounds good, General, but I need to stay with Nestor," interjected Deirdre. "I have his medicines and..."

"Just leave the bag with me, dear," said Glades, adjusting his sunglasses. "I should be able to take care of myself for an hour or two. I'll be just fine."

Bao Trung briefly spoke into a handheld radio and then commented, "We have a field hospital established in the main terminal. I have notified a physician there. He can be here in two minutes if there's a need."

"We'll take good care of him, Deirdre," said Ourecky, smiling as he shifted a blue nylon satchel from one hand to the other. "I promise. Why don't you go look around? I'm sure that Bea would appreciate your company."

"Well, I suppose," conceded Deirdre. She swung her daypack from her shoulder, unzipped it, removed a plastic shopping bag filled with medicine bottles, and handed it to Glades. "Don't forget the big white pills, Nestor. Two of them, with plenty of water, every half hour."

"I won't forget, *ionúin*. For once, don't fret about me. I'll be fine."

Ourecky nodded to a Secret Service agent standing close by. The agent shadowed Bea and Deirdre as they boarded the van, and then climbed in after them.

"Please come this way," said Bao Trung, gesturing toward the operations building as the van pulled away. "I think we would all be much more comfortable once we have some shelter from this dreadful sun. Nestor, your friend—Matthew Henson—is waiting inside."

As they walked toward the single-story metal structure, they compared notes as old men often do, sweeping aside decades of separation to focus on the truly important issues. "I brought you something," said Glades. He pulled some photographs from an envelope and handed them to the Vietnamese officer. "I have grandchildren."

Bao Trung examined the photographs. "I do, also," he said, grinning. He reached into his pocket, extracted a photo of his extended family and held it towards Glades. In the picture, he and his wife were surrounded by their three sons, their daughters-in-law, and ten grandchildren.

Adjusting his bifocals, Glades smiled weakly. "Your wife looks much happier now. As I recall, she didn't look too happy in the last picture I saw of her."

"Thank you for allowing me to meet my grandchildren, Glades," said Bao Trung solemnly. He slipped the cellophane-laminated photograph back into his pocket. "I am indebted."

"As I am." Glades twisted the cap from a plastic bottle and took a long drink. He swallowed, obviously with some difficulty, and added in a raspy voice, "I doubt that I would have made it home if I hadn't followed your advice."

The three men entered the building, where Matt Henson waited in the foyer. He hugged Glades and asked, "I heard about the diagnosis. I'm really sorry. How are you holding up?"

Grimacing, Glades replied, "Not so good. Ask me again in a few weeks, after I'm dead."

"Don't be so negative, Nestor," said Henson. "Miracles happen all the time. Don't be so quick to count yourself out."

"I'm not looking for a miracle," replied Glades. "I'm just looking for it to be over."

Henson turned to Ourecky, shook his hand and said, "It's good to see you again, sir. Thanks for your help with the K-Max's. They're worth their weight in gold."

"I'm just glad that I could of assistance," replied Ourecky. "Look, Matt, I hate to sound impatient, but I really want to know where Carson is. We'll have time to catch up later."

Bao Trung nodded at a stocky Vietnamese officer, who guided the four men into a small conference room. On one wall, a whiteboard displayed a rough sketch of the airport grounds, detailing the locations of different staging areas and loading sites. A wood-bladed ceiling fan spun slowly, hardly stirring the oppressively humid air.

As the four men took their seats at a table, a young Vietnamese soldier passed through the room, bearing a lacquered serving tray heavily laden with soft drinks and bottled water. Slowly shaking his head, Ourecky politely waved him away as he removed a bulky satellite phone from a nylon satchel. Glades took a liter-sized bottle of water from the soldier's tray. Bao Trung and the soldier spoke briefly in Vietnamese; afterwards, the soldier adjusted the dial of a wall-mounted thermostat.

Ourecky switched on a small voice recorder and said, "Colonel, your message said that you had information about Major Drew Carson, who was shot down northeast of Haiphong in December of 1972. Do you know what happened to him?"

"I do, General. But please be patient with me. I really want to start at the beginning." Bao Trung looked at Glades and said, "Before I can tell you what happened to Carson, I have a confession for you."

"A confession?" asked Glades. With trembling hands, he screwed the cap back on the water bottle.

Bao Trung nodded. "I know it's probably difficult to remember, but just before you left me to be found by our forces, you concealed some of your excess equipment in the swamp, so you could move faster. Do you remember?"

Glades closed his eyes momentarily, frowned, and then replied, "I do. What of it?"

"Although you thought I was unconscious, I overheard you and your men discussing what would be cached and what you would carry with you. The soldier with blonde hair was reluctant to abandon some gold coins, but you ordered him to leave them in the swamp."

"That's right. There were gold coins in a bartering kit. Ulf Finn wanted to hang on to them. You have a good memory."

"Yes, Glades, but if you knew the things that I am compelled to remember, then you would understand that my good memory is not a blessing but a curse."

"You said that you had a *confession*?" interjected Ourecky in an urgent tone. "Does it have to do with Carson?"

"Indirectly, yes." Ashamed, Bao Trung confided, "Here's my confession. Four months after we went our separate ways, I was authorized to go on furlough to see my family. As much as I longed to see my wife and son, I did not travel directly home."

"So you went after that gold?" chided Ourecky. Behind him, a wall-mounted air conditioner screeched loudly as it started running.

"I did, but let me explain. Do you believe in God, General?"

"I do."

"Do you ever have doubts about God?"

"Of course. That's the nature of faith. We all have our doubts sometimes."

"Well, General, when I was a younger man, I fervently believed in Socialism, probably in the same manner that you believe in God. But I also had my doubts, so the prospect of that gold was very alluring. I decided to find it and recover it for myself. I justified my actions by swearing to myself that I would dutifully labor on as a good Socialist, but I would hide the gold away, perhaps bury it somewhere for safekeeping."

Bao Trung continued. "You must remember that at that point, the war was progressing well for us. Your nation had grown weary of conflict and death, so your imperialist forces were finally leaving. It was only a matter of time before we defeated the South. I was optimistic about the future, but if our fortunes had changed or if you Americans had elected to return to the fight…"

"You wanted to hedge your bets," interjected Glades. He shivered slightly, zipped up his khaki poplin windbreaker, and then reached out to adjust the air conditioner's vents to blow away from him.

"*Hedge my bets?* I'm not familiar with that expression," said Bao Trung. "But I assume that it accurately describes my actions. Sincerely, I didn't act out of greed; I just wanted to protect my family if the South prevailed, particularly if we were compelled to escape to another country."

Bao Trung continued. "So when I departed on furlough, I returned to that swamp. It took me three days of grubbing in the muck, but I found the bag that contained the gold coins. The bag also held a uniform, boots, a compass, a medical kit, an American flag, a rubber ground cloth, and some other things. I can't explain why, but I kept them. I concealed the bag with the gold and other items not too far from our camp. Afterwards, I hitchhiked to Hanoi and then caught a ride to my hometown with a convoy of engineering equipment."

"So you kept the gold," noted Ourecky. "I can appreciate your motivations for doing that, but it doesn't answer my question. What happened to Drew Carson?"

Without sparing any painful details, Bao Trung described Carson's ordeal. He told them how Carson was tortured and otherwise mistreated during his first weeks in incarceration, and that he never broke under even the most barbaric treatment and brutal deprivations.

He explained how Carson was kept in the reeducation camp, isolated from other Americans, and the dilemma that the camp commander faced when American prisoners were being processed for repatriation during Operation Homecoming. He described going home on furlough, returning to discover that Carson had fallen deathly ill with malaria.

"The camp commander, Major Thanh, and I sincerely wanted him to go home," claimed Bao Trung. "But he kept refusing to correctly identify himself. All that we asked of him was to divulge his true name, so we could correlate him to a name on the reconciliation list that the Americans had provided. That's all."

"Did he?" implored Ourecky. "Did he give his correct name?

"He did, but then he just would not be quiet," muttered Bao Trung, slowly shaking his head. "He would have immediately gone home had he not claimed to be an astronaut."

"An *astronaut?!* Did you believe him?" asked Ourecky. "Don't you suppose that it was the malaria that caused him to say that?"

Bao Trung was silent for a moment, and then said, "Did I believe him? Rather than answer that directly, General, I would rather tell you the rest of the story."

Bao Trung continued. "We were waiting for transportation to take him to Hanoi when the Soviets passed word that they wanted to interrogate him. They were sending a special detachment of *spetsgruppa*, under the command of a general directly from Moscow, along with a specially qualified doctor to treat his malaria. They were adamant that he not be released back to the Americans, and they demanded that we withhold confirmation that we had captured him. I was concerned that they would almost assuredly take him to the Soviet Union. I felt responsible for causing this unfortunate chain of events, so I decided to take action to thwart the Russians."

"*What did you do?*" demanded Ourecky, slapping his hand on the table. "What happened?"

"Here's what happened to Carson," said Bao Trung. As he recounted the details of that fateful night, Bao Trung was transported back in his thoughts:

After he finished writing the name on the wall of Carson's cell, Bao Trung jammed the shard of dried clay in his pocket. As he admired his handiwork in the flickering light of the candle, he heard thunder in the distance, and saw a faint flash of lightning through the cell's small window.

As he groggily struggled to button the unfamiliar uniform, Carson looked back over his right shoulder, obviously curious about was Bao Trung was doing. "Who is that?" he asked.

"I ordered you to face the *door*," snapped Bao Trung, dumping the contents of the rubberized bundle onto the coarse concrete floor. As he rummaged through the items, he added, "Can you not follow instructions?"

Bao Trung neatly folded Carson's threadbare blanket and placed it at the head of the straw mat. He did the same with his prisoner uniform, placing the carefully folded bundle atop the blanket. Then he took some other items from the bag and arranged them beside the maroon and gray striped uniform.

He then helped Carson finish dressing. The American pilot had shed so much weight in the past few weeks that his trousers would not stay up, so Bao Trung used a strand of baling twine to hitch them around his narrow waist.

Finally, he squatted down, stuck the black canvas Bata boots on Carson's feet and laced them up. He quickly crammed the remainder of the items back into the waterproof sack. He checked his watch; it was nearly midnight. He blew out the candle and stuck it in his pocket. He drew in a deep breath, thought of his wife and son, and then resolved himself to do what had to be done. It was time to go.

The sky grew darker as clouds obscured the moon. Drizzling rain pelted the dusty ground as they left the cell. Whispering, Bao Trung directed Carson to wait in the shadows as he slipped the lock back on the door hasp. Less than an hour ago, he had completed his final walk-through inspection of the guard posts, so the oncoming shift of guards had just come on duty and the outgoing shift had already returned to their barracks or left the camp.

Except for a two-man patrol that circulated through the area where the Vietnamese prisoners were housed, no one should be wandering about in the camp. So long as the debilitated prisoners were locked in their cells, the camp's security was rather lax. It was a habitual discrepancy that Bao Trung intended to correct, provided that he survived this night.

The unexpected shower was fortuitous; the rainfall would blot out their tracks. Moreover, the lazy night shift guards would be less prone to emerge from overhead cover. Consequently, slipping Carson away from the camp was a simple matter of walking about a hundred meters to the sector that contained the guards' barracks. Beyond that, there was a "secret" break in the fence behind the barracks that the guards occasionally used to sneak off the camp. Besides, with Carson dressed as he was, he looked like an NVA soldier in the darkness.

In mere minutes, the two men had squeezed through the narrow gap in the barbed wire, and were outside the compound, trudging along a narrow dirt track used mostly by farmers. Although he cajoled Carson to move faster, he was compelled to adjust his pace to match the American's pained shuffle. After all, besides being terribly ill with malaria, Carson hadn't walked for any significant distance since arriving at the camp and was also clad in stiff new boots.

As they walked, Bao Trung reviewed his plan. He knew of a Buddhist monastery located twelve kilometers west of here. The secretive monks who resided there practiced the ancient traditions; they did not ascribe to the newer and politically correct version of Buddhism officially sanctioned by the central government. Although their religious customs were not condoned, the central government did not overtly interfere with the monks, provided that they kept to themselves within the confines of their monastery. They were tolerated so long as they did not become a nuisance like the monks in the South who doused their bodies with gasoline and immolated themselves in protest of discrimination.

Bao Trung intended to use the monastery as a temporary way station for Carson, until he could devise some scheme to spirit the American pilot out of the country. He stuck his hand in his pocket and hefted the pouch of gold coins. If nothing else, despite their contrary beliefs, the monks were practical men; one or two of the Americans' coins would go a long way towards underwriting the operating expenses of their monastery.

In exchange for the gold, all Bao Trung desired was that the monks take care of Carson and keep their silence until he was able to return. That might take weeks or even months, but Carson would be safe there, much safer than if he fell into the hands of the Soviet *spetsgruppa* intent on swooping him away.

Although he was confident that the monks would be accommodating and discreet, the rest of Bao Trung's scheme was much more tentative and considerably more risky. Supposedly, in the coastal villages not too far away, there were wily fishermen who could be hired to smuggle people and goods to South Vietnam, and vice versa. If he was successful in contacting one of those pirates, certainly the remainder of the gold would purchase Carson's passage to freedom.

At Carson's lagging pace, it took them over an hour to cover just five kilometers. At least the rain had slackened and the clouds were starting to slowly scud away. Consumed with malaria, the American was weak, teetering on the edge of unconsciousness.

Staggering, he struggled to remain upright, constantly pleading for Bao Trung to stop so that they could rest. Bao Trung was terrified that they would not make it to the monastery in sufficient time so that he could return to the camp before the sun rose. He anxiously contemplated finding a place to hide Carson, perhaps in the undergrowth beside the track, so that he could return to him tomorrow.

Barely conscious, Carson slumped against Bao Trung's side and then toppled to the ground. Kneeling to assist him, Bao Trung was startled when three men, obviously soldiers, suddenly approached on the narrow road. It was a very disconcerting development; he knew where all the military units were garrisoned in the area and had painstakingly planned this route to dodge them. But as the men drew closer and he listened to their slurred words and boisterous voices, he immediately knew that they were not where they were supposed to be.

"Halt," ordered Bao Trung in a low but authoritative voice. The three men obediently stopped. He issued the day's challenge and the men responded with the proper reply.

"Out late, also?" sneered the tallest of the three carousers. Bao Trung could not glimpse their faces in the darkness, and he hoped that they couldn't discern his. "We're headed back to our camp from a…patrol."

"A patrol?" asked Bao Trung, slipping his hand to his side to unfasten the flap on his holster. He quietly tugged out his Tokarev sidearm and rested it against his thigh. "Sure. Perhaps I should accompany you to your camp and have a word with your officers."

"Uh, you're not an *officer*, are you?" asked the man hesitantly.

"I am. A lieutenant. I'm escorting this man back to his outpost. He's a good soldier when he's in the field, so I'm trying to avoid disciplinary action against him. We're going South soon, so I need every man on my roster. Especially those with strong backs and sharp eyes."

"You're headed south? You must be Infantry, sir," said the man. "The 250th Regiment is quartered near here, to the northwest. Your outfit, Comrade Lieutenant?"

"Very astute assessment. Undoubtedly, you must be in Intelligence."

"Air Defense, sir. We work at a radar site." In a shamed voice, he added, "Our equipment is being repaired. Otherwise, we would not be out tonight. Honestly, sir, I swear."

"What's wrong with him, sir?" asked one of the men. In the woods, an owl hooted.

"He's drunk," explained Bao Trung, lightly kicking Carson's shadowy form. He looked up; a few stars twinkled and the moon was beginning to peek through the departing clouds. With just a hint more illumination, the wayward soldiers would realize that Carson was not Vietnamese and the furtive charade would be over.

"I don't smell alcohol, Comrade Lieutenant," offered the third soldier, shifting the weight of the submachine gun slung over his shoulder. "Are you sure he's drunk?"

"This fool roamed onto a dispersal airfield, siphoned de-icing fluid from a storage tank and drank it," replied Bao Trung. "It's alcohol, all right, but it's methanol. It's poisonous. He'll be lucky if he's not blind in the morning. If he is, I could care less if he lands in the disciplinary barracks."

"I have a flashlight," offered the first soldier, pulling an object from his pocket. "I can check him over, Comrade Lieutenant, if you like."

"You have a pocket torch?" asked Bao Trung. "Good. I need to write your names in my notebook, so I can report that I encountered you tonight. Shine it here, soldier."

"Uh, it doesn't seem to be working, Comrade Lieutenant," stated the soldier, fiddling with a switch. "Perhaps the batteries are dead."

"Perhaps. Anyway, I'll take care of this drunken dimwit. With any luck, your radar will have been repaired and you'll be much busier tomorrow. I trust that you three will have clear heads when the sun rises?"

"We will, Comrade Lieutenant," averred the tall man.

"Then I see no need to make a report. Be on your way."

As the trio padded off in the darkness, Bao Trung holstered his pistol and breathed a sigh of relief; the chance encounter could have been disastrous. He coaxed Carson to his feet, and they continued their trek.

Moving on wobbly legs, the American limped less than five hundred meters before collapsing again. Sprawled on the damp ground, he wheezed and shuddered. In a weak, quavering voice, he begged, "Please, Bao Trung. Just let me rest. Let me sleep."

"I can let you rest here for a moment, but there is no time for you to sleep. You should be quiet now. You need to conserve your strength. We still have a long journey ahead of us."

In moments, Carson was snoring. Bao Trung nudged him, but he barely woke. He crouched down, awkwardly scooped up the drowsy American, draped him over his shoulders, and slogged on into the dark stillness, toward the sanctuary of the monastery.

Descended from the sturdy stock of mountain dwellers, Bao Trung was well accustomed to bearing burdens. Over three years ago, on his way south to fight the imperialist invaders and their South Vietnamese lackeys, he had humped a thirty-kilo case of mortar ammunition. The cumbersome box was in addition to his personal gear and weapon, so the sickly and emaciated American pilot wasn't much of a hindrance. Even with his crippled arm, he scarcely had to exert himself. Now, without having to adjust his pace to Carson's, he made good headway, steadily covering ground almost twice as fast as before.

But he soon discovered that although his resolve was firm, his endurance fell woefully short. Struggling, he marched on for another kilometer. Panting, he paused, painfully realizing that he was not in the same physical condition that he enjoyed when he led his platoon down the Ho Chi Minh Trail.

As he caught his breath, he oriented himself to his surroundings. He was familiar with this area; the dirt road ran parallel to an abandoned agricultural collective. The farm, once the temporary site of a surface-to-air missile battery, had been heavily damaged by American bombing in December. By his reckoning, the monastery was still slightly more than four kilometers away. Tired and frustrated, he was beginning to realize that this escapade was much more than he had bargained for.

He saw an opening in the woods on the right side of the trail, walked to it, and gently placed Carson in the tall grass. His stomach churned and his back ached as he considered his options. At the rate he was traveling, it would take him at least another hour to reach the monastery.

Once he reached it, it would take him at least two hours to get back to camp before dawn broke, and even then he would have to run most of the way. More importantly, he had to consider how long it would take to negotiate with the monks. They would likely be in a foul disposition after being awakened from a sound slumber, and probably would be in no mood to haggle, particularly with an Army officer seeking safe harbor for a woefully sick Caucasian.

He groaned quietly as he came to grips with his situation. The monastery was an unrealistic goal at this point. He decided that he would conceal Carson in the ruins of the secluded farm, and then come back for him when and if things settled at the camp. He would leave the gold coins and the bag with the American flag. At least he would have food and water for a few days. If nothing else, if he was unable to return, Carson might be able to find his own way home with the compass and gold coins.

Bao Trung pulled the rubber poncho from the bag and wrapped it around Carson to keep him warm. Kneeling over him, he explained his plan. In conclusion, he whispered, "You stay here. I'll scout the area to find a hiding place. Be quiet and don't move."

"Okay," replied Carson weakly.

Searching in the dim light of the moon, Bao Trung found a partially collapsed shed. It was big enough to accommodate Carson and shelter him from the elements. It wasn't much, but would have to suffice for at least a few days.

When he returned, Carson was still fragile, but seemed slightly more alert and coherent. Overhead, the clouds had cleared completely and the stars were absolutely brilliant against a stark black sky. Carson was obviously engrossed with the glimmering stars; Bao Trung realized that the American had not seen the night sky in its entirety since he was captured. He had never seen any man so mesmerized by the heavens.

"I used to go there," said Carson weakly, holding his hand up to the tranquil sky.

"I believe you," said Bao Trung. "Honestly, I do."

"I was happy there," added Carson. "It's where I belong."

"Maybe someday you will go back." A light wind rustled the tall grass.

"I doubt it. Hey...that's the Pleiades," muttered Carson, pointing toward a tiny shimmering cluster of stars. "That was Scott's favorite."

"I know," whispered Bao Trung, leaning close to him. "Scott was your friend. You told me that."

"Yeah. Scott Ourecky," said Carson. His words were slurred and he shook violently as a chill passed over him. He sighed and added, "We used to fly together."

"I hope that you see your friend again soon. Right now, I need to hide you, like we talked about. I will leave food and water for you. If I don't…"

Carson interjected, "I like it here. This is peaceful. Can we stay just a little while longer?"

"*No,*" replied Bao Trung impatiently. "I need to hide you, and then I must return to the camp before they discover that you are gone."

"Please let me rest for just a little while. I'm really tired, and I want to sleep."

Bao Trung prodded the American, to no avail. He wasn't moving. He decided to let him rest just a few minutes longer. Just a few minutes wouldn't matter.

Carson stared up at the sky like a blind man who had momentarily found his sight. After a few minutes passed, his eyes closed and he breathed softly, like an infant falling asleep. His breath became ragged and erratic, progressively growing fainter and fainter. Then he exhaled softly and made no other sound. Bao Trung leaned toward him and listened for him to inhale, but heard nothing. He felt for a pulse, but Carson's heart had apparently stopped beating. He tried to revive the American, but his efforts were futile.

Distraught and almost physically spent, Bao Trung agonized until he realized that there was nothing more he could do. Weeping, he swathed Carson's body in the rubber poncho and bound the corpse like a parcel with baling twine. He slowly dragged the body to the derelict shed. Clawing with his bare hands, he hastily dug a shallow trench in the soft ground next to the shed. He scooped the loose earth onto Carson's body and then used pieces of the shed and debris to conceal the impromptu grave. After a brief prayer, he returned to the road and ran as fast as his legs would carry him.

Birds were chirping in the trees and the morning fog was dissipating as he arrived at the camp. His entrance coincided with the dawn muster. He strolled purposely into the courtyard as the audit was in progress, as if nothing was amiss. He took his station, received the morning report from the senior guard, marched up to Major Thanh, saluted and then dutifully recited the counts.

Nodding, Thanh verified the accounting and transcribed the numbers into his notebook. "You look like you haven't slept," he observed.

"I haven't, sir," replied Bao Trung as he adjusted his sweat-damp tunic. "My son is suffering from an ear infection. My wife and I were up all night with him. He finally went to sleep just before I left."

"Hopefully, you can grab a quick catnap before our Russian visitors arrive. After muster, have your guards lock the prisoners back in their cells. I don't want anyone moving about while our guests are here."

"Comrade Major, you know that locking them up in the daytime will likely kill at least ten of them today, don't you? As hot as it is, we might even lose double that."

"You're right. We'll grant them an extra water ration for the heat," replied Thanh. "Have you checked on the American this morning?"

"Not yet," answered Bao Trung. "I looked in on him last night when I did my last rounds. He didn't look very well. He's probably still asleep right now. I'll go wake him for his breakfast."

"Don't," said Thanh, gently polishing the scuffed toe of his boot on the back of his trousers leg. "Let's wait until the Russians tell us what they want to do. I don't want his belly stuffed with rice if they intend to work him over right away."

Bao Trung heard horns honking in the distance. "So much for my nap," he grumbled, adjusting his cap and straightening his belt. "I think our exalted guests have arrived." He waved for a pair of guards to swing up the bars at the truck checkpoint.

The checkpoint guards had just barely opened the gates when a small convoy—two small trucks, a field ambulance and a staff car—raced through the entrance and straight into the courtyard. Aghast at the idiotic behavior of their allies, Bao Trung cringed; the bare ground was still soggy from last night's intermittent downpours and was slick as a provincial commissar's political rhetoric.

As the vehicles bore down upon them, frantic prisoners broke ranks and scattered. Some reacted too slowly. The field ambulance swerved as its brakes squealed, skidding on the slippery mire and plowing directly into a row of prisoners. The hapless captives toppled over like human dominoes. Several were killed instantly; screaming and groaning, the others lay broken and bleeding in the stinking muck. In his mind, Bao Trung amended his estimates for the day; they would probably lose fifty or more if this lunacy continued at its present pace.

Bellowing directives, a large Caucasian woman in a starched white dress clambered down from the ambulance. Responding to her instructions, two men in white uniforms jumped from the rear of the vehicle and quickly began unloading boxes and medical equipment.

A pudgy Russian man in civilian clothes walked up and jabbered in faltering Vietnamese. The Russian's skin was pale, like he lived in an underground bunker, but his eyes were unnaturally bright, like he bore some secret obsession. He fidgeted anxiously with his hands, like a toddler waiting for a long-promised piece of Tet candy.

Understanding little of what the stranger was trying to convey, Bao Trung shrugged his shoulders and shook his head.

The man tried Russian, which Bao Trung partially understood, and then English.

"I speak English," replied Bao Trung. "Fluently."

"Good," exclaimed the Russian. "Then we'll speak English. I am Major Morozov of the Soviet GRU. He waved toward a huge, redheaded man, also dressed in mufti. "That is my boss, General Federov. We're here to collect our American."

Collect their American? thought Bao Trung. Yesterday, their message indicated that they wanted to interrogate Carson. *Collecting* was a far cry from *interrogating.* "We know. He's very sick. Weren't you supposed to bring a doctor with you?"

Morozov pointed at the woman in white. "That's our doctor. She's a specialist in treating tropical diseases," he boasted. "She will cure the American's malaria. Where is your clinic?"

"That building," replied Bao Trung, indicating the infirmary.

Morozov shouted at the doctor in Russian. Ignoring the moans and pleas of the badly injured prisoners wallowing in the mud, the aloof doctor and her two orderlies carried the boxes and equipment to the infirmary.

"Where are they going?" asked Bao Trung. "The American is not there."

"If he's not in the clinic, then where is he?" demanded the abrasive Russian.

"We keep him in an isolation cell in that next building," answered Bao Trung. "As I told you, he's very ill, but sticking him the infirmary would likely only make him sicker. Anyway, it's early and we weren't expecting you yet. I'm sure that he's still asleep."

Morozov sniffed arrogantly, swiveled around, and walked away from Bao Trung. He then proudly guided Federov and his entourage like he personally owned the camp. They strolled to Carson's cell, chatting excitedly along the way. Nervously wringing his hands, Morozov waited impatiently as Bao Trung selected the proper key. The Russian officer behaved as though he was a renowned archeologist preparing to unseal a pyramid's burial vault reputed to be overflowing with fabled treasures.

As he slipped the key into the brass lock, fear gnawed at Bao Trung's gut; he dreaded that his treasonous acts would soon be discovered. He turned the key, slipped the lock off the hasp and then stepped to the side as Federov, Morozov and two other Russians rushed into the cell.

He heard a collective gasp. Holding his breath, he looked within and saw that the unoccupied cell was exactly as he had left it hours earlier. A small American flag was spread out on Carson's straw sleeping mat. Beside it, his prisoner's uniform and threadbare blanket were neatly folded. The deceptive scene was precisely as Bao Trung had staged it; it was like Carson had just miraculously vanished—naked—into thin air. The Russians were obviously taunted by the name on the wall, boldly chalked in white block letters.

"*Nestor Glades?*" blurted Morozov, reading the name aloud and then pivoting towards Bao Trung. "The dispatches claimed that you had *Carson!*"

"We *did* have Carson."

"You did?" growled Morozov. "Well, then who the hell is *Nestor Glades?*"

"A *ghost!*" snarled Federov furiously. "Glades is a damned specter! A phantom with mystical powers! Uzbekistan, the Congo, Berlin, Cuba and now *here!*" Flying into a livid rage, he roared vile curses as he repeatedly bashed his massive fist into the wall. The whole building shook as if suddenly jarred by an earthquake. Fragments of chipped concrete fell at his feet.

In his entire life, Bao Trung had never seen a man so angry. He resisted the urge to smile; his ruse had worked far better than he could have ever hoped. The giant Russian hothead obviously had some sort of history with Glades, and it clearly was not a pleasant one. It took minutes for Federov's tantrum to subside. He had pounded the wall so intensely that bits of flesh were shredded from the knuckles of his right hand. As he stood with his fists at his sides, bright red blood dripped to the floor and spattered his shoes. Cowering, most of the Vietnamese officers slinked towards safety.

"We had this one right in our grasp and the Americans sent their damned ghost soldier! Poof, and he's gone!" ranted Federov. "How could they possibly know?"

"Comrade General, perhaps there is a simpler explanation," offered Morozov in Russian, looking furtively towards Bao Trung. "I suspect that the guards here may have assisted the American to escape."

Ignoring him, the surly Soviet general stepped out into the sunlight and glanced up into the sky. His ire seemed to be gradually dispelled by fear. Speaking to an aide, he looked to the center of the camp and said grimly, "We are extremely vulnerable at this moment. Our vehicles are exposed where they are, out in the open. Move them under tree cover and conceal them under tarps. See if there are camouflage nets available."

"*Da*, Comrade General," replied the aide. He saluted, turned away, and scrambled toward the vehicles in the courtyard.

Federov stepped back into the cell and addressed Bao Trung in English, "You seem to be the only one here with your wits about you. Tell me: did you hear any helicopters last night?"

Bao Trung shook his head, yawned, and solemnly attested, "No, Comrade General." He polled the other Vietnamese officers; they shook their heads as well.

"As I thought," replied Federov. Turning toward the heavyset *spetsnaz* officer who oversaw his security detail, he growled, "Ensure that your goons are on full alert. We must be especially vigilant right now." He drew his Makarov automatic and racked back its slide to check its chamber. Quaking with a mixture of anger and fear, he scowled at the name written on the wall, and added, "Fetch me an AK, a bandolier of magazines and a sack of grenades. This damned peashooter won't be worth a tinker's damn if Glades is still lurking about."

"But, Comrade General," said Morozov. "I think we should question the Vietnamese. If there is some sort of plot behind Carson's disappearance, then maybe we can ferret it out."

Federov gritted his teeth and frowned. Bao Trung had never seen such a sinister-looking man; it seemed as if evil seeped from his every pore.

Stammering, Morozov spoke again. "Comrade General, I swear to you that Carson was here. Shouldn't we expend every effort to bring him back under our control? After all, he's…" Morozov flinched as Federov stepped toward him, nudging him into a corner of the dank cell.

Uncharacteristically, Federov's demeanor suddenly changed as he spoke softly in Russian: "Anatoly Nikolayevich, listen to me. I believe that you were right that Carson was here, but we must accept that he has been wrested from our grasp. The Americans probably parachuted Glades and a team here last night. If they knew your alleged astronaut was here, then they likely knew that we would be coming as well. I have tracked Glades long enough to know that he would not leave a ripe fruit still dangling on the tree. Believe me, if he is still in the vicinity, all of us are in mortal danger. I am angry, but I don't blame you. This whole episode might have just been a trick to lure me here."

"Perhaps so," replied Morozov. "What can I do now?"

"Make yourself useful," replied Federov. "Instruct the *spetsnaz* wireless operator to assemble his kit. Prepare a coded message to immediately request a helicopter from Hanoi, highest emergency priority. Start formulating a contingency plan to get us out of here by road."

Morozov nodded, pulling topographic maps of the area from his canvas valise. "Should I prepare our *spetsnaz* soldiers to search outside the camp?" he implored.

"*Nyet!*" replied Federov in an adamant whisper. "Under *no* circumstances will our men leave this camp on foot. Our Vietnamese allies have special troops to locate and kill commandos. This is their problem now, so we'll leave the dirty work and dying to them."

"And that's what happened," said Bao Trung, concluding his recollections of the fateful night when Carson disappeared from Reeducation Camp # 4. "I swear it."

"They actually believed that I came into that camp and took him that night?" asked Glades incredulously.

"They did," answered Bao Trung, nodding solemnly. "Without question."

"What happened to Carson's body?" demanded Ourecky. "You *left* him at that farm to rot?"

"He was dead, but I did not abandon him. I was not able to return until a week later. I went to the Buddhists and asked them to care for his body. They wanted to cremate him, in their custom, but I convinced them to bury him in the Western tradition instead. I offered them a gold coin for their services, but they refused to accept it."

"So you kept the gold for yourself?" asked Henson.

"I did," admitted Bao Trung. He slid a leather pouch across the table to Ourecky. "But I never spent it. There are ten coins in there, exactly like the day I dug them out of the swamp."

Ourecky gently shoved the faded pouch back to Bao Trung.

"Thank you, but I cannot accept this. I feel guilty for keeping it so long. I feel guilty for everything that has happened."

"So where is Carson's body?" asked Ourecky. "Still at the monastery?"

"No. The monastery closed years ago. Last week, I sent some specialists to recover his remains. They located an old monk who had studied at the monastery. He showed them where Carson was buried. His memory was failing, but he still felt very confident that it was Carson."

"But you're not *absolutely* sure?"

Bao Trung shook his head. "No, I am not. After the monastery closed, some of the villagers also buried their loved ones in the cemetery, but the monk seemed sure that he remembered where Carson was buried." He looked toward an officer; they spoke briefly in Vietnamese, and then the officer nodded. Standing to his feet, Bao Trung said, "Your friend is in the next room. I'm sorry for the delay, but my specialists were preparing him so he could be viewed."

The four men walked deliberately into the next room to find a skeleton displayed in an open wooden casket. The scene could have been taken from a page in a medical textbook; the discolored bones were arrayed with meticulous care. Carson's personal effects, including the items confiscated from him when he was captured, as well as the contents of the rubberized bag that Bao Trung had salvaged from the swamp, were in a pasteboard box next to the casket.

For several minutes, they stood silently and contemplated the skeletal remains. Ourecky closed his eyes and said nothing. After a while, he reached into the box that contained Carson's belongings and pulled out a tarnished Breitling chronograph. He examined it, smiled, and then replaced the watch in the box.

"There's no rush, General," said Bao Trung quietly. "I know that you want to mourn your friend. You can stay here as long as you desire."

"I guess that I've already done all my grieving long before I came here," replied Ourecky. "I suppose that I've long since accepted the notion that he was dead, but I've always wanted to know where he died and how it happened. Thank you for answering those questions."

Glades coughed, cleared his throat and apologized. "I guess I'm ultimately responsible for this. If I hadn't botched the mission, we would have rescued him. I let him down. I let you down."

Ourecky shook his head and replied, "Nestor, I'm sure that you did your best. I don't think that anyone could possibly find fault with you. Please don't think that you're responsible."

"Then *I* must apologize," interjected Bao Trung. "I cannot tell you how sorry I am. If I had not taken your friend away from the camp, he would have received care from the Soviet doctor. I am sure that he would have survived his malaria. He would probably be alive today."

Ourecky shook his head and said, "Yeah, Bao Trung, you're right. He probably could be alive today, but I think we both know that there are some fates that are worse than death. I know that if I had been in his shoes that night, and I was conscious of what was taking place, I would have gone with you. If it meant the difference between dying or possibly spending the rest of my days hidden somewhere in Siberia, I would have willingly gone with you."

"Damned right," commented Glades, shuddering. "In an instant."

"Me too," added Henson.

There was a knock at the door, and the forensic specialist entered. "I don't mean to interrupt, General, but..."

"I know that you have work to do," replied Ourecky.

The forensic specialist took a quick glance at the skeleton and noted, "All here."

"What?" asked Ourecky.

"This is an intact set of remains, General," asserted the forensic specialist. "It's a complete skeleton. In fact, I've never seen a skeleton this well preserved, at least not a set of remains recovered in the field."

"You got that from just one look?"

The specialist nodded. "Someone obviously took a lot of care when they buried him, and someone else took just as much care when they disinterred him."

"Is it Carson?" asked Ourecky.

"General, these remains appear to closely match the physical description in his records, but I have to caution you that a positive identification is still going to take at least a week or so, maybe longer, once we transport the remains back to the States."

"I requested that you bring equipment for DNA analysis," said Ourecky. "Did you bring it?"

"I did, General. I have a portable kit."

Ourecky reached into his bag, pulled out a small plastic vial, and handed it to the forensic specialist. The vial had a white identification label that bore a reference number but no name. "This sample was properly collected, exactly by regulation, and documented," he explained. "Can you compare it to Carson's DNA to confirm that they're related?"

"I can, sir. If you permit me, I'll need to take a bone to extract a sample of mitochondrial DNA. But there's still something I'm curious about."

"What?"

As a quizzical look passed over his face, the forensic specialist referred to the first page in Carson's records. "Forgive me if it's not my place to ask, General," he said, closing the dossier. "This man was classified as missing in action in 1972 and presumed dead back in 1973. That was *long* before the military started collecting DNA specimens for identification purposes."

"And your point?"

"Sir, his records indicate that he had no living relatives. If that's the case, where did that DNA sample come from and why do you want a match?"

"That's none of your concern," replied Ourecky. "Just *do* it."

The specialist took a bone and left the room. When they were alone, Ourecky turned to Bao Trung and asked quietly, "Let's go back to something you mentioned earlier. Did you believe Carson when he claimed he was an astronaut?"

Bao Trung swallowed. A nervous expression crossed his face as he somberly replied. "After the last time he was interrogated, when he finally identified himself correctly, I escorted Carson back to his cell. I'll admit that I was very angry with him, because I knew there would be a delay when the interrogation team rendered their report to Hanoi. Certainly, I also had no idea that Hanoi or Moscow would take his claims seriously."

Bao Trung continued. "When we returned to his cell, Carson obviously knew that I was disappointed with him, because he kept apologizing to me. He kept trying to explain that he wasn't lying, and started telling me things. At first I thought he was just rambling, that his mind was addled by the malaria. But…" He looked at the floor and fell silent.

"But *what?*" asked Ourecky.

"The stories he told me were just too detailed. It was not something that he could have arbitrarily invented. I am confident that it happened exactly as he described it. With that said, I must offer another confession, General."

"What?"

"At that point, I knew that if the Russians came and discovered even a fragment of what I knew, they would take Carson away and he would never return. I was also certain that if the Soviets ever realized what I knew, then I would have been swept away as well."

Bao Trung continued. "I was afraid for Carson, but I was afraid for myself as well. So while I took Carson from the camp to protect him, that wasn't my only motive. I was deathly concerned that the Soviets would eventually find me out. I have lived the past forty years under that fear, and I have lived the past forty years knowing that I was responsible for Carson's death."

"Listen to me," said Ourecky. He looked toward Glades and then toward the skeleton on the table. "All of you. Carson chose to come here. What happened here was tragic, but these circumstances were the result of Carson's choices, not yours. The three of you are entirely without blame."

"Perhaps," said Bao Trung. "But I still feel guilty."

"You shouldn't."

"I do. We Buddhists don't believe in God the same way that you do," observed Bao Trung, intertwining his fingers. "But I still think that I was punished for my sins."

"How so?" asked Ourecky.

"I think that for my transgressions, I was forced to remain at that camp for ten more years. I watched hundreds of men die, mostly by starvation. After the South collapsed, we became even busier, and I was eventually placed in charge. Thousands of men and women passed through my camp. Most died, but some were reformed and went back into society. I'll tell you; with the burden that I carried on my soul, there were many days that I would have gladly joined their ranks as they marched into their graves."

"Well, I may never change your mind, Bao Trung, but I am still grateful for what you did. If nothing else, for his last few hours on earth, Carson was *free*."

The four men were silent for several moments, contemplating the skeleton on the table. With Henson's assistance, Glades opened a jar of pills, took two, and chased them with a swallow of water. He cleared his throat, and in a gravelly voice said, "I want to know something. There were a lot of other American POWs who didn't make it home. What happened to them?"

"Honestly, I don't know," answered Bao Trung. "Besides those that I killed in combat, you, your team and Carson were the only Americans that I ever had contact with. Except for Carson, I never had direct contact with any American prisoners, so I cannot tell you what happened after the Americans were released in 1973. But I've always been curious about something…"

"What's that?" asked Henson, twisting the lid back on the medicine bottle.

"I could never understand why your country was so swift to close the books about the prisoners. With so many men unaccounted for, I'm sure that questions still lingered. If I were in your shoes, I would certainly have doubts."

"How so?" asked Ourecky.

"Well, consider this: I know just enough about jet aircraft to be aware that an emergency ejection is a traumatic event. Men sometimes lose a limb or two when they are forced to eject even during routine training missions. You have to assume that for all of your pilots who ejected in combat during the war, at least a few would have suffered a traumatic amputation. Yet for all of the 591 American prisoners who were repatriated, none were missing limbs. Do you *really* suppose that's possible?"

Glades started to reply, but there was a knock at the door. The forensic specialist entered the room and declared, "It's a match, General."

"Are you *sure?*" asked Ourecky.

"I am. They are definitely related. I have absolutely no doubt."

Ourecky nodded and said, "Bao Trung, I want to finish this conversation, but I have some urgent business to attend to. Would you mind?"

Bao Trung shook his head. "Please do whatever it is that needs to be done."

"Henson, I need a favor from you. I need to make a couple of calls on the satphone. In the meantime, would you jump on the radio and call the security detail accompanying Bea and Deirdre? I would like to get them back here as soon as possible."

"Will do, sir."

Ourecky tapped in a string of numbers on the satellite phone, waited for a minute to hear an answer, and then said, "Sorry, son, I know it's really late there, but it's *him*. I just thought that you'd want to know.

He terminated the call and said, "Henson, one more thing: bring Major Smith back in here, please."

Accompanied by her co-pilot, Major Smith arrived. "You called, sir?" she asked.

"I did. Your C-17 is taking up a lot of valuable real estate, and since we've accomplished our mission here, I want to get it back into the air ASAP," said Ourecky. "To be frank, Major, my friend has been waiting a very long time, and I want to get him home as quickly as possible. Do you understand that?"

"I do, sir," she answered. "I lost friends in Afghanistan, so I know exactly what that feels like. I guess that you're aware that we came in here with an extra flight crew aboard, so crew rest is not a constraint. I've already inquired about take-off spots, and the locals have told me that they can slot us to depart in about three hours."

"Three hours?" he asked, glancing at his watch. "I would have guessed they could get us out of here sooner."

"This airport is a very busy place," she replied. "That's the nature of humanitarian relief. Full planes in, empty planes out, all day, every day, until the work is finally done."

"Then so be it," he said. "Please make the necessary arrangements. I'll work with Bao Trung and the Vietnamese to have Major Carson's remains ready for movement."

"I'll have my loadmaster unload the transfer case and bring it over here," said Smith. She spoke quickly in a hand-held radio, and then said, "Begging your pardon, General, but there's no need for you and your party to be uncomfortable on the way back."

"How so?"

Turning toward the window, she pointed to a Gulfstream executive jet parked next to the C-17. "The State Department diverted that airframe from a trade mission in China, on orders from the White House," she explained. "It's earmarked for your party, General. You'll be here a few extra hours, maybe overnight, and make a layover in Hawaii, but you'll travel back in comfort."

Accompanied by Glades and Henson, Bea and Deirdre walked into the room. "Scott, dear, what's going on?"

"The State Department has dispatched an executive jet for us, but I intend to go back on the C-17 with Drew. I'll be leaving in a few hours, but there's no need for the rest of you to be in a rush or be uncomfortable. You, Nestor and Deirdre can stay here for a day or so and then ride back in comfort on the Gulfstream."

"How about the funeral?" asked Bea. "Won't that be at Arlington? Will we make it in time?"

"The funeral won't be for another couple of weeks, at a minimum, so you'll have plenty of time."

"General, are you absolutely sure you want to do it this way?" asked Major Smith.

"I will fly back with Major Carson's remains," declared Ourecky. "On the C-17."

"Sir, it's going to be a long haul," counseled Major Smith. "Are you *certain* that you don't want to just ride back in the Gulfstream with the others?"

"I *will* fly back with Major Carson's remains," reiterated Ourecky, obviously becoming impatient. "I'm staying with Drew."

"I understand,"

"Sir, I'll go with you also," said Glades. "Please."

"Nestor, I'm a hardheaded old Czech, but there's certainly no need for you to be uncomfortable," observed Ourecky. "Especially in your current state. You and the ladies can fly back in the Gulfstream."

"General, if it's all the same to you, I will accompany Major Carson's remains as well. This was my mission, and I intend to finish it."

Deirdre shook her head. "I go where Nestor goes. And there's no sense arguing about it."

"I'm flying back with Drew, dear," added Bea. "And you. It's only fitting that we take him home together."

Less than an hour later, Carson's remains had been carefully shifted from the wooden casket to the aluminum casket, the "human remains transfer case," to be loaded on the C-17.

"General, uh, we have a problem," confessed Major Smith. "There's a specified protocol for transporting the remains of deceased military personnel, and in our rush to get here, and the humanitarian relief mission, uh, uh, we neglected to…"

"Just spit it out, Major," said Ourecky. "What's the problem?"

"Sir, we don't have a flag to cover the remains."

Ourecky shook his head. "Well, Major, that's a pretty significant oversight, but one way or another, we need to get in the air. Why don't you check with the Gulfstream's crew to see if they have one?"

As Smith rushed away, Bao Trung stepped forward. "Perhaps this might suffice," he offered, pulling a carefully folded American flag from the box that contained Carson's personal effects. He handed it to Henson, who unfolded it and stretched it out for Ourecky to see. It bore a few dark stains, and its colors were dulled by time, but it was clearly recognizable as the Stars and Stripes.

"Where did *that* come from?" asked Ourecky.

"It was in the bag that your team cached in the swamp," explained Bao Trung, looking toward Glades. "It was the same flag that I left in Carson's cell the night that I took him away."

The four men carefully draped the small, faded banner on the shiny casket. In unison, they stepped back and saluted.

After slowly lowering his salute, Henson commented, "You know, General, under any other circumstances, that would sure look pathetic, but I couldn't think of anything more appropriate."

"That's *perfect*," commented Glades, as he dropped his salute.

"You're right, Nestor," noted Ourecky, smiling at the notion that his friend was going home at last. "*Absolutely* perfect."

30

TWO FUNERALS

Faith Methodist Church, Perryton, West Virginia
Present Day

As much as Nestor Glades desired to be at Carson's funeral, he just missed it. Just two weeks after he had returned from his final journey to Vietnam, his pancreatic cancer finally caught up with him. Like everything he had done in his working life, his funeral was planned in meticulous detail. All that remained was to execute the plan.

In accordance with his wishes, Deirdre Glades took him back to West Virginia to be buried in a graveyard beside a Methodist church that had long since been boarded up and abandoned as a house of worship.

Some of his former comrades had arrived the day prior, to make the place more presentable. They mowed the grass, pulled weeds, trimmed the hedges, straightened toppled memorials and touched up the black paint on the metal picket fence surrounding the cemetery. After a long day of it, they retired to a pub in nearby Morgantown for an impromptu wake.

Although she expected at least a few old soldiers to make an appearance, Deirdre was absolutely astonished by the turn-out. Hundreds of former comrades and acquaintances came, pouring in from all over the country. An official Honor Guard was sent from Fort Bragg, North Carolina.

Lying in his coffin, Nestor wore an Army dress uniform, resplendent in awards. He wasn't empty-handed. In one hand, for his parents, he clasped a single .22 caliber bullet wrapped in an ancient scrap of possum fur. In the other hand, for his father-in law, was a sand-colored beret of the British Special Air Service, presented to him many decades before.

On any other day it would have taken a mere twenty minutes to drive from the funeral home to the cemetery, but today it took the funeral procession more than two hours to make the transition. Although they were effectively in the middle of nowhere, in a deserted coal mining community, the tiny cemetery and surrounding grounds were absolutely filled—and then some—with scores of men who had known and fought with Nestor Glades.

Ourecky and Bea sat beside Deirdre and comforted her as the bugles sounded Taps and the folded flag was gently handed to her in memory of her husband's service. After presenting her with the flag, the ceremonial leader of the Honor Guard, a former Special Forces colonel who had served with Nestor, leaned forward and took her hands. Whispering in her ear as he surreptitiously slipped an object in her palm, he said, "Your father's Regiment sends their regards."

As grief-stricken as she was, Deirdre could not help but smile when she later opened her hand to glimpse the SAS cap badge, a winged dagger above the inscription "Who Dares Wins."

Nestor Glades was buried next to his mother. His father was not buried here; he and two of Nestor's brothers were entombed a few miles away, thousands of feet below ground, trapped in a collapsed coal mine.

Nestor's grave was marked with the simple stone that marks a soldier's passing: a GI slab of marble, provided by the Veterans Administration, marked with his name, rank, highest decoration and some of the places he had fought.

When the official ceremony was concluded, one of Nestor's MACV-SOG teammates—an elderly Nung mercenary who had escaped Vietnam in 1978 and later settled in Cincinnati—used a stamp-sized P-38 can opener to methodically slice the top from a Number Ten can of peaches. In remembrance of him, Nestor's former teammates gathered in around his coffin to participate in the quirky communion that used to mark his departure from Vietnam. As the men stood over his yet unclosed grave, they passed the big can and each ate a peach slice in his memory. While they were sad at his passing, none cried in his memory, just as Nestor had not shed a tear at his own mother's funeral. They saved the next-to-last slice for Deirdre, and then placed the final slice—dripping with sweet syrup—atop the coffin of Nestor Glades.

Deirdre lingered there long after they left. Tomorrow, she would return home to Florida, to dote on grandchildren and do those things that elderly widows do, but she knew that in time she would come back here, to this forlorn cemetery by an abandoned church in a gritty little town that had long ago ceased to exist. She would come back here to be with Nestor always.

Arlington National Cemetery, Arlington, Virginia
Present Day

Ourecky didn't venture to Arlington very often, but when he did, he set aside plenty of time to walk the hallowed ground and pay homage to the men and women who rested here. Consequently, on the day prior to Carson's funeral, he and Bea had spent hours roaming amongst the stones and monuments, visiting old friends and acquaintances.

Legions of heroes rested here, as did more than a few cowards. In the gentle hills were thousands of fathers never granted the opportunity to meet their children, resting in the midst of multitudes of teenagers whose lives were erased even before they were written. It was a place where rank and social standing mattered little; in these lawns were the mighty and powerful—Generals, Admirals, Ambassadors, Senators, Congressmen—eternally taking their places amongst common Soldiers, Sailors, Marines and Airmen.

Here on these grounds, an eternal torch burned for a young President, taken away in his prime, whose stirring words set the Nation on a course to the Moon. Victims of a horrific fire that occurred as we stumbled along that path, a pair of Apollo 1 astronauts—Virgil "Gus" Grissom and Roger Chaffee—were not too far away, buried beside each other. Several other astronauts were also buried in Arlington, including men and women who had perished in the Challenger and Columbia space shuttle disasters, as well as two brave explorers—Charles "Pete" Conrad and James Irwin—who had long ago strolled on lunar dust.

Mark Tew was here, just a stone's throw from the graceful sky-reaching curved spires of the Air Force Monument. Leon Tarbox and Virgil Wolcott were not. Admiral Tarbox had died just five years after Carson was shot down near Haiphong. At his request, he had been buried at sea in the Pacific, at the site of a naval battle where his older brother had died in 1944.

As he had always intended, Wolcott retired to his family's ranch in Oklahoma shortly after the Project folded. He lived to be ninety-two years old, and succumbed to a heart attack while mending fences. With his favorite roan hitched to a nearby fencepost, he died with his boots on, a Stetson on his head, a hammer in his hand and an unfiltered Camel cigarette between his lips. He was buried near the family's homestead, under the spreading branches of a grand oak he had planted as a child.

A month had elapsed since they had returned from Vietnam. The day had come to pay respects to Carson. Although long retired, Ourecky donned his Air Force dress uniform. For only the second time in his life, with special authorization from the President, he wore the ribbon that signified the Medal of Honor, as well as his astronaut badge.

In fact, even though he had earned a stack of ribbons during a long and illustrious career, his astronaut wings and the pale blue ribbon were the only awards that adorned his uniform on this bright morning, and everyone present knew that he wore those decorations not to honor himself, but rather to honor his friend.

Although Carson's funeral wasn't a large ceremony, it wasn't without auspicious attendees. The current President, Secretary of the Air Force, and Chief of Staff of the Air Force were present to honor Carson. Five of his shipmates from the Vietnam era—including "Badger," his former squadron commander, and "Beans" Leesma—were guests as well.

Of the seven pilots—other than Ourecky—who had worked on the Project, only two—Tim Agnew and Mike Sigler—were still alive. Both men were in attendance. Ed Russo had succumbed to cancer less than five years after being rescued from orbit. Parch Jackson had been killed while flying a vintage racing plane at an air show in 1995; his last conscious act was to steer the stricken aircraft away from the crowd.

With Bea at his side, Ourecky took his place beside Rebecca, Carson's daughter. Although the field DNA test in Vietnam showed that they were related, a more sophisticated analysis definitively proved that Carson was her biological father. As his sole surviving relative, she would receive his military insurance and back pay—compounded with interest—which came to a fairly substantial sum since Carson had continued to be administratively promoted with his peers, eventually achieving the rank of full colonel.

But Rebecca really had no desire for the windfall; Ourecky was quite aware that she had married a good man, and that they had no cause for want. They planned to set aside enough for their children's education, and then contribute the remainder to service organizations that cared for wounded veterans.

The service was short and very simple: Ourecky read a heartfelt but suitably vague eulogy, a prayer was said by an Air Force chaplain, three volleys of seven rounds were fired by an Honor Guard, and the flag covering Carson's casket was painstakingly folded thirteen times and tucked into a tight triangle. The Honor Guard paused as four fighter aircraft—-two from the Air Force and two from the Navy—screamed overhead, with one suddenly breaking away to form the "Missing Man" formation.

After the jets roared off over the horizon and their noise dissipated, the leader of the Honor Guard came forward and carefully handed Rebecca the folded flag. He saluted it in painfully slow motion, and then quietly declared, "We present this on behalf of a grateful nation."

After Taps had been sounded and the Honor Guard departed, Ourecky presented Rebecca with a blue leatherette case, approximately the size of a hardbound book, and explained, "Your father earned the Medal of Honor."

"That's been in a vault all these years," commented the President. "We want you to have it."

Ourecky and Bea remained long after the President and other official guests had departed. He looked toward Rebecca, surrounded by her three children, and smiled. Long ago, after he had graduated from MIT, he and Bea had returned to Ohio, where he was stationed at Wright-Patterson for almost ten years. In no time, their son became reacquainted with his best friend from childhood; the two took up right where they had left off, became fast friends—virtually inseparable—and eventually married after he graduated from the Air Force Academy in Colorado.

"I'm really sorry that Andy couldn't be here," said Rebecca. "He wanted to be, but they're in the middle of a big flight test. It's all very hush-hush."

"I know *that* feeling," commented Bea, gazing towards Ourecky.

"Sometimes I wish he was back at NASA, flying the Shuttle again," said Rebecca. "But at least he's home more often now. And he's happy."

"That's good," said Ourecky.

Rebecca's youngest child, a seven-year-old boy, scampered up. The child took Ourecky's hand and led him to Carson's still open grave. Ourecky knelt down, and the two of them read the marker stone's inscription together. It said simply:

ANDREW M.
CARSON
COLONEL
U.S. AIR FORCE
VIETNAM
MARCH 18 1935
APRIL 1973
MEDAL OF HONOR
COMMAND PILOT

"So, Pops, the man they buried here, he was my other Pops?" asked the boy, pointing at the stone.

"He was," answered Ourecky. "He died a long time before you were born."

"Really? Was he brave like you?"

"Oh, *much* braver than me."

"Then was he a hero like you?" The boy reached out and touched the ribbon on Ourecky's chest. In the bright sunlight, the child's eyes almost exactly matched the shade of the ribbon; they were crystalline blue, almost unnaturally so, as if they had been snatched from the sky.

"No. I'm not a hero. *He* was a hero." Ourecky felt a familiar twinge of pain in his shoulder, something he hadn't felt in decades. He rubbed it, thought of Carson, and smiled to himself. Gazing at the bright eyes of his grandson, he knew that a part of his friend would be with him always.

31

EPILOGUE:
A VISIT WITH DOCTOR SOPHIE

Office of Doctor Sophie Dubuission
Grande-Rivière-du-Nord, Haiti
Present Day

Doctor Sophie Dubuission paused for a moment's respite. One of the few doctors in Grande-Rivière-du-Nord, and the only pediatrician, her services were in great demand.

Her medical practice was housed in a simple one-story building consisting of three rooms. One served as a reception area, the middle room—the largest of the three—functioned as her office and examination area, and a small room towards the back was her living quarters. There was a little porch at the rear of the building, where Sophie periodically came for a break. The sun had set less than thirty minutes ago; she particularly enjoyed this time of twilight.

Gazing up at the darkening sky, she glimpsed a bright spot of light tracing swiftly through the stars. It did not flicker, so she decided that it must be a satellite rather than an airplane. Curious, she admired it for a moment and watched it as vanished over the horizon. Then she remembered that children were still waiting to be seen, so she could not linger here for long.

With striking features that always seemed to make men weak at the knees, a gift from her mother, Sophie was a sprite, hardly bigger than some of her patients, weighing barely more than a shadow. She had not always practiced medicine here; she had returned recently after spending several years as a doctor in Africa. In that capacity, on that bleak continent, she had been one of the busiest pediatricians on the planet, saving tens of thousands of children from disease, starvation, wounds and other trauma. In fact, she probably would not have returned to Haiti if her father had not been deathly ill, and she probably would not have stayed if he hadn't specifically asked her to come back to care for the destitute children of Grande-Rivière-du-Nord.

Caring for children was her life and her passion. Although she was nearly forty years old, she had never married. She came close a few times, but she had yet to find a man who could truly understand or appreciate her obsession, and had scarce time to spend looking for a mate.

Her mother was always concerned that Sophie did not have a husband and family. *Life is short*, she insisted, *you must find yourself a good man.* As it was, her parents had passed away in their late fifties; longevity was not a common trait in Haiti. Her mother was right: *Life is short.* Now, as her life gradually settled back into normalcy, Sophie lived a very ascetic existence. This evening, after she had seen her last patient, she would sit down to a simple meal of rice and beans. Afterwards, she would curl up with a copy of *Harrison's Principles of Internal Medicine*, read until her eyes could bear no more, and then lie down to sleep until dawn.

Sophie looked back up at the stars one more time and then walked back into reception area. She was surprised to see that no patients remained. Her nurse looked up, handed her a folder, and explained, "Most of the last parents decided that their children's complaints were minor, so they will be back in the morning. I've arranged for them to immediately go to the front of the queue. But there's only one more patient yet to see."

"*Really?* Just one?"

"*Wi*. It's *Madan* Madabat's son. They're waiting outside."

"Ear infection *again?*" asked Sophie, quietly clicking her tongue and rolling her eyes.

"*Wi.*"

Sophie sighed, shook her head and frowned. She wasn't particularly fond of *Madan* Madabat. First, her six-year-old son was particularly prone to ear problems, and to make matters worse, he seemed especially susceptible to some sort of microorganism—which Sophie had yet to isolate and identify—that resided in the river. She had repeatedly advised *Madan* Madabat to keep the little rascal out of the river, or at least caution him to keep his head out of the water, but every two weeks or so they showed up at the office, with the child presenting with yellow pus oozing from an ear canal swollen nearly shut.

Second, *Madan* Madabat was a woman of means, at least by Haiti's humble standards. She made an obnoxious practice of putting on airs and acting superior to the impoverished mothers who frequented the clinic. But even though *Madan* Madabat was certainly capable of paying for a visit, such an act was obviously beneath her, because she always seemed to leave the office without opening her purse. She always insisted on receiving a bill from the nurse as she departed, but the invoices were never paid.

Sophie found it ironic, if not also aggravating. She would see any child, regardless of their parents' capacity to pay, and often received compensation in the form of a chicken, vegetables, bananas or mangos. And although she appreciated anything she was given, it was just as likely that she would send the offerings out the door with the next patient rather than eat them herself.

She took a deep breath and reminded herself of her pledge to be cheerful and encouraging with every child and every parent. Regardless of their complaint, whether minor or most severe, she resolved herself to be patient and kind.

Sophie adjusted the stethoscope draped around her neck and sniffed a delightful odor in the air. "What on *earth* is that wonderful smell?" she asked, almost immediately salivating.

"Remember that boy you stitched up this morning?" asked her nurse. "His father left you some dinner." The nurse reached under her desk and brought out a lunch carrier consisting of several nested aluminum pans held together by a metal bale. "I was going to surprise you."

Sophie's first thought was how much more time she would have to study without the annoyance of cooking dinner. The scent of the food was almost overwhelmingly enticing; at this instant, after tending to over a hundred suffering children over the course of thirteen hours, she desired nothing more than to immediately sit down and eat.

The nurse sniffed at the lid of the uppermost pan and excitedly declared: "There's *roast chicken* in here! It smells delicious. There's rice, beans and fried plantains as well! And *pudding*! A feast! I really think that man is sweet on you, Doctor Sophie."

Sophie's stomach growled audibly. She held her hand over it, and said, "If it's a feast, then have some for yourself, Marie. Just leave a little bit for me, *souple*."

"*Mèsi anpil*, Sophie."

Sophie yawned as she reviewed the boy's medical records. Almost assuredly, the child's ear infection would probably be just a simple case that could be resolved by irrigating the ear canal with diluted vinegar. But if it were worse, perhaps a bacterial infection spreading from the ear to the soft tissues of the face, it might require an extensive regimen of antibiotics. She frowned; she tried to use the potent medicines as sparingly as possible.

She closed the folder, looked up and said, "Send them in."

Nibbling on a greasy drumstick, the nurse left her desk and stepped outside. She beckoned for *Madan* Madabat and the child to follow her to Sophie's office.

As they entered the room, Sophie lifted the boy and sat him upright on her exam table. The child flinched as she playfully wiggled his outer ear. Although that simple gesture was sufficient to tell her most of what she needed to know, she spent several minutes examining him, peering into his ear with an otoscope as he fussed and fidgeted.

She irrigated the inflamed ear canal with sterile saline, dried it, and then used a cotton swab to gingerly apply a topical antibiotic. Afterwards, handing *Madan* Madabat a sample bottle of medicinal eardrops, she recited the same litany of treatment and precautionary measures that she had told her so many times in the past.

As she stood to leave, *Madan* Madabat spied two pictures—a photograph and a child's drawing—hanging in simple frames beside the door. "I've not seen these pictures before," she commented.

Sophie grinned and replied, "I've just put them up recently, since your last visit. That photograph is of my parents."

"They *were* a handsome couple," said *Madan* Madabat, haughtily turning up her nose. "I'm not one to judge, Doctor Sophie, but wasn't your mother a prostitute? I've heard stories…"

"Well, her past was somewhat checkered, but she was *never* a prostitute." Forcing a smile, Sophie could not believe the woman possessed the gall to speak ill of the deceased.

Wrinkling her nose and squinting, *Madan* Madabat studied the framed drawing. "And this? It's very crude and odd-looking. Did one of your patients draw it?"

"*Non.* I did, when I was as a little girl. It shows a story that my father used to tell us when we were young. See those dark spots near the bottom? Those are bloodstains. *My* blood."

"You *bled* on this picture?"

"I did. When I was seven," explained Sophie. "My parents enrolled me in the missionary school in Hinche. The other children were curious about my upbringing, so I drew that picture and told them the story. One of the nuns heard about it, and she was furious. She snatched the picture away from me and stuck it in her desk. She rapped my knuckles with a ruler trying to force me to recant the story. She beat my knuckles until they bled, but I would not give in. *Ever.*"

As Sophie took the batteries out of her otoscope so they would not corrode the interior of the expensive instrument, she added, "Last year, that ugly old nun died. I went back to the school and they let me look in her desk. I found my picture and brought it home."

"And it was a picture of a story? What on earth could have made a nun so upset?" asked *Madan* Madabat, holding her son's shoulder. "What was the story about?"

"*Yon zanj nwa,*" replied Sophie. "A black angel."

"A black angel? I *must* hear this. Tell me about it."

Sophie smiled, closed her eyes, and said, "Every year on March 13th, my father and my uncle would gather their families together. Before they moved closer to town, they used to live on a hilltop, all by themselves. The townspeople would not allow my father to come down from the hill, because he was a leper."

"*Lèp?*" said *Madan* Madabat, shuddering as if a chill passed over her. "The leper on the hill? I vaguely remember hearing that story, but I thought it was just a rumor, like a ghost story to scare misbehaving children. It was true? He was your father? He lived there with your uncle?"

Sophie nodded, opened her eyes and continued. "They lived on that hill together. They were content with themselves, but they also knew that they could never marry and have families of their own. But one night, a flaming chariot fell from the sky, right into their sugarcane patch."

"A flaming chariot?" scoffed *Madan* Madabat. Her young son seemed entranced by the story, tilting his head so that his good ear was pointing towards Sophie.

"*Wi.* Then later, angels came from the sky and descended on their hilltop. Most were *zanj blan,* but those white angels were preceded by a black angel. According to the story, the angels cured my father of leprosy, and later the black angel took him to Cap-Haïtian where he introduced him to my mother. And that's how I came to be here."

Gasping, *Madan* Madabat snatched her enthralled son and roughly pushed him through the doorway into the reception area. Frowning, she slammed the door shut and exclaimed "Doctor, *please* don't fill my child's ears with such silly nonsense! Flaming chariots? Angels descending from heaven? A black angel? What kind of *idiot* would believe such a ridiculous thing?"

Stifling an urge to sigh, Sophie smiled politely as she gazed over the many diplomas that declared her to be an educated woman. For all of her learned knowledge, she knew that the world was a mysterious place, with many aspects that simply could not be explained with logic and reason. Looking at the contented smiles of her parents, Henri and Lydie, she knew that love—even more so than hope and faith—was the most powerful force in the world. And as her father had so often told her, the real lesson of the *zanj nwa* was that the greatest thing we can do on this earth was simply to be kind to one another.

Madan Madabat's question reverberated in her ears: *What kind of idiot would believe such a ridiculous thing?* "*Mwen. Mwen kwè ke,*" replied Sophie. "Me. I believe."

ACKNOWLEDGEMENTS

I am indebted to Dr. John Charles, for kindly and consistently contributing his extensive knowledge of manned spaceflight, timely advice and sage counsel.

For their encouragement, friendship and enthusiastic support of the Blue Gemini series, I would like to extend my sincere appreciation to Bo Canning, Frankie Fisher, Ronald G. Purviance, Dennis L. Rodgrick, Karen Weldon, Ken Klein, Tonja Klein, Dr. James Busby, James Pacheo, John Gresham, Tim Gagnon, Michael Mastin, Dan Bramos, Steve Schultz, Shannon Pettigrew, Taylor Robinson, Emily Carney, Stuart Sentell, Elliot Lounsbury, Andreas P. Bergweiler, Roy Houchin, Jose Clavell, Donald Franck, Byron Craig Russell, David Jacob Heino, Gregg Correll, Richard Johns, Paul Ryan, Christopher Rottiers, Robert Hawthorne, and BG John Scales.

I would be remiss if I failed to mention the contributions of my brother Ed, who was responsible for some of the key technology described in the story. Ed has also executed an extensive collection of technical illustrations that depict some of the key equipment and facilities, like the fictional "Pacific Departure Facility" launch complex at Johnston Island, to help readers in visualizing these aspects of the story. I would also like to extend my most heartfelt appreciation to three gentlemen—Mike Pierce, James F. Rosencrans, and John McGinley—who kindly assisted Ed in his efforts.

Last but definitely not least, I am indebted to my family and especially my wife, Adele, for their love and infinite patience.